Warnings and Risk Communication

Warnings and Risk Communication

EDITED BY

MICHAEL S. WOGALTER
North Carolina State University
Raleigh, North Carolina

DAVID M. DEJOY
University of Georgia
Athens, Georgia

KENNETH R. LAUGHERY
Rice University
Houston, Texas

UK Taylor & Francis Ltd, 11 New Fetter Lane, London EC4P 4EE
USA Taylor & Francis Inc., 325 Chestnut Street, Philadelphia, PA 19106

Copyright © Taylor & Francis 1999, except for Chapter 12 which is a work of the US Government.

All rights reserved. No part of this publication may be reproduced, stored in a retrieval system, or transmitted in any form or by any means, electronic, electrostatic, magnetic tape, mechanical, photocopying, recording or otherwise, without the prior permission of the copyright owner.

British Library Cataloguing in Publication Data

A catalogue record for this book is available from the British Library.
ISBN 0–7484–0266–7 (cased)

Library of Congress Cataloging in Publication Data are available

Cover design by Rob Steen

Contents

Editor Biographies	*page* ix
Foreword	xi
Mark S. Sanders California State University at Northridge	
Preface	xiii
Michael S. Wogalter North Carolina State University	
David M. DeJoy University of Georgia	
Kenneth R. Laughery Rice University	
Contributors	xvii

PART ONE Introduction   1

This first section introduces the area of warnings and risk communication, describing its importance and providing a brief history of the field. An overview of a communication-human information processing (C-HIP) framework is described that organizes the research literature presented in subsequent chapters.

1 Overview
Kenneth R. Laughery and Amy Hammond Rice University   3

2 Organizing Theoretical Framework: A Consolidated Communication-Human Information Processing (C-HIP) Model   15
Michael S. Wogalter North Carolina State University
David M. DeJoy University of Georgia
Kenneth R. Laughery Rice University

PART TWO Methods/Techniques   25

The second section describes the different methods of investigating warning effectiveness. Behavioral compliance and measures of intermediate processing stages are discussed.

3 Intermediate Processing Stages: Methodological Considerations for Research on Warnings 27
Stephen L. Young Liberty Mutual Group
David R. Lovvoll Rice University

4 Methodological Techniques for Evaluating Behavioral Intentions and Compliance 53
Michael S. Wogalter North Carolina State University
Thomas A. Dingus Virginia Polytechnic Institute and State University

PART THREE Research on Warnings: Stages of the Model 83

This section reviews and summarizes research on warnings. The research is organized around the stages of the communication-information processing (C-HIP) framework. In addition to providing an organizing framework, the model has utility in explaining research findings as well as why warnings may succeed or fail in application.

5 Source 85
Eli P. Cox III University of Texas at Austin

6 Channel 99
Michael B. Mazis American University
Louis A. Morris PRR Research Division

7 Attention Capture and Maintenance 123
Michael S. Wogalter North Carolina State University
S. David Leonard University of Georgia

8 Comprehension and Memory 149
S. David Leonard University of Georgia
Hajime Otani Central Michigan University
Michael S. Wogalter North Carolina State University

9 Attitudes and Beliefs 189
David M. DeJoy University of Georgia

10 Motivation 221
David M. DeJoy University of Georgia

11 Behavior 245
N. Clayton Silver University of Nevada at Las Vegas
Curt C. Braun University of Idaho

PART FOUR Practical Issues of Warning Design 263

This section provides practical guidance on warning development. Government regulations, industry standards, and general guidelines can serve as a basis for the design of warnings, but these rules may not be adequate by themselves. Procedures for developing warnings and testing them on appropriate target populations are described.

12 Standards and Government Regulations in the USA 265
Belinda L. Collins National Institute of Standards and Technology

13 Practical Considerations Regarding the Design and Evaluation of Product Warnings 291
J. Paul Frantz and Timothy P. Rhoades Applied Safety and Ergonomics, Inc.
Mark R. Lehto Purdue University

PART FIVE Forensics 313

The last section describes the legal aspects of warnings in the USA. Implications of statutory and case law for warning design, including potential consequences of failure to warn, are discussed. The chapter on the human factors expert witness provides insight on how litigation is aided by research and analysis.

14 The Law Relating to Warnings 315
M. Stuart Madden Pace University School of Law

15 The Expert Witness 331
Kenneth R. Laughery Rice University

Subject Index 351
Author Index 357

Editor Biographies

Michael S. Wogalter, Ph.D.
Michael S. Wogalter is an associate professor in the Department of Psychology at North Carolina State University (Raleigh). He received a B.A. from the University of Virginia, an M.A. from the University of South Florida, and a Ph.D. from Rice University. He has held faculty appointments at the University of Richmond and Rensselaer Polytechnic Institute. Most of his research focuses on the factors that influence the effectiveness of warnings and people's hazard and risk perceptions. He also has interests in human-technology interaction and information displays. He is a Fellow of the Human Factors and Ergonomics Society and has been HFES Secretary-Treasurer and a member of the Executive Council. He has also held the offices of Chair and Technical Program Chair of the HFES Safety Technical and Forensic Professional Groups, and the General Sessions Chair and Special Sessions Chair of the Technical Program Committee. He is a member of a number of other professional associations including the American Psychological Association, American Psychological Society, Psychonomic Society, and Sigma Xi. He is on the editorial boards of *Human Factors*, *Ergonomics*, *Occupational Ergonomics*, *Psychology & Marketing*, and *Theoretical Issues in Ergonomic Science*.

David M. DeJoy, Ph.D.
David M. DeJoy is a professor in the Department of Health Promotion and Behavior and Faculty Administrator for Research in the College of Education at the University of Georgia. He received his B.A. and M.A. from the State University of New York College at Geneseo and his Ph.D. from the Pennsylvania State University. Most of this research focuses on the behavioral aspects of injury prevention and control, especially risk perception/communication and the analysis of self-protective behavior. He has published over 50 articles and has served as Chair and Technical Program Chair of the Safety Technical Group of the Human Factors and Ergonomics Society. He is on the editorial boards of *Safety Science* and the *Journal of Safety Research*. He also is a member of the American Psychological Association and the American Public Health Association.

Kenneth R. Laughery, Ph.D.
Kenneth R. Laughery is the Herbert S. Autrey Professor of Psychology at Rice University. He has a B.S. in metallurgical engineering and an M.S. and Ph.D. in psychology, all

from Carnegie-Mellon University. Past employment has included the State University of New York at Buffalo (1963–72) and the University of Houston (1972–84). He is a Fellow of both the American Psychological Association and the American Psychological Society, and a Fellow and Past President of the Human Factors and Ergonomics Society. He has published approximately 120 papers, including more than 40 on warnings and risk perception. His research interests focus on consumer safety, particularly the topics of risk perception and warnings.

Foreword

It seems that warning and risk communications are everywhere. Just within my nice, safe (or so I thought) office, my little bottle of correction fluid warns me that I could kill myself if I deliberately concentrate and inhale the stuff. A desk lamp warns me that I could start another Chicago fire if I use a bulb bigger than 75 watts. My label maker warns me not to pull on the labels to get them out or I can kiss the little puppy goodbye. The cell phone charger dares me to remove the base plate knowing that there are 'hazardous voltages present.' A quick perusal of my weekly news magazines tells me that there are new doubts (risks?) about an old heart monitoring procedure; a certain mutual fund went up 28% last year, but 'past performance does not guarantee future results' (is there a risk?); and a certain drug reduces deaths from heart disease by 42%, but there is a 1% chance (risk?) it will cause liver dysfunction.

Questions regarding how best to communicate warnings and risk information, whether such communications are likely to be effective, and what factors influence the communication process are of importance to a wide range of players in society today. Often these players have different perspectives, values, and vested interests. Manufacturers, consumers, consumer advocacy groups, government organizations, plaintiff attorneys, defense attorneys, and behavior scientists are just some players involved. Also, decisions regarding how, when, and where to warn about a hazard may be based on little or no scientific information. Frequently, politics, expediency, self-interest, and litigation influence such decisions. Probably we shall never eliminate the influence of such factors, but the challenge is to incorporate the best scientific information into the process to ensure the most effective results.

Although there have been books and standards addressing how to design warnings, there has really not been a comprehensive, well organized book summarizing the empirical scientific literature on warnings and risk communication. The area of warnings and risk communication is not a narrow field, but rather encompasses the entire field of experimental psychology, including perception, information processing, decision making, attention, memory, motivation, and personality. Synthesizing this vast field to extract the salient theories and principles as they apply to warnings and risk perception is quite a task, perhaps rather a lot to expect in a single book. This book, however, attempts that feat. The strengths of the book include: a distinguished group of authors, an organizing theoretical model, a review and critique of research methodologies, and a preference for behavioral studies over subjective evaluations as the basis for conclusions.

This book is not a design guide or a 'how to' book. This is a book for those who want to know the whys, whats, and wherefores of warning and risk communication. It is an excellent source of ideas for researchers, or for a graduate student thesis or dissertation. It is a comprehensive source of information for attorneys and expert witnesses who need to explain the effectiveness or lack of effectiveness of a warning, or who wish to analyze the factors in a situation that affect whether a warning will be noticed, received, understood, and heeded. However, in fairness to all, the reader should note:

> **WARNING**
> **This Book May Be Hazardous to Your Ignorance**

MARK S. SANDERS
Northridge, California

Preface

As products, equipment, and environments become more technologically complex, many potential hazards associated with them have become less apparent. One of the ways to prevent accidents involving personal injury and property damage is to warn about them. The purposes of warnings are to inform persons at risk about hazards and to promote safe behavior.

Over the past 15 years or so there has been increasing interest in warnings and in research on the topic of risk communication. One probable reason for the earlier scarcity of warning studies is that research having applicability to real-world hazards is difficult to conduct. The foremost problem is that it is unethical to actually expose participants to hazards while manipulating different kinds of warning systems to see whether they comply. More recently, methodologies have been developed to measure compliance under realistic and safe conditions (without exposing participants to real hazards). Research has also examined aspects of the mental processes that precede compliance or, in other words, the intermediate stages of information processing between exposure to a warning and behavior.

Some of this research has been fueled by three other concurrent concerns. First there has been increasing interest in safety and health, in part because of rising health care costs and the pain and suffering that generally accompany injury and disease. The second interest in warnings derives from legal concerns. In the USA, the adequacy of warnings can play a large role in the initiation and outcome of a law suit. Third, governments and standards organizations around the world have mandated rules or issued guidelines on the design of warnings for a variety of situations.

As a consequence of the newly developed research methodologies, increased interest in safety and health, and the contemporary litigation milieu, there has been an upsurge in warning research. This book reviews, organizes and synthesizes theory of and research into warnings, including applications and applicable law. The broad coverage of warnings in a single volume should make this book of interest to a wide audience.

ORGANIZATION AND CONTENT

The book itself is a multi-authored edited volume with 15 chapters. The chapters are divided into five sections: Introduction, Methods/Techniques, Research on Warnings: Stages of the Model, Practical Issues of Warning Design, and Forensics. A short synopsis of each section is given in the Contents section describing the general purpose of the subsequent chapters. In the following paragraphs, we offer a brief overview of these sections and chapters.

The first section, Introduction, contains two chapters. Chapter 1 by Laughery and Hammond introduces the area of warnings and risk communication, describing its importance and providing a short history of the field. The chapter also discusses the role of warnings in the hazard-control hierarchy. Chapter 2 by Wogalter, DeJoy and Laughery outlines the basic theoretical framework that the book adopts to organize the warnings research literature. The framework combines basic parts of both communication and human information processing models which we have labeled as the C-HIP model. The chapter describes how processing bottlenecks can prevent a warning from producing the desired safe behavior. Then the chapter discusses some of the limitations of a simple, linear model and extends this framework by suggesting that feedback from later stages influences earlier stages and that, under certain conditions, stages may be skipped entirely.

The second section, Methods/Techniques, introduces how research in this area is conducted. Chapter 3, by Young and Lovvoll, describes methods for investigating aspects of the intermediate stages of processing following exposure to a warning. Included are methods of measuring subjective impressions, memory, and eye movements. The limitations of these techniques are also discussed. Chapter 4, by Wogalter and Dingus, describes methods used for measuring behavioral compliance/adherence in a variety of situations including laboratory experiments and field evaluations.

The third section, Research on Warnings: Stages of the Model, comprises roughly half of the book. Each of the seven chapters in this section reviews research pertinent to a stage of the communication-information processing (C-HIP) framework. The first two chapters in this section are taken from the basic communication model, source and channel. The source concerns the originator/transmitter of risk information. There is limited research on this topic, which is rather surprising when one considers that warnings emanate from many sources, such as government, industry, trade associations and nonprofit public service organizations. The perceived credibility (or lack thereof) of the source could add to (or detract from) the impact of the message. Because of the scarcity of research on this topic with respect to warnings, Cox, in Chapter 5, extracts theory and research from social-persuasion theory in discussing potentially relevant factors such as expertise, likeability, trustworthiness, and others.

The channel concerns the way the message is transmitted from a source to the receiver. Warnings can be transmitted through one of several sensory modalities, but usually vision and audition are the central focuses for warnings. Each of the senses has its own characteristic advantages and disadvantages. Consideration of the channel also concerns the kinds of media presentation that are used, delivering information through one or more sensory modalities. Different media can be more or less effective in different situations. Morris and Mazis, in Chapter 6, discuss these issues using research from both the warning and nonwarning domains.

The remaining group of chapters in the third section of the book concerns the processes that occur within the third part of the basic communication model, the receiver. Within the receiver, an effective warning undergoes a series of mental operations as described by a human information processing framework. At the receiver, processing starts with the information's arrival at the senses, and the processing may continue through intervening stages to produce changes in behavior.

Wogalter and Leonard, in Chapter 7, describe the factors important for capturing and maintaining attention. These factors include the characteristics of the message and its surrounding environment. An effective warning will stand out (i.e., be salient) in cluttered and noisy environments. Once attention has been gained it must be maintained at least briefly so that information is transferred.

The next processing stage is comprehension and memory. In Chapter 8, Leonard, Otani and Wogalter describe the factors that facilitate understanding and retention of warning messages. Whether the text and symbols can be understood by the targeted groups are among the issues discussed. Additionally, strategies useful in developing prototype warnings are described. Emphasis is given to comprehension testing as a necessary step in the production of warnings. In addition, factors that influence the encoding, storage and retrieval of warnings in memory are presented.

In Chapters 9 and 10, DeJoy describes the next two stages of the model. The chapter on attitudes and beliefs reviews the literature on topics such as perceived hazard and familiarity. The chapter on motivation describes factors that energize users to comply with the warning-directed behavior, and these include costs of complying and anticipated severity of injury. In both chapters, various individual difference factors are described.

The last stage of the sequence of stages is behavior. Correct, safe behavior is the ultimate desired outcome of a warning. Silver and Braun, in Chapter 11, review the factors that have been shown to influence behavioral intentions and compliance, both positively and negatively.

The last two sections of the book address specific areas of application: one is the development of real-world warnings and the other is legal challenges. The two chapters in the section called Practical Issues of Warning Design give practical guidance on developing warnings. Collins, in Chapter 12, describes the content and process involved in forming standards and guidelines on warnings, and includes an extensive description of selected US government warning-related regulations. Guidelines, standards, and rules do not always provide adequate specification, but they can serve as a basis for initial design prototypes. Generally, testing is needed to verify the effectiveness of prototype warnings. Frantz, Rhoades, and Lehto, in Chapter 13, outline some of the practical methods of producing and evaluating warnings for use in real settings. This information should be particularly helpful to individuals who develop hazard communications for actual applications.

The last section of the book, on Forensics, describes the litigation aspects of warnings in the USA. Chapter 14 by Madden gives relevant US case law for warning design, including potential consequences of failure to warn. This chapter is also formatted in the style often used in legal writings. Laughery, in Chapter 15, describes the role and activities of the expert witness in warning-related litigation, and provides insight into how testimony is aided by a combination of research and analysis.

READERSHIP

We believe that this book will be of interest to several groups of people. One major group will be human factors professionals (ergonomists) who are involved in research, consulting, or expert witness work in legal cases concerned with warnings. Many of these individuals, both academics and practitioners, have been trained in psychology or in industrial engineering, and many of them hold memberships in the following professional organizations: Human Factors and Ergonomics Society, Division 21 of the American Psychological Association (Applied Experimental and Engineering Psychology), Ergonomics Society (UK), Canadian Human Factors Society, International Ergonomics Association, and other country-specific ergonomics organizations. Additionally, we expect the book to be relevant and of interest to: (a) safety professionals, (b) technical communication professionals and documentation writers working with product manufacturers, (c) product designers, (d) persons involved in consumer marketing, and (d) attorneys involved in product liability and personal injury cases. Moreover, there are individuals in other specific areas (e.g., government agencies responsible for labeling and signage for specific products, equipment, and environments) that will find the information in the book useful. We believe the book could be a text in college seminar-type classes (special or advanced topics courses) for graduate students and advanced undergraduates in human factors and ergonomics and in other allied fields.

Readers will achieve maximum benefit if they have had some exposure to basic behavioral science research and associated methodology. However, we have tried to ensure that readers without this background will understand most of the conceptual content. Because some of our target audience will not have knowledge of some technical jargon, we have tried to limit its use and where it inevitably occurs, we have tried to include additional explanation and examples.

ACKNOWLEDGMENTS

We thank Richard Steele, Tony Moore, Luke Hacker, and Rachel Brazear of Taylor & Francis for their support of this project. We are extremely grateful for the efforts and insights of the following colleagues who graciously served as reviewers. They helped make the book better.

John W. Brelsford	Arthur D. Fisk	Dieter W. Jahns
Blair M. Brewster	Alan Heaton	John Lust
Kevin Celuch	Henriette C. Hoonhout	Russell J. Sojourner
Robert E. Dewar	Richard J. Hornick	Harm J.G. Zwaga

Finally, we hope this book will stimulate critical discussion on the current state-of-the-art of warnings, and that it will help generate new and better ideas concerning warning design and effectiveness as well as the methods used for research and application. Most importantly, such progress should produce better methods of communicating risk information that ultimately will reduce the likelihood and extent of personal injury and property damage.

Contributors

Curt C. Braun
Department of Psychology, University of Idaho, Moscow, Idaho 83844-3043.

Belinda L. Collins
Director, Office of Standards Services, National Institute of Standards and Technology, Room A313, Building 226, Gaithersburg, MD 20899.

Eli P. Cox III
College and Graduate School of Business, The University of Texas at Austin, Austin, TX 78712.

David M. DeJoy
Department of Health Promotion and Behavior, University of Georgia, Athens, GA 30602-6522.

Thomas A. Dingus
Department of Industrial and Systems Engineering, 302 Whittemore Hall, Virginia Polytechnic Institute and State University, Blacksburg, VA 24061.

J. Paul Frantz
Applied Safety & Ergonomics, Inc., 3909 Research Park Dr, Suite 300, Ann Arbor, MI 48108.

Amy Hammond
Department of Psychology, Rice University, Sewall Hall, P.O. Box 1892, Houston, TX 77251.

Kenneth R. Laughery
Department of Psychology, Rice University, Sewall Hall, P.O. Box 1892, Houston, TX 77251.

Mark R. Lehto
Department of Industrial Engineering, Grissom Hall, Purdue University, West Lafayette, IN 47907.

S. David Leonard
Department of Psychology, University of Georgia, Athens, GA 30602-3013.

David R. Lovvoll
Department of Psychology, Sewall Hall, Rice University, P.O. Box 1892, Houston, TX 77251.

M. Stuart Madden
School of Law, Pace University, 78 North Broadway, White Plains, NY 10603.

Michael B. Mazis
Department of Marketing, Kogod College of Business Administration, American University, Washington, DC 20016-8044.

Louis A. Morris
PRR Research Division, 17 Prospect Street, Huntington, NY 11743.

Hajime Otani
Department of Psychology, Central Michigan University, Mount Pleasant, MI 48859.

Timothy P. Rhoades
Applied Safety & Ergonomics, Inc., 3909 Research Park Dr, Suite 300, Ann Arbor, MI 48108.

N. Clayton Silver
Department of Psychology, University of Nevada at Las Vegas, Las Vegas, NV 89154.

Michael S. Wogalter
Department of Psychology, North Carolina State University, 640 Poe Hall, Campus Box 7801, Raleigh, NC 27695-7801.

Stephen L. Young
Applied Safety & Ergonomics, Inc., 3909 Research Park Dr, Suite 300, Ann Arbor, MI 48108.

PART ONE

Introduction

This first section introduces the area of warnings and risk communication, describing its importance and providing a brief history of the field. An overview of a communication-human information processing (C-HIP) framework is described that organizes the research literature presented in subsequent chapters.

CHAPTER ONE

Overview

KENNETH R. LAUGHERY AND AMY HAMMOND

Rice University

Recent years have witnessed increased attention to the use of warnings in addressing environmental hazards as well as hazards associated with products. Warnings are considered a third line of defense against hazards, behind design alternatives and guarding. At a general level, warnings are intended to improve safety. More specifically, they are intended to influence people's behavior and to enable more informed judgments and decisions. Two theoretical frameworks have been fundamental to warnings research and design principles. A communications model emphasizes the sender, receiver, channel, and message as factors to be considered. An information processing model focuses on the receiver and defines a series of stages through which warning information must pass successfully in order to be effective. A warning system may consist of several components (messages and media). A number of principles exist for deciding when, what and who to warn as well as the basis for prioritizing warnings.

1.1 INTRODUCTION

In recent years there has been increasing concern for public safety in the USA. This growing concern has been manifested in a variety of ways. Laws at local, state and federal levels have been passed to address safety issues. US government agencies such as the Occupational Safety and Health Administration (OSHA), the Consumer Product Safety Commission (CPSC), the Food and Drug Administration (FDA), and the Environmental Protection Agency (EPA) have taken on responsibilities for public safety in various domains. Regulations and guidelines have been promulgated by public and private institutions to influence environmental and product safety. Still another manifestation of increasing concern or demand for public safety, perhaps of a different type, is the growth in product liability and personal injury litigation. Clearly one of the results of litigation is increased attention to safety in the design of environments to which the public will be exposed and of products that the public will use.

At a somewhat different level, another outcome of greater safety concern is the growing use of safety communications and warnings to inform people of hazards and to provide instructions as to how to deal with them so as to avoid or minimize undesirable consequences. Warnings are used to address environmental hazards as well as hazards associated with the use of products.

A topic that is closely associated with warnings is risk perception; that is, people's knowledge and/or understanding of hazards and their consequences. Risk perception and warnings are closely related, since obviously when and how to warn is a function of the knowledge people have about hazards and the factors that influence this knowledge.

The purpose of this chapter is to provide an introduction or overview of some of the issues, ideas and facts/data that will be presented and discussed in greater detail in the chapters that follow.

1.2 HAZARDS AND RISKS

In this section some terms will be defined and the role of warnings in the broader context of hazard control will be discussed.

1.2.1 Definitions

It is useful to define the terms hazard, danger and risk. These terms are commonly used in the warnings and risk perception literature. There is some variation in how these terms are defined and used, both in the technical literature and by the lay public. Indeed, at times the terms are used synonymously. Nevertheless, the following definitions are typical.

Hazard is defined as a set of circumstances that may result in injury, illness or property damage. These circumstances may include characteristics of an environment, of a product, and/or of a task. They may also include the abilities, limitations, experience, perceptions and knowledge of the person(s) in the system.

Danger is a term that is used in a variety of ways. If something is dangerous, it is believed to have some degree of hazard. One common definition is to view it as a multiplicative function of hazard and likelihood. If one has quantified values for hazard and likelihood, a value for danger would be obtained by multiplying the two quantities. Note, an implication of this definition is that if either hazard or likelihood is valued at zero, there is no danger. If the hazard is serious but will not occur, there is no danger. Conversely, if the probability of the event occurring is high, but there will be no resulting undesirable outcome, there is no danger.

Risk is a term that has had many definitions and has been used in a variety of contexts. For example, often it is used to refer to the probability or likelihood that an undesirable event will occur. *Risk perception* encompasses a broad notion of safety awareness. It concerns the overall awareness and knowledge regarding the hazards, likelihoods, and potential outcomes of a situation or set of circumstances. In the warnings literature, frequently the term *hazard perception* is used in referring to this notion of safety awareness.

1.2.2 Hierarchy of Hazard Control

In the field of safety there is a concept of hazard control that includes the notion of a hierarchy or priority scheme (Sanders and McCormick, 1993). This scheme or hierarchy consists of an ordered set of approaches to or procedures for dealing with hazards. The key elements in this sequence are (1) *design out* the hazard, (2) *guard* against the hazard, and (3) *warn*. The first preference is to eliminate the hazard by an alternative design. If a nonflammable solvent can be used for some cleaning task, such a solution is preferable to wearing protective equipment or warning against using a flammable solvent in a situation where a possible ignition source exists. Obviously it is not always possible to eliminate hazards. Physical or procedural guarding is a second line of defense, and has the purpose of preventing contact between people and the hazard. Barriers and protective equipment are examples of physical guards, while designing tasks in ways that keep people out of the hazard zone is an example of a procedural guard. Like alternative designs, however, guarding is not always possible. The third line of defense is warnings. Warnings are third in the priority scheme because influencing behavior is sometimes difficult, and seldom foolproof. This priority scheme has another important implication; namely, warnings are properly viewed as a supplement to, not a substitute for, other approaches to safety (Lehto and Salvendy, 1995).

The distinction between active and passive approaches to injury control is related to the above hierarchy. Active approaches in this context refers to situations where some knowledge and/or action is required on the part of the person(s) in the system, while the passive approach does not. Three-point manual seat belts in automobiles are an example of an active safety approach in that the occupant is required to fasten the belt. Air bags, on the other hand, are an example of a passive approach. As a general principle, passive approaches to injury control are preferred.

In addition to the three-part hierarchy, there are other approaches that may be effective in dealing with hazards. Generally they are similar to warnings in the sense that they are means of influencing the behavior of people. Training and selection are examples. Supervisor control would be another example. These approaches are especially applicable to hazards in the context of job performance. They are, of course, much more difficult to implement with regard to consumer products.

1.2.3 Perspectives of Hazard Control

The history of hazard control is not so much one of periods in time as it is of different perspectives. For example, in her book *Read the Label*, Susan Hadden (1986) noted that for centuries the implicit doctrine governing consumer products was caveat emptor, or 'let the buyer beware.' This doctrine imputed that consumers would use their intelligence and experience to protect themselves. Perhaps when the world was simpler such a view was more tolerable than it is with today's complex environments and products.

A related issue is that in our culture we often seem to have a predisposition to assume that when an accident occurs it is because someone made a mistake, usually the injured person. In the arena of industrial safety, Heinrich's (1941) work in the 1930s proclaimed that 85% of industrial accidents are caused at least in part by human error. This work had enormous influence on thinking about industrial safety in that a major emphasis in attempting to improve safety in industrial settings focused on influencing and controlling the behavior of employees. While such efforts obviously are an important ingredient in industrial safety programs, it is equally if not more important to design safe work environments and safe equipment. Certainly such a view is consistent with the design-it-out/guard/warn hierarchy discussed above. But even in the context of this hierarchy, one must be careful not to adopt the overly simplified perspective that safety can be viewed as unsafe conditions versus unsafe acts. Circumstances are inevitably more complex. For example, events that appear to be the result of unsafe acts often are the product of poorly designed systems. The concept of 'induced error' refers to such situations. In the USA and other countries, government bodies such as OSHA have played a role in broadening and changing perspectives about work safety.

The point of the above comments is simply to note that different perspectives have characterized people's thinking about hazards and how to deal with them. The perspective here is that influencing people's behavior through warnings has an important place in coping with hazards but, where hazards can be eliminated or guarded against, such approaches should take precedence.

1.3 HAZARD ANALYSIS AND IDENTIFICATION

How do we know what hazards to warn about? Clearly before we can design or evaluate a warning we must know what hazards are associated with an environment or a product. The identification of hazards encompasses a variety of methodologies, many of which are discussed in Chapter 13 (Frantz, Rhoades, and Lehto). Generally, these methodologies fall into two broad categories. The first category consists of a set of analytical tools or procedures that are useful throughout the lifecycle of an environment or product, including the design and development phase. These hazard analysis procedures vary along a number of dimensions such as formality (how quantitative or qualitative they are), logic (inductive or deductive) and focus (product or user oriented). As a rule, the procedures have in common noting and analyzing various circumstances that will or might arise and identifying hazards associated with these circumstances. In the case of a product, the circumstances would encompass not only intended uses, but foreseeable misuses as well. For example, a chair may not be intended as a device to stand on, but clearly such uses occur and should be considered. The various analytical techniques have different names or labels, some of the more common being fault-tree analysis and failure-mode analysis.

The second category of hazard analysis methods consists of analyzing actual environment or product use outcomes, especially accident data. Much has been written on the subject of accident analysis, and it is not within the scope of this chapter or this book to address the topic. The point is that such data provide important opportunities to identify and understand hazards, often beyond those that may be identifiable through the analytic procedures noted above. Numerous accident data bases exist. One example is the National Electronic Injury Surveillance System (NEISS) compiled by the Consumer Product Safety Commission. This database provides information about injuries associated with the use of a large number of consumer products. A second example is the Fatal Accident Reporting System (FARS) compiled by the Department of Transportation, which contains information regarding highway accidents that involve fatal injuries.

From a warnings perspective, it is important to keep in mind what the questions and issues are that one is trying to address in identifying and analyzing hazards. The key question is 'What information or knowledge does the person(s) in the system need to function safely?' Other relevant questions include: 'What are the characteristics of the person(s) in the system relevant to this information and knowledge?' 'What do they already know?' 'What are they capable of understanding?' 'How can the information needed best be displayed or transmitted?'

1.4 HAZARDS AND RISKS: PERCEPTION AND KNOWLEDGE

Risk perception is a term that refers to people's perception, awareness and knowledge of hazards, including potential consequences, associated with a situation or set of circumstances. For good reviews of this topic see Fischhoff (1989) and Slovic *et al.* (1982). The concern here is to note how risk perception considerations enter into decisions regarding the design, implementation and effectiveness of warnings.

An important factor in considering the hazards associated with any situation or product is the perception or knowledge of the people involved. Obviously the information people have from past experience or that they derive from the existing situation or circumstances is relevant to decisions about when, where, what and how to warn. However, an understanding of people's knowledge about a particular hazard or situation is not always an adequate basis for making warning decisions. An important consideration is what knowledge they have available at the time it is needed. This distinction concerns knowledge versus awareness.

1.4.1 Knowledge Versus Awareness

The distinction between knowledge and awareness is important in understanding issues of risk perception and how they map on to the design and effectiveness of warnings. The difference is analogous to a distinction made in cognitive psychology between short term memory (this may be thought of as what is in consciousness) and long term memory (one's more permanent knowledge of the world). The point is simply that people may have information or experiences in their overall knowledge base that at a given point in time is not what they are thinking about (i.e., they are not aware of or conscious of that information). In the context of dealing with hazards, it is not enough to say that people know something. Rather, it is critical that people be aware of (thinking about) the relevant information at the right time. This distinction has significant implications for one of the important functions of warnings: they serve as *reminders* or cues which help access that information stored in memory.

The issue of awareness is probably more important than it may seem at first glance. One of the important characteristics of people that is relevant to their role in systems concerns limited attention capacities. We are not capable of attending to many things simultaneously. Indeed, a reasonable working assumption for systems design is that people should not be required to attend to more than one thing at a time. Without dwelling on what is the appropriate number in this regard, the implication is that even though people may have knowledge of some hazard, warnings may be necessary to draw their attention to the hazard at the critical time. This requirement may be especially important when there are other factors simultaneously vying for their attention. Anyone who has analyzed industrial accident reports probably has encountered numerous accounts or causal explanations that refer to 'not paying attention'. Often, a detailed analysis of the circumstances reveals that the problem was quite the opposite: specifically, the person had too many things to attend to and his/her attentional capacity was exceeded.

1.4.2 Sources of Hazard Knowledge

There are many ways in which people become aware of and knowledgeable about hazards, consequences, and appropriate procedures or behavior. Warnings are among them. There are others. Experience, of course, is one way people acquire safety knowledge. Certainly 'learning the hard way' by having experienced an accident or by knowing someone else who has can result in such knowledge. Such experiences, however, do not necessarily lead to accurate knowledge, as they may result in overestimating the degree of danger associated with some situation or product. Similarly, the lack of such experiences may lead to underestimating such dangers or not thinking about them at all. Nevertheless, experience clearly plays an important role in risk perception.

Another source of information about dangers is the situation or product itself. In law there is a concept of 'open and obvious.' This concept refers to the notion that the appearance of a situation or product or the manner in which it functions may communicate the nature of the hazard. Moving mechanical parts such as chain-driven sprockets may be an example of an open and obvious pinch point hazard. Even more obvious may be the hazard and consequence of a fall from a height in a construction setting. Of course many safety problems are not open and obvious. Examples here would be chemical based products like solvents and pesticides. Ingestion, inhalation and skin contact hazards often associated with such products will seldom be in the category of open and obvious.

A final point regarding risk perception concerns the problem of overestimating what people know or are aware of. To the extent that it is incorrectly assumed that people have information and knowledge, there may be a tendency to provide inadequate warnings. Thus, it is an important aspect of job, environment and product design to take into account people's understanding and knowledge of hazards and their consequences. Indeed, as noted in the earlier section on hazard analysis, an important factor to consider in such analyses is what people know. A further discussion of this issue can be found in a paper by Laughery (1993).

1.5 SAFETY COMMUNICATIONS: WARNINGS

1.5.1 Purpose of Warnings

The purpose of warnings can be addressed at several levels. Most generally, warnings are intended to improve safety, i.e., to eliminate or reduce incidents that result in injury, illness or property damage (e.g., warnings not to use a medication in circumstances where alcohol has been consumed). At a different level, warnings are intended to influence people's behavior in ways that will improve safety (e.g., warnings to wear protective equipment when playing contact sports or handling a toxic chemical). At still a third level, warnings are intended to provide information that enables people to understand hazards, consequences and appropriate/inappropriate behavior which, in turn, enables them to make *informed* decisions (e.g., warnings on cigarette packages).

The emphasis in the third point above is on informed decisions or informed choice. People may opt not to follow the instructions provided by the warning, but rather 'take the risk.' However, if the warning is effective, the decision will be made on the basis of adequate information. In this regard, a warning is successful if the information is properly transmitted and received. This point places warnings squarely in the category of a communication, which of course they are.

There are two additional points to be noted regarding the purpose of warnings that are related to warnings as communications. First, warnings are a means of shifting or assigning safety responsibility to the people in the system, the product user, the worker, etc., in situations where hazards cannot be designed out or adequately guarded. This point is not intended to imply that people do not have responsibility for safety independent of warnings, of course they do. Rather, the purpose of warnings is to provide the information necessary for them to carry out such responsibilities. The second point regarding the communication purpose of warnings concerns an issue that has received little attention in the technical literature, namely, people's right to know. The notion is that even in situations where the likelihood of warnings being effective may not be high, people have the right to be informed about safety problems confronting them. The hazard communication standard promulgated by OSHA places considerable emphasis on informing workers about hazards they may encounter in the workplace. Obviously, this right-to-know aspect of warnings has personal, societal and legal dimensions. Nevertheless it is a matter related to the overall purpose of warnings.

1.5.2 Theoretical Frameworks

Historically, the research, analysis, and principles of design regarding warnings have been driven by one of two theoretical frameworks, a *communications model*, and an *information processing model*. A typical communications model includes a sender, a receiver, a channel or medium through which a message is transmitted, and the message. The receiver is the user of the product, the worker, or any other person to whom the safety information must be communicated. The message, of course, is the safety information to be communicated. The medium refers to the channels or routes through which information gets there.

The information processing model focuses on the receiver. It defines a series of stages through which the warning information must pass successfully in order for it to be effective. Generally these stages include attention, comprehension, attitudes and beliefs, motivation and behavior. The usual logic of this approach is that the information flow and effect are serial, and a failure at any one stage results in a failure of the warning. If the warning is not noticed, it fails; if it is not understood, it fails; and so forth. Obviously, such logic is incorrect in that it ignores considerations such as feedback loops and the possibility that some stages simply may be skipped in some situations.

In Chapter 2 by Wogalter, DeJoy, and Laughery the communication model and information processing model are integrated into a single theoretical framework. This framework in turn serves as an organizing context for most of the remainder of the book.

1.5.3 Concept of a Warning System

The notion of a warning being a sign or a portion of a label is much too narrow a view of how such safety information gets transmitted. The concept of a warning system is that a communication for a particular setting or product may consist of several components. These components may include a variety of media and messages. An example or two can help make the point.

A warning system for a product off the drug store shelf, such as an antihistamine, may consist of a number of components; a printed statement on the box, a printed statement on the container (bottle), and a printed package insert. It may also include warnings in print ads and verbal warnings in television commercials about the product. Similarly, a warning system for tires and rims that may be mismatched during mounting with a resulting potential explosion might have a number of components. Examples here are: warnings in raised lettering on the sidewall of the tire, tread labels on new tires, stickers or stamping on the rim, statements on wall posters in shops or stations where tires are mounted, statements in tire and rim product catalogs and manuals, statements in handouts that accompany sales of tires and rims, verbal statements by employers of people who mount tires, etc. A third example would be warnings for a pesticide used in agricultural settings. The components here might include printed on-product labels, printed flyers that accompany the product, statements in advertisements about the product, verbal statements from the salesperson to the purchaser, and material safety data sheets provided to the employer.

An important point regarding warning systems is that the components may not be identical in terms of content or in terms of purpose. Some components may be intended to capture attention and direct the person to another component where more information is presented. An example is 'See the owner's manual for more information'—a common statement on visor warnings about restraint systems in vehicles. Similarly, different components may be intended for different target audiences. In the example of the pesticide given above, the label on the product package may be intended for everyone associated with the use of the product including the end user (farm worker), while the information in the material safety data sheet may be directed more to the safety professional working for the employer. A final example concerns prescription medicines where one warning may be directed to the prescribing physician (such as the information in the Physicians Desk Reference) while another is intended for the patient (such as the flyer accompanying the medicine). Clearly these components may differ in content due to their different purposes and the different characteristics of the two target audiences.

1.5.4 General Criteria for Warnings

The most important rule or criterion for warnings is that their design should be viewed as an integral part of the overall system design process. Although, when dealing with hazards, warnings are a third line of defense behind design and guarding, they should not be considered for the first time after the design (including guards) of the environment or product is fixed. Too many warnings are developed at this stage, the afterthought phenomenon, and often their quality and effectiveness reflect it.

In this section, criteria or guidelines are presented addressing three warnings issues: (1) when/what to warn; (2) how to prioritize warnings; and (3) who to warn.

When/What to Warn. There are several principles or rules that guide when a warning should be employed. They include: (i) a significant hazard exists; (ii) the hazard, consequences and appropriate safe modes of behavior are not known by the people exposed to the hazard; (iii) the hazards are not open and obvious; and (iv) a reminder is needed to assure awareness of the hazard at the proper time.

Prioritizing Warnings. The concern here is what hazards to warn about when multiple hazards exist. How are priorities determined in deciding what to include/exclude, how to sequence them, or how much relative emphasis to give them? To some extent the criteria overlap with the above rules about when/what to warn. Clearly, when the hazard is known and understood or when it is open and obvious, warnings may not be needed. Other considerations include likelihood, severity, and practicality. (a) With likelihood, the more likely an undesirable event, the higher the priority for warning. (b) With severity, the more severe the potential consequences of a hazard, the greater the priority that it should be warned. (c) With practicality, there are circumstances when limited space (a small label) or limited time (a television commercial) does not permit all hazards to be addressed in a single or primary component of the warning system. In such situations, hazards with lower priority often are addressed in secondary components such as package inserts or manuals.

Who to Warn. The general principle regarding who should be warned is that it should include everyone who may be exposed to the hazard and everyone who may be able to do something about it. There are occasions when people in the latter category may not themselves be exposed to the hazard. An example would be the industrial toxicologist who receives warning information about a product to be used by employees and uses the information to define job procedures and/or protective equipment to be employed in handling the material. The physician who prescribes medications that have contraindication and side-effect hazards is another example. Other warnings may be directed to a very specific audience. Warnings about toxic shock syndrome in the use of tampons would be directed primarily to women of child bearing age. Warnings about contraindications associated with prescription medications, as noted above, may be directed to physicians. There are, of course, situations and products where the target audience is the general public. Equipment in a public playground or most products on the shelf of a drugstore or hardware store are examples. If warnings are to be effective, it is imperative that the characteristics of the target audience be taken into account.

Clearly, target audiences, the receivers of warnings, may differ. Laughery and Brelsford (1991) discussed several dimensions along which intended receivers may differ. One of the most relevant factors is familiarity/experience with the situation or product. This factor has been researched extensively in the context of warnings, and will be discussed in later chapters. Another factor is competence. Competence may be general or specific. General competence includes considerations such as intellectual abilities and language or reading abilities. Specific competence would include knowledge about a technical area such as a physician and medications or a mechanic and engines. A common problem in warnings is technical information being provided to audiences that do not have the specific technical competence to comprehend it. There are also demographic factors such as age and gender that may be relevant to the design of warnings.

There are four general principles that apply when taking receiver characteristics into account in the design of warnings.

Principle 1. Know thy receiver. Gathering information and data about relevant receiver characteristics may require time, effort and money, but without it the warning designer, and ultimately the receiver, will be at a serious disadvantage.

Principle 2. When variability exists in the target audience, design warnings for the low-end extreme. Do not design for the average. If you design for the average, you may miss half the audience.

Principle 3. When the target audience consists of subgroups that differ in relevant characteristics, consider employing a warning system that includes different components for the different subgroups. Do not try to accomplish too much with a single component.

Principle 4. Market test the warning system. Despite the designer's knowledge of receiver characteristics and efforts to apply that knowledge, warnings generally should be market tested. Such tests may consist of 'trying it out' on a target audience sample to assess comprehension and behavioral intentions. This principle will be addressed in Chapter 4 by Wogalter and Dingus.

1.6 A FEW HISTORICAL COMMENTS

In this chapter we have attempted to provide a brief overview regarding topics and issues associated with warnings. In this regard, the chapter has been intended to set the stage for the remainder of the book. It seems appropriate in this final section to provide some brief comments on a broader historical perspective of warnings.

No one knows when the first warning sign appeared, but it is likely that the occasion was during ancient times. Publications concerning warnings can be dated to the early part of the 20th Century, and guidelines or recommendations were in print at least by the 1920s. However, much of the body of technical literature reporting research and theory about warnings has appeared during the past two decades. Our own analysis of the literature indicates that the late 1970s witnessed a noteworthy upsurge in such publications.

The types of issues and questions addressed also have broadened. During the late 1970s and early 1980s much of the research focused on issues of design. Where to put them, how big, what color, what signal word, and what reading level are typical of the kinds of questions studied. Dependent measures such as noticeability and comprehension were commonly employed. In the mid 1980s the issues being addressed broadened to encompass concerns about effectiveness, that is, the conditions or circumstances in which warnings do or do not make a difference. Dependent measures included behavioral intentions and actual behavior. Also, the work began to be placed into theoretical contexts such as communications theory and information processing theory. Still another development has been an increasing interest in the past decade or so on the use of pictorials in warnings, and the research has clearly reflected this evolving interest. In the following chapters these issues and others are presented along with the many research outcomes.

There have been increased efforts in recent years to establish regulations, guidelines and recommendations for warnings. These efforts have included government regulations that carry the force of law. Examples of agencies that have been involved include the Food and Drug Administration, the Environmental Protection Agency and the Consumer Product Safety Commission. Nongovernment organizations such as the American National Standards Institute, Underwriters Laboratories, and a variety of trade associations have promulgated guidelines. Chapter 12 by Collins presents a discussion of these efforts and the existing standards.

A final development to be noted concerns the litigation process in the USA. Personal injury and product liability litigation has increasingly involved warnings as an issue in lawsuits. Human factors specialists, psychologists, and communications specialists have been involved in such litigation as expert witnesses in increasing numbers. Clearly, this development has resulted in greater attention to issues of warnings in at least two ways. First, obviously it has led designers and manufacturers to be more concerned about warnings. Second, it has highlighted the need for more and better scientific research and information regarding the various issues associated with warnings. In part, the increased research activity has been in response to this need. The manner in which warnings fit into the litigation process and the role of the warnings expert are discussed in Chapter 14 by Madden and Chapter 15 by Laughery.

REFERENCES

FISCHHOFF, B. (1989) Risk: A guide to controversy. In *Improving Risk Communication*. Washington DC: National Research Council.
HADDEN, S.G. (1986) *Read the label*. Boulder, CO: Westview Press.
HEINRICH, H.W. (1941) *Industrial Accident Prevention*, 2nd Edn. New York: McGraw-Hill.
LAUGHERY, K.R. (1993) Everybody knows: or do they? *Ergonomics in Design*, July, 8–13.

LAUGHERY, K.R. and BRELSFORD, J.W. (1991) Receiver characteristics in safety communications. In *Proceedings of the Human Factors Society 35th Annual Meeting*. Santa Monica, CA: Human Factors and Ergonomics Society, pp. 1068–1072.

LEHTO, M.R. and SALVENDY, G. (1995) Warnings: a supplement not a substitute for other applications to safety. *Ergonomics*, 38, 2155–2163.

SANDERS, M.S. and MCCORMICK, E.J. (1993) *Human factors in engineering and design*, 7th Edn. New York: McGraw-Hill.

SLOVIC, P., FISCHHOFF, B., and LICHTENSTEIN, S. (1982) In KAHNEMAN, D., SLOVIC, P., and TVERSKY, A. (eds), *Judgment Under Uncertainty: Heuristics and Biases*. Cambridge: Cambridge University Press, pp. 463–489.

CHAPTER TWO

Organizing Theoretical Framework: A Consolidated Communication-Human Information Processing (C-HIP) Model

MICHAEL S. WOGALTER
North Carolina State University

DAVID M. DEJOY
University of Georgia

KENNETH R. LAUGHERY
Rice University

Much of this book is organized around a sequential or stage model of the warning process that incorporates aspects of two existing models: the communication and the human information processing frameworks. Processing begins with the presence of warning (or other) information. From communication theory, the model takes the concepts of source, channel, and receiver. From information processing theory, the model decomposes the receiver component into the stages of attention, comprehension, attitudes and beliefs, motivation, and behavior. The receiver must notice the information and understand it. The message must be consistent with the person's attitudes and beliefs, or sufficiently persuasive to change them and to motivate the person to carry out the directed behavior (i.e., comply with the warning). This model is useful in (a) organizing the substantial amount of warning-related research that has been generated in the last 15 or so years, and (b) pinpointing the reason or reasons why a specific warning failed to produce adequate levels of behavioral compliance. Although earlier models describe the information processing stages as an invariant linear sequence, this chapter puts forth the proposition that later stages can influence earlier stages through feedback loops, and that in some instances entire stages can be skipped.

2.1 INTRODUCTION

When hazards are associated with products, equipment, and/or environments, steps must be taken to produce a system that will minimize injuries to people and damage to property. Chapter 1 by Laughery and Hammond describes several basic steps that should be carried out. These steps, in order of priority, are (a) eliminate the hazard through design changes or other modifications, (b) physically or procedurally guard against the hazard, and (c) warn those at risk about the hazard. Warnings, therefore, carry a heavy burden in situations where a hazard cannot be eliminated or adequately guarded at its source. Warnings are intended to keep people from engaging in unsafe behavior, and often this involves rerouting or stopping people from doing what they would otherwise do. The complexity and difficulty of modifying human behavior are substantial, but a considerable amount of psychological research shows that safety related behavior can be changed by warnings and that there are a number of factors that influence the success or failure (effectiveness) of warnings. In this chapter, we describe a model or theoretical framework for classifying and exploring the various factors influencing warning effectiveness. This chapter describes the model in general terms while subsequent chapters provide detailed discussions of each stage.

2.2 ORGANIZING THE LITERATURE

A considerable body of warning-related research has been reported in the last 15 years or so. This research has made use of a broad array of techniques and performance measures, and because of its diversity, organizing this work is a challenging task. We employ a hybrid or composite model involving multiple stages to help pull this literature together. This model combines the basic communication model with the human information processing framework. A representation of this communication-human information processing (C-HIP) model is shown in Figure 2.1.

From the communication model, C-HIP takes three major components: source, channel, and receiver. The first two of these components are reviewed in Chapter 5 by Cox and Chapter 6 by Mazis and Morris, respectively, and the third stage of the communication model, the receiver, is the connecting point for the human information processing model. In other words, the receiver stage of the communication model is the superordinate category that incorporates a number of information processing stages: attention/noticeability, comprehension and memory, attitudes and beliefs, motivation, and behavior. These stages are discussed in detail in Chapters 7 through 11. Although these two frameworks are fairly standard and derive from the well established disciplines of communications and cognitive psychology, we do not know of any theoretical treatment that has combined them into a single consolidated model.

It should be pointed out that existing, extensive research and theory associated with the two frameworks has produced many refinements. The model depicted in Figure 2.1 is somewhat simplified and idealized for heuristic purposes. For example, the model in the figure does not show the basic concept of 'noise' and how it affects the communication process. Noise (random changes to the message) can affect any of the stages and is an important element of the communication process. We did not include noise and other potential elements in the model because it would make the figure unnecessarily complex. We have chosen instead to address these nuances and refinements in the discussions of

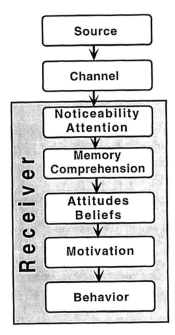

Figure 2.1 Communication-human information processing (C-HIP) model.

the specific stages of the model. For example, the effects of background noise coincident with auditory warnings and cluttered surroundings frequently associated with visual warnings is discussed in Chapter 7 by Wogalter and Leonard. Likewise, many details about human information processing are omitted from the figure but are discussed within the chapters. We believe the C-HIP model captures a broad range of relevant warning-related processes in a simple and straightforward representation, and it is useful in organizing the diverse factors that influence warning effectiveness.

2.3 THE C-HIP STAGES

In the following paragraphs, we offer a brief overview of each stage of the model. Also, these overviews provide a preview of the upcoming chapters (5–11).

The *source* is the originator or initial transmitter of hazard and risk information. Characteristics of the source influence the effectiveness of the warning. There are many possible sources, such as manufacturers, the federal government, nonprofit public service organizations, and industry trade organizations. The perceived credibility (or lack thereof) of the source may add to (or detract from) the impact of the message. Because there is so little source-related research in the warnings domain, Cox in Chapter 5 extracts theory and research from the communication and social-persuasion literatures in discussing potentially relevant factors such as expertise, likeability and trustworthiness, among others.

The *channel* concerns the way the message is transmitted from the source to receivers. Warnings can be transmitted through one or more sensory modalities: visual, auditory, kinesthetic, olfactory, and so forth. The channel also involves the media used to present the material. Depending on the medium, more than one sensory modality might be involved. A video warning, for example, could relay information to both the auditory (nonverbal alarms, speech) and visual (alphanumeric text, pictorials) senses. Each of the senses has its own characteristics that could be considered to be beneficial or disadvantageous depending on the message, the environment, and tasks involved. In other words, different media might be more or less effective in different situations. Long complex messages are not conveyed well via the auditory channel because they may overwhelm attentional capacity and memory. Long warning messages are not a good idea in the first place but, if used, they can be conveyed more effectively through the visual print medium (assuming attention is given to them). Short, easy-to-understand messages are quite effectively conveyed by voice. Mazis and Morris in Chapter 6 discuss these issues using research from both the warning and non-warning domains.

The next group of chapters focuses on the processes that occur within the receiver. A sequence of mental operations starts with the information's arrival at the senses. The receiver's first operation is *attention*. Wogalter and Leonard in Chapter 7 discuss the factors important for capturing and maintaining attention, which include the characteristics of the message itself and its immediate surroundings. Context or background factors are important because they enable the warning to stand out (i.e., be salient, prominent, conspicuous). In addition to the composition of the warning itself, other situational or environmental variables can influence *noticeability*, including physical location, stress level, and ambient noise conditions, among others. Once attention is captured, it needs to be maintained to extract information. Factors that facilitate the maintenance of attention include legibility and brevity.

The next processing stage is *comprehension and memory*. Chapter 8 by Leonard, Otani and Wogalter describes the factors that facilitate understanding and retention of warning messages. Issues such as whether warning message text and pictorial symbols can be understood by the targeted group are examined. Strategies that can be useful in developing prototype warnings are described. Comprehension testing as a necessary step in the development of warning messages is emphasized. In addition, factors that influence comprehension, including storage and retrieval of warnings from memory are presented.

The next two stages of the model are *attitudes and beliefs* (Chapter 9) and *motivation* (Chapter 10). These two chapters, discussed by DeJoy, describe various potentially relevant individual-differences factors. Research on the highly influential factors of perceived hazard and familiarity is described. The motivation chapter describes factors that energize users to comply with warnings, and these include cost of compliance, explicit consequences, and anticipated injury severity.

The last stage is *behavior*. Correct, safe behavior is the ultimate desired outcome. Silver and Braun in Chapter 11 review the factors that have been shown to influence behavioral intention and compliance, both positively and negatively.

2.4 BOTTLENECKS AT THE STAGES

At the outset of this chapter, we noted that other kinds of hazard control techniques are preferred over warnings. If the hazard can be eliminated or guarded, then these measures ought to be incorporated into the system before warnings are considered. Generally, these more direct control strategies are more reliable than warnings in preventing harm. Warnings are best used to handle residual risk, or that which remains after reasonable design and engineering measures have been taken. Warnings are inherently less reliable because of inherent limitations and complexities of human beings.

Another major purpose of the C-HIP model is that it helps to identify potential points of failure. The model can help explain how a warning message might fail to promote safe behavior. Before safe behavior can occur, the warning information must pass through several points or stages. This path is traced by the linear route from the source stage to the behavior stage as shown by the downward arrows in Figure 2.1.

Ignoring feedback for a moment, for a warning to be effective in influencing behavior, information must pass through each of the preceding stages. In a nutshell, the process starts with the warning information moving from the source through some channel to arrive at the receiver. The receiver must then notice and attend to the warning. Once it has been attended to, it must be understood, and the information must, in turn, be consistent with the person's attitudes and beliefs. Motivation is the last stage before behavior is achieved. Sufficient motivation must be present or induced to produce the appropriate behavior.

As described thus far, the C-HIP model proceeds in a linear, temporal sequence. However, each stage of the model is a potential 'bottleneck' that could prevent the process from being completed. If the source does not communicate a warning about a hazard, then clearly, persons at risk will not receive the information and there will be no subsequent behavior change (assuming there are no other opportunities to acquire information about the hazard from other sources). Even if the source attempts to convey a warning, the warning could be ineffective if the channel used to transmit the message is inappropriate or inadequate. Again, hazard information transmitted but not received produces little or no processing in persons at risk. Suppose the source does transmit the warning information and it moves successfully through one or more appropriate channels. The warning could be unsuccessful if the receiver does not attend to it. This end result is the same as a warning that was never transmitted by the source or one that was sent using an inappropriate channel. As a consequence, the information will not move forward to any subsequent information processing stages in the receiver.

To be effective, a warning needs to capture and maintain attention. But even if the warning is attended to, it may not be effective if the message is not understood. Merely examining and reading the warning does not necessarily mean that people comprehend it. People must understand the meaning of the printed words and symbols (i.e., properly interpret the printed language and graphics) comprising the message. Of course, we are assuming that the basic content of the warning message itself is adequate for the task at hand. However, even if the information is understood, the process will go no further if the message does not fit with the person's current beliefs and attitudes. For processing to continue in the face of antagonistic attitudes and beliefs, the warning itself must be sufficiently persuasive to change or overcome those beliefs and attitudes. Failing this, the processing stops prematurely before behavior change. But even if the person believes the message, the message still may be inadequate if it does not motivate or energize the user to perform the appropriate safe actions.

Thus, the C-HIP model shows that each stage in the sequence is a potential bottleneck that could cause processing to stop, thus hindering the warning from ultimately modifying behavior. Chapters 5–11 describe in greater detail the factors that influence warning effectiveness both positively and negatively at each stage of the model.

2.5 MODEL AS AN INVESTIGATIVE TOOL

The C-HIP model is useful also as an investigative or diagnostic tool for discovering why a particular warning does not fulfill the goal of promoting safe behavior. For example, when a warning for some consumer product, piece of industrial equipment, or hazardous environment fails to produce adequate levels of safe behavior, it could be that it lacks sufficient salience (i.e., it fails to be noticed and attended to). One solution might be to add or change features that increase the warning's conspicuousness. The warning might also have failed because people did not understand it. Making the warning more understandable to the target audience might remedy the low compliance rate.

The model can help differentiate which of the stages is causing a bottleneck. For example, it might be noted by a manufacturer that a warning is failing to influence behavior. The manufacturer might assume that the failure is due to a lack of warning conspicuousness resulting in a decision to enhance its prominence. However, this change might not solve the problem. Using measurements assessing attention to the warning, it might be found that virtually all people noticed the warning (so therefore the lack of conspicuousness is not the root of the problem), but rather the warning failed because people did not understand the message. Another example is that people might see and understand the message (as assessed by attention and comprehension measures), but just do not believe the message. Through systematic testing one can find out why a warning is not working. Thus the C-HIP model provides a framework for systematically analyzing why a particular warning application failed to produce its intended effects.

Typically, after a warning is found not to work in the field, most attempts to remedy the problem involve either adding prominence-type features or altering the content of the warning message. These particular fixes will be helpful only to the extent that the limited effectiveness is related to the warning not being noticed or to some critical piece of information not being present. However, as noted above, it is possible that the warning is adequate in terms of both salience and comprehension, and the reason for the low rate of compliance is traceable to discordant attitudes and beliefs with respect to the message being conveyed or inadequate motivation to carry out the directed behavior. In such cases, the obstacle is at the beliefs and attitudes stage or the motivation stage. For example, a person may ignore or discount the warning message because they believe that it does not apply to them personally. This perception might arise from being highly familiar with the task, activity, or environment in question and confident that any related hazards pose very little personal risk. When such discordances in beliefs and attitudes exist, the warning needs to be sufficiently persuasive to convince these individuals to take note of and heed the warning.

Finally, a warning may be physically apparent, understandable, and consistent with beliefs and attitudes, but it still might not be behaviorally effective if it does not motivate people to exert the effort to comply with it. In such situations, the warning might be inadequate in terms of conveying how badly they could be hurt or the effort required to comply may be greater than people are willing to expend in this particular situation. Beliefs or expectations about threat provide much of the initial motivation for compliance, but compliance might ultimately be a cost–benefit decision, in which the benefits of compliance (typically injury prevention) are weighed against the costs or barriers associated with performing the indicated precautions.

Thus, the model can help pinpoint the reasons for the failure of a warning to produce the desired end result: safe behavior. This model can be particularly useful in applied settings where determining the cause of the failure and then rectifying it needs to be targeted precisely and cost-effectively. With knowledge of the factors that influence each stage of the model, and a little detective work, the model can be used retroactively to diagnose and remedy failures. It can also be used proactively to guide the design of new warnings.

2.6 FEEDBACK PROCESSES: INFLUENCE OF LATER STAGES ON EARLIER STAGES

Up to this point, we have described warning information as flowing through a linear sequence of stages. Original conceptions of the warning process have advanced the simplistic view that for warnings to influence behavior the information must go successfully through each of the stages. For the most part, this early version of the model was a logical perspective. In order to read a product label, the person needs to have noticed it in the first place. To understand and remember the warning, one must have examined (read) it. In other words, certain types of processing must logically occur before others. Indeed there is early research (e.g., Strawbridge, 1986; Friedmann, 1988; Otsubo, 1988) reviewed by DeJoy (1989) that shows systematic declines in the percentage of people seeing the warning, reading and remembering the warning, and behaviorally complying with the warning. These results support the simple, linear model with bottlenecks. The decrements are caused by processing being impeded at different stages, which decreases the percentage of people who ultimately comply with the warning. While this model appears to concur with some data, the simple, linear conception of the warning process is almost certainly not true for several reasons.

First, some of the data used to support the simple linear model may not have measured what they purported to measure. Although behavioral compliance and memory were measured objectively in the above cited studies, the measures of seeing and reading the warning were derived from data collected by post-task questionnaires. Post-experimental questionnaires such as these can sometimes be inaccurate in reflecting what actually occurred during exposure to the warning. Did people who said they saw and/or read the warning really do so? Extraneous factors such as participants' interpretation of the questions, whether they actually can remember what they did and when they did it, social desirability, and other demand characteristics can all affect how people answer the questions. Objective measures such as eye movement recording or looking behavior would provide more objective assessments of seeing and reading behavior. With objective measures of performance, we would be more confident of these stage-related decrements, and would have a firmer handle on what actually occurred once the research participants were exposed to the warning.

A second problem with the simple linear model is that it assumes that the perceptual and cognitive processing of warnings occurs within a single (or short duration) point in time upon initial exposure. The individual was essentially viewed as a passive recipient of the warning. We take a broader and more interactive view of the processing that occurs, including the fact that people have different levels of preexisting knowledge and experience. Prior to being exposed to a particular warning, people may have varying levels of familiarity with the tasks and environment involved and may have been exposed to information related to the hazard from multiple sources. These factors (and others) enter into the equation of the warning process, and consequently make it more complex than the simple linear model would suggest.

A third problem with the simple linear model is that later stages can influence how warning information is processed at earlier stages. This is illustrated in Figure 2.2 by the arrows pointing back from the later stages to the earlier stages of the model. These pathways are feedback loops. These additions to the model are related to the second problem cited above—that people's preexisting knowledge and experience often influence how warning information is processed at a given point in time. Two examples of this feedback mechanism will serve to illustrate this. First, repeated exposure to a particular

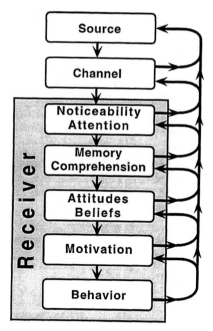

Figure 2.2 Communication-human information processing (C-HIP) model with feedback loops illustrating that later stages influence earlier stages.

warning creates memory. With enough exposure, the warning stimulus becomes habituated, and this reduces the likelihood that a person will look at the warning in the future. Here, preexisting knowledge affects attention or, in other words, a later stage (comprehension and memory) influences processing of an earlier stage (attention). A second example of feedback concerns the effects of beliefs and attitudes on attention. Individuals who assume something is safe may not look for a warning. Even if they notice one, they may not examine it further. Here again, preexisting knowledge, a later stage, influences attention, an earlier stage of processing. Although we have noted only two examples on how later stages of processing affect earlier stages, we believe that all of the stages probably influence each other. The major point is that in most instances the information flow through the model's stages is neither simple nor linear.

2.7 BYPASSING STAGES

The simple linear model is limited also in that it requires the warning information to be processed at each stage before compliance is achieved. It is possible that not all of the stages are needed for safe behavior to occur. As we noted in the above section, people have different levels of preexisting knowledge about the hazard and the warning material. They might have heard about the hazard in the media or from their work supervisor prior to coming into contact with a particular warning. For someone who has some knowledge of the hazard and/or the warning message, the warning stimulus itself might serve simply as a timely cue that elicits safe behavior without going through much further processing. That is, complete processing of the warning is not necessary to produce the desired end result. For example, highly knowledgeable individuals might only need to catch a glimpse of a pictorial symbol (and no other parts of the warning) and know what they need to do.

This is true for a sign containing a directional arrow on the roadway or 'slippery when wet' self-standing floor placard. In the first case, you go in the direction that the arrow is pointing and in the other case you avoid the area. While there may be printed material on the sign accompanying the pictorial symbol, one does not have to read it, or read it completely, to know what to do. Therefore, in some situations some people might not fully examine a warning but still engage in the safe behavior. This analysis also suggests the possibility that the actual rate of compliance with a warning could be greater than the number of people who actually read the warning. Thus the 'funneling down' process suggested by the traditional linear model may not always occur.

2.8 SUMMARY AND IMPLICATIONS

This chapter has provided a brief overview of the C-HIP model. The model is useful in organizing the diverse warnings research literature that will be described in more detail in Chapters 5 through 11. We have shown how the model is useful in determining why a warning might fail to achieve the goal of changing behavior, and how cost-effective corrections may be made by pinpointing the stage(s) where the compliance process breaks down. We have also described some of the problems and limitations of the simple linear model, specifically that later stages might affect earlier stages of processing and that some stages might be skipped altogether. Although the chapters are organized around the basic or traditional model, it will become obvious that the processes involved are complex and that the model displayed in Figure 2.1 provides only a simplified heuristic view of the warning process. Nevertheless, we believe the linear model possesses many positive aspects. Among them is that it provides a useful tool for organizing the literature, for making predictions about effectiveness, and for tracking down why a warning has failed to influence behavior.

REFERENCES

DEJOY, D.M. (1989) Consumer product warnings: review and analysis of effectiveness research. In *Proceedings of the Human Factors Society 33rd Annual Meeting*. Santa Monica, CA: Human Factors Society, pp. 936–940.

FRIEDMANN, K. (1988) The effect of adding symbols to written warning labels on user behavior and recall. *Human Factors*, 30, 507–515.

OTSUBO, S.M. (1988) A behavioral study of warning labels for consumer products: perceived danger and use of pictographs. In *Proceedings of the Human Factors Society 32nd Annual Meeting*. Santa Monica, CA: Human Factors Society, pp. 536–540.

STRAWBRIDGE, J.A. (1986) The influence of position, highlighting, and imbedding on warning effectiveness. In *Proceedings of the Human Factors Society 30th Annual Meeting*. Santa Monica, CA: Human Factors Society, pp. 716–720.

PART TWO

Methods/Techniques

The second section describes the different methods of investigating warning effectiveness. Behavioral compliance and measures of intermediate processing stages are discussed.

CHAPTER THREE

Intermediate Processing Stages: Methodological Considerations for Research on Warnings

STEPHEN L. YOUNG
Liberty Mutual Group

DAVID R. LOVVOLL
Rice University

Warning information must pass successfully through a series of intermediate processing stages (attention, comprehension/memory and attitudes/beliefs) in order for it potentially to influence user behavior. Research on warnings tends either to deal with these stages explicitly while ignoring the effect of later stages on behavioral compliance or to examine behavior while ignoring these intermediate stages. Both types of omission are costly and reduce the utility of research findings. This chapter examines how warning information traverses these serial stages and demonstrates how the flow of information can be measured at each point. A better understanding of how warning information is processed should lead to a better conception of where warning failures occur and how they might be remedied.

3.1 INTRODUCTION

We know that warning information must traverse a maze of potential distortions and roadblocks in order for complete, understandable, and useful information to affect motivation and behavior. In particular, a failure of information to pass successfully through a given stage reduces the quantity and/or quality of the information that advances to subsequent stages. In some cases, information may not be passed through a stage at all, producing a short-circuit of the warning's processing. Given this situation, a great deal of research has been devoted to assessing the effect of warning manipulations on these intermediate stages of processing. We review this literature and point out the relevant methodological issues related to each study. It is hoped that a better understanding of the research and the ways that different studies fit into a larger model of information processing will help guide practitioners in the development of relevant studies.

3.2 ATTENTION

Attention is an important consideration of warnings for the simple reason that, if a warning is not noticed, it can have no direct effect on behavior. Thus, a great deal of research has been devoted to examining manipulations that attract attention to warnings (i.e., salience features). The ability of warnings to 'attract' attention is based on some basic psychological principles. The first, and most basic, involves the notion of 'figure-ground' from Gestalt psychology. A warning is a 'figure' (an object) embedded in some visual 'background' (e.g., on a product, in an advertisement), and it can be detected or recognized only if it is sufficiently different from its background. For example, a blue dot on a similarly colored blue background will not be detected because there is no discriminating feature to distinguish it from the surrounding information. Thus, a minimal criterion for attention is that a warning be distinguishable from its background. The second basic principle involves not only the ability of information to be distinguishable, but also to attract attention to itself. The principle of 'pre-attentive processing' (see Treisman, 1986, 1989) suggests that information (e.g., a warning), if designed properly with respect to the background, can 'pop out' and attract attention without conscious control on the part of the individual. For example, if an individual is shown a random configuration of 100 equal size dots, a single red dot embedded in 99 blue dots will 'leap off the page' and be detected prior to any conscious, attentive processing on the part of the individual.

These psychological principles have implications for research on attention to warnings. It is almost taken for granted that salience features help distinguish warnings ('figure') from their background (or 'ground'). In most studies, it is assumed also (although not always explicitly) that salience features can produce a 'pop-out' effect, such that the warning draws attention to itself regardless of attentional control on the part of the individual. While the latter issue is the more important from a practical standpoint, it is not always easy to determine which features direct attention on the part of users. When one is measuring something overt (e.g., behavior), it can be observed directly. However, a psychological construct, like attention, must be inferred from other measures. Some methods of assessing attention are more direct than others, but none of them measures attention directly. The following sections detail the different ways that attention has been and can be measured. Each method has advantages and disadvantages that must be considered carefully before implementation.

3.2.1 Attention Measured More-or-Less Directly

Three relatively common methodologies have been used to measure attention more-or-less directly: eye movements, detection or reaction time (D-RT), and self-reports. These methods are examples of the most direct techniques for assessing attention. Each will be discussed in some detail, along with the benefits and drawbacks of each method.

Eye movements

Of all the attentional measures, eye movements are probably the most direct method of assessing visual attention. Eye movements can be used in several different ways, depending on the interest of the researcher. For example, Laughery and Young (1991) had participants scan alcohol labels (presented on a computer screen) to determine whether or not a warning was present. The warning, when present, was either plain, or it was manipulated with different salience features (border, icon, color, and/or signal word). Eye-tracking equipment was used to determine the path participants took to find the warning (a more direct path indicated that the warning 'drew' attention), the amount of time participants took to find it, and the amount of time participants spent looking at the warning itself in order to determine that it was the target (less time indicated that the viewed item was easily identifiable as a warning). Since this was a timed search task, the authors were interested in the pre-attentive aspect of the salience manipulations. Specifically, they were interested in the ability of salience features to 'pop out' and draw the attention of the participants. Eye movements are good measures for this type of attention.

Fischer, Richards, Berman, and Krugman (1989) used eye-tracking equipment in a slightly different manner. They gave participants a fixed amount of time to view advertisements for tobacco. Data were collected on whether adolescents viewed the standard Surgeon General's warning (with no salience manipulations), and if so, what percentage of time participants spent examining the warning as a percentage of total time viewing the ad. Since timed search was not of interest, there was no interest in or measure of pre-attentive processing. With this method, Fischer et al. were able to demonstrate the attention-demanding nature of the background (an attractive cigarette advertisement) compared to a relatively undemanding figure (a nonsalient warning). The use of eye-tracking equipment to determine where people look (Mori and Abdel-Halim, 1981) or what people read (Galluscio and Fjelde, 1993) is more common than testing specific hypotheses regarding visual attention (e.g., Laughery and Young, 1991). While eye-movements are an attractive dependent measure, it should be noted that eye-tracking equipment is very expensive and data collection can be time-consuming. Unless a great deal of sensitivity is required (as when attempting to measure pre-attentive processes), other measures, such as detection or reaction time, may be sufficient.

Detection or reaction time

Detection or reaction time (D-RT) is a measure of how quickly an individual can search for and detect a pre-defined target. Lower D-RTs indicate that the 'figure' or target is more easily distinguished from its background. In the warning's domain, much of the research using D-RT as a dependent measure is based on research from other domains. These include: (i) D-RT to highway signs or markers of different configurations and external conditions (Asper, 1972; Plummer, Minarch, and King, 1974; Loo, 1978; Testin and Dewar, 1981); (ii) D-RT to auditory alarms (Adams and Trucks, 1976; Fidell, 1978), especially in aircraft cockpits (Simpson and Williams, 1980; Wheale, 1983); and (iii) D-RT to color (Christ, 1974), symbolic (Samet, Geiselman, and Landee, 1982), and alphanumeric displays (Snyder and Taylor, 1979).

As a dependent measure, D-RT is conceptually quite similar to eye movements—the more salient the 'figure' (warning), the more easily and more quickly it will be detected in a background. While it is much easier and cheaper to use D-RT, this measure does have some drawbacks. First, the experimenter does not gain information about the visual path that participants take in order to locate the warning. This information can be important to determine what background information is competing for attention with the warning. Second, and more importantly, participants can falsely report having detected the warning. Participants, responding quite naturally to demand characteristics (the desire to look good in the eyes of the experimenter or to give the experimenter 'good' data), may report locating the warning without actually having detected it. The only real way to deal with the first issue is to ask participants for self-reports of what they scanned when searching for the warning. This method provides some useful information, but it is prone to demand characteristics as well. With regard to the second problem, there is one major check or remedy—signal detection theory or SDT (see Wickens, 1984).

SDT can be used to analyze participants' responses to determine the extent to which they respond correctly (hits and correct rejections) or incorrectly (misses or false alarms). There are two important issues from SDT that must be considered before using D-RT in a study. First, if the proportion of non-targets (or distractors) in a study is too high or too low, participants will be able to employ probabilistic guessing strategies which can distort the results. If this type of guessing is not desired then the proportion of distractors should be in the range of 50%. Second, analysis needs to be conducted regarding errors. By looking simply at response times in D-RT data, there is no indication of how accurate participants are in their detection. Participants who respond immediately to all stimuli (either affirmatively, negatively, or randomly) will have very low D-RTs for the targets, but they will also have a large number of errors. Analysis of errors can provide an indication of the quality of participant's D-RT data. In addition, participants can be induced, through instruction or the use of rewards or punishments, to respond with fewer errors. Manipulation of instructions technique should be considered in conjunction with error analysis when using D-RT.

Considerations of guessing and accuracy have been incorporated into research using D-RT as a dependent measure. For example, Godfrey *et al.* (1991) had participants search for an alcohol warning on 100 alcohol beverage containers. Half of the containers had the US government-mandated warning and half did not. Young (1991) used the government-mandated alcohol warning label, but manipulated its appearance with different salience features (pictorial, color, border, signal word). In this study, half the labels did not contain a warning, and the information in the non-warning labels was manipulated in the same manner as the warnings were in the target labels. For example, in one target label there was a warning that was printed in red. There was a corresponding non-target label that had non-warning information printed in red. Thus, participants had to evaluate the information to some extent before responding. In both studies, participants were also instructed to answer correctly and to be fairly sure of their answer prior to responding. Instructions can have a major impact on response patterns.

Self-reports of attention

Self-reports involve asking participants to respond whether or not they saw the warning, independent of whether they can recall its content. It simply entails a response to the question, 'Did you see a warning?' Several studies have used this method (Goldhaber and deTurck, 1988; Otsubo, 1988; Jaynes and Boles, 1990; Kaskutas and Greenfield, 1991; Duffy, Kalsher, and Wogalter, 1993), and almost all of these studies subsequently asked participants to recall the content of the warning. Some of the more relevant uses for this method have been with 'interactive' warnings—warnings that force the user to interact with them before product use (Gill, Barbera, and Precht, 1987; Wogalter and Young, 1994)—and with 'active' environmental warnings—warnings that direct attention with an auditory or active signal (Wogalter, Rashid, Clarke, and Kalsher, 1991; Wogalter, Racicot, Kalsher and Simpson, 1993).

While self-reports provide important information in themselves, really they are used only in the context of memory studies with incidental exposure. For example, Young and Wogalter (1990) found differences in the rate of recall for information imbedded in an instruction manual. However, the findings could have been biased by the degree to which participants saw or did not see the warning in the first place. Since self-report data on attention were collected, the authors were able to determine the effect of the independent variables on memory. However, as a measure of pure attention, self-reports are not recommended since they are subject to various demand characteristics. Goldhaber and deTurck (1988) demonstrated this effect in a study which asked middle-school students if they saw a 'No Diving' sign that had been posted for one month: 8% of the students reported seeing the sign where no sign was posted (with another 44% saying that they were not sure whether one was posted). The problems of demand and falsification can be ameliorated, to some extent, by informing participants that they later will be asked to recall the sign's information.

Summary: Attention measured more or less directly

These three methods of measuring attention have their benefits and drawbacks. The most obvious benefit is that they are direct and require less inference than other dependent measures (e.g., for memory). Another attractive aspect of these measures is their lack of dependence on or sensitivity to individual differences. There is little reason to think that gender, racial, or other differences will be exhibited in measures of pure attention. Most humans are quite similar in this regard, making results more generalizable. The primary drawback is that these measures are somewhat contrived. Eye movements and D-RT data are collected in controlled environments, with high levels of intrusion compared to the real-world acquisition of information. People do not normally scan labels, ads, etc. as quickly as possible for a particular item. In addition, people are not usually self-conscious about their eye movements when scanning a visual field. Thus, while these measures are very useful for assessing the effects of various design factors on warning noticeability, they do not tell us much about the likelihood a warning will be noticed.

3.2.2 Attention Measured Through Comprehension/Memory

Although comprehension and memory are grouped together in a single processing stage, they are independent in some very important ways. First, no study was found that used comprehension as a dependent measure in a study of attention. The reason for this is obvious: asking someone about the extent to which they comprehend a given message necessitates that they, at the very least, have seen it. Thus, memory is the only component of this stage which is a practical measure of attention. However, comprehension does play a role when using memory as a dependent measure, since generally one must comprehend the message to some extent before remembering it.

Memory as a measure of attention

All the studies in this category use incidental exposure techniques. The incidental aspect of the studies can take two forms: a purely incidental exposure field study or an incidental exposure lab experiment. The primary difference between the two is whether or not they recruit participants. As an example of a field study, Goldhaber and deTurck (1988) placed a sign around a swimming pool and left it there for one month. After that time interval, students that frequented the pool were given a questionnaire about the content of the sign, including whether or not they ever saw it. Another example is Kalsher, Clarke and Wogalter's (1991) survey of college students measuring their knowledge of alcohol facts and other information from a sign that was posted in a fraternity house. Such field experiments are useful for assessing the impact of situational or environmental warnings, but they are less useful for assessing the impact of warning information for consumer products. Survey research assessing a specific warning intervention (see Gallup Canada, Inc., 1989; Canadian Inter-Mark, 1972a,b) for a specific consumer product may be an exception. For example, Kaskutas and Greenfield (1991) performed a survey of people 12 months after the US government-mandated alcohol beverage warning appeared. They asked people whether or not they had seen the warning and asked them to recall its content. Likewise, Godfrey and Laughery (1984) surveyed women on their knowledge of existing tampon warnings.

In most other cases, the incidental exposure lab experiment has been used to assess the impact of attention on memory. In lab studies, the recruited participants are not told of the warning-related nature of the study prior to stimulus exposure. Usually there is some ruse which has participants examining instructions (Young and Wogalter, 1990; Wogalter, Kalsher, and Racicot, 1992), advertisements (Barlow and Wogalter, 1991; Loken and Howard-Pitney, 1988), labels (Wogalter and Barlow, 1990), or actually using products (Duffy, Kalsher, and Wogalter, 1993; Gill et al., 1987; Strawbridge, 1986) under conditions which do not draw undue attention to the warning. After exposure, participants are given a recall test which assesses their memory of the relevant information. Some of the studies have asked participants to recall additional information about the warning such as its location (Fischer et al., 1989; Barlow and Wogalter, 1991), and configuration (Barlow and Wogalter, 1991).

Conclusions: Attention measured through comprehension/memory

The idea behind using memory as a dependent measure in this type of research is that people cannot remember and recall information that they do not notice and comprehend. Under incidental exposure, and assuming the information being tested is not generic or well known (which is no small assumption), any difference in retention for warning information is due to the level of the warning's noticeability. The extent to which a warning's information is known can be assessed by a control group of participants not exposed to the warning information. These participants could serve as a baseline of prior knowledge for the information contained in the warning(s). However, important participant differences would not be revealed by comparing knowledge scores across groups. For this reason (and others), memory is not the best dependent measure of attention.

While there are significant drawbacks to using memory as a dependent measure, there are some important benefits associated with it. First, it is relatively easy and inexpensive to collect memory data—it requires little (or no) equipment or other resources. Second, because memory assesses a combination of attention, comprehension, and memory, it provides an indication of how the warning is likely to perform in 'real world' situations (considering the three stages as a whole). While attention may be interesting at a basic level, we are interested ultimately in how attentional factors affect the ability of information to pass from attention to subsequent stages. A memory measure allows one to evaluate the end-result of this process of information transmission.

The drawbacks associated with this method are numerous. First, this method is noisy, because it relies on attention, comprehension, and memory. As such, this method may not reveal actual deficiencies in the attentional component of warnings (e.g., that the warnings were not really salient), since the results may be attributable to the comprehension and memory components. Second, using memory to assess attention does not reveal much about the construct of attention itself. For example, a salience manipulation, such as color, could be used to draw attention to some information that is printed in a foreign language. If the targeted individual does not comprehend the message in that language, then their failure to recall the content of the message does not reflect a failure of the salience manipulation to attract attention. In general, failures or disruptions at higher levels in the processing chain do not necessarily reflect on any of the individual components at the lower levels.

3.2.3 Summary: Attention

Attention is an important consideration in warning research since warnings must be noticeable if they are to have any potential impact on behavior. However, research on attention offers limited information about warning effectiveness, since high scores on attention do not necessarily translate into increased positive behavior (see Strawbridge, 1986; Friedmann, 1988; Otsubo, 1988; Jaynes and Boles, 1990; Wogalter *et al.*, 1992). Thus, attention is a necessary but not sufficient condition for warning effectiveness. Because of this, we might begin to think of attention as an independent criterion for warnings—one which can and should be tested independently of the behavioral criterion. As this section demonstrates, the best way to determine if a warning meets the attentional criterion is to test it as directly as possible. If attention is not measured directly, steps need to be taken to ensure that the contribution of attention can be accounted for by other means (e.g., the use of control groups in memory studies).

3.3 COMPREHENSION

The attentional component of a warning is important, since a warning must be noticed before any subsequent processing may occur. However, in most situations getting users to notice warnings is not an end in itself. Users must not only notice a warning; they should also gain some information from it. Thus, warnings, and any icons or pictorials that accompany them, must be understandable to the target population. If the warning information is understood, then it has the potential to be processed further and remembered.

3.3.1 Comprehension of a Verbal Warning Message

The primary consideration of a warning, after it is noticed, is that it be understood. There are three levels in the comprehension process to consider. At the most basic level, a verbal message must be written in a language that is read and spoken by the target population. At the next level, the message must be written in words and sentence structures that are familiar and can be understood by most people in the target population (the information must be readable from a grammatical standpoint). At the next level, warnings must be coherent—they must convey a message which can be understood by individuals in the target audience. Related to this third level, the content of the warning must be unambiguous in conveying the relevant information, it must contain words which are understood, and it must be relatively brief. Researchers have used different methods to evaluate warnings on all three levels. In this chapter, we will not deal with language skill *per se* other than to stipulate that warnings should be presented in the desired language of the target population (or, at the very least, multiple translations should be provided).

Readability

There are over 100 indices of readability, reading level, or complexity of written material (Duffy, 1985), and all of them are designed to quantify the degree to which messages are comprehensible. Anyone interested in a summary of the major readability indices should obtain a copy of Klare (1974–1975). While readability metrics are popular, there are important questions regarding their usefulness and applicability (Lehto and Papastavrou, 1993). First, in general these metrics are designed to evaluate longer passages of prose text for books, instructions, etc. Hazard sign statements often are terse statements designed to convey information in a relatively efficient manner. Thus, readability formulas may not evaluate warning statements properly. Second, readability indices must take into account the nature of the expected audience. Certain technical terms may be well understood by people in one context, but not in others (Duffy, 1985). Third, some of the rules employed by readability formulas are not valid in all contexts. For example, simplifying vocabulary and sentence structures does not always increase comprehension (Duffy and Kabance, 1982). Powell (1981) suggested that readability formulas have been abused, but that they can provide valuable information if used properly.

Some authors essentially have dismissed simple readability criteria (e.g., letters-per-word, words-per-sentence), and have concentrated on other ways to evaluate information and increase comprehension. For example, Siegel, Lambert, and Burkett (1974) focused on mental load and suggested, among other things, that the number of letters in a word was not as important as the number of morphemes—the smallest unit of speech sound that has meaning. Duffy and Kabance (1982) suggested that comprehension could be increased according to the 'transformer' concept—having an individual or group of people ('transformers') evaluate the comprehensibility of information on behalf of the target audience. While a bit more difficult to implement, procedures like mental load and transformers may be more powerful in the hands of qualified individuals.

Several studies have used notions of readability to evaluate warnings and safety instructions. Pyrczak and Roth (1976) used the Dale–Chall formula to analyze directions on non-prescription drugs. Silver, Leonard, Ponsi, and Wogalter (1991) used four indices of readability to evaluate pest-control product warnings: number of words (word count), number of sentences (sentence count), the Flesch (1948) index as modified by Gray (1975), and the Coleman and Liau (1975) index. Participants in this study reported that they would be more likely to read warnings which were more complex (e.g., with more sentences/statements and warnings which were written at a higher grade level). This finding may be a reflection on the participant population (e.g., college students), suggesting, as above, that the characteristics of a potential audience are important.

Coherence

In addition to the words and sentences needing to be understandable, the message being communicated must be coherent. One way to evaluate the coherence of warning messages is to test them in the target population. Much of this testing involves exposing participants to the message and collecting some data on the degree to which they understand it. Funkhouser (1984) demonstrated the importance of such testing by showing that drastic differences in comprehension could be attributed to relatively minute wording differences in affirmative disclosure statements. Another method is to conduct phone surveys regarding people's use of certain products in response to warnings (Morris and Klimberg, 1986). Focus groups have also been used to pre-test knowledge of hazards in the target population prior to development of warnings (see Eberhard and Green, 1989).

Eye movements are another way to test the comprehensibility of verbal information in warning signs. For example, Galluscio and Fjelde (1993) altered the sequence of presented information (consequence statement followed by an action statement or vice versa) and examined eye movements of participants who read the signs. Eye movements were a good measure in this instance because they allowed the experimenters to determine the extent to which people could understand the content presented in a certain order. Saccadic 'backtracking' was an indication that people had to go back and re-read some information in order to understand the message.

Explicitness

Warning and safety information must not only convey understandable information but also it must convey this information explicitly. Kreifeldt and Rao (1986) suggested that often 'fuzzy' terms were used in instructions to convey concrete information. Words like 'squeeze hard' or 'squeeze gently' or 'push firmly,' etc. are members of fuzzy sets because they convey ambiguous information. If warnings are pre-tested for comprehension (as they should be), emphasis should be placed on comprehension of fuzzy words, especially in important instructions. In general, ambiguity is a quality that warnings should not possess. Another area of explicitness which has received attention is on the consequence side—warnings should be presented with explicit information about the potential results of exposure to the hazard (Laughery et al., 1993). Previous research has concentrated on two major aspects of explicit consequence information: (a) people's preference for or aversion to such information, and (b) the extent to which people might not purchase products with explicit consequence information. In general, both aspects of explicitness have been measured with Likert-type rating scales (e.g., 0–7), indicating degrees of aversion or preference (e.g., Vaubel, 1990; Langlois et al., 1991; Vaubel and Brelsford, 1991).

Comprehension of terms

When describing a hazard, certain terms (e.g., flammable, combustible, corrosive, irritant, etc.) must often be used in order to convey information about the nature of the hazard and its consequences. Researchers have demonstrated, by asking participants to provide definitions, that some terms are not well understood by the general population (see Leonard, Creel, and Karnes, 1990, 1991a,b). If certain technical terms are to be used in a sign, research should be conducted regarding the target population's understanding of those terms.

Brevity

While warnings must be explicit in their content, they must also be brief. Long warnings may take too much time to read and they may induce people to ignore them. While explicitness and brevity seem contradictory goals, several studies have examined ways to communicate important information while remaining brief. For example, Young et al.

(1995) examined the relative importance of different sign components (signal word, hazard statement, consequence statement, instructions, and pictorials) by having participants construct signs from scratch. From the finished signs, the relative size of individual components and their ordering within the sign indicated what information participants thought was most important. Pre-testing of warnings might use a similar technique to determine what information needs to be present and/or emphasized within a sign or label. The importance of pre-testing was demonstrated by Polzella, Gravelle, and Klauer (1992), who found different levels of comprehension for signs with different components.

Summary: Text comprehension

Very subtle differences in the wording of safety messages or warnings can affect the information conveyed to people (Funkhouser, 1984; MacGregor, 1989). Therefore, it is recommended that warnings be pre-tested in the target population, preferably using an interview or focus-group technique (see Eberhard and Green, 1989). Without pre-testing, comprehension cannot be guaranteed, and it should not be assumed. With pre-testing, the relevance of certain warning components and the quality of information they convey can be evaluated.

3.3.2 Comprehension of Pictorial Symbols

Several researchers have discussed the differences between pictorial symbols and verbal material on different stages of information processing (Childers and Houston, 1984; Childers, Heckler, and Houston, 1986). One of the most important differences is that pictorials convey information in the absence of words—a feature which is a benefit to both non-English speakers and those with low literacy skills. Pictorials have an advantage over verbal statements wherever and whenever the verbal message is unlikely to be read (because of length) or understood (because of the population characteristics). Of course, this is true for pictorials only to the extent that they can be understood or comprehended by the target population (see Lerner and Collins, 1980).

The most common method of testing comprehension for a pictorial is to show it to participants and have them define it in the absence of written, environmental or other external cues. Some pictorials (i.e., biohazard or radiation) have almost 0% comprehension, while others (i.e., skull-and-crossbones) are almost universally understood. As such, many studies have attempted to determine which pictorials are comprehensible in the absence of environmental context and which are not (Krampen, 1965; Dreyfuss, 1970; Keller, 1972; McCarthy and Hoffmann, 1977; Mackett-Stout and Dewar, 1981; Collins, Lerner, and Pierman, 1982; Hodgkinson and Hughes, 1982; Laux, Mayer, and Thompson, 1989; Wolff and Wogalter, 1998). These studies demonstrate clearly that the rate of comprehension for pictorials is highly variable and should never be assumed.

Another method of testing the comprehensibility of pictorials is to show them to participants with contextual information (Cahill, 1975, 1976). In general, including context provides an ecologically valid test of pictorial comprehension, since pictorials are rarely viewed or interpreted in the absence of context in the 'real world'. However, Frantz, Miller, and Lehto (1991) demonstrated that context does not always enhance comprehension. While environmental cues are one piece of additional information which can make pictorials more comprehensible, researchers have studied ways to increase comprehension in other ways. For example, Wogalter, Sojourner, and Brelsford (1997) suggested that comprehension for pictorials can be enhanced with brief training (presenting the meaning of the pictorial once). Baber and Wankling (1992) and Booher (1975) tested combinations of pictorials and verbal messages. The provision of context and the nature of the context should be considered carefully before conducting research on pictorial comprehension (Wolff and Wogalter, 1998).

One additional method of studying pictorial comprehension warrants mention. While most of the studies discussed so far simply evaluate existing pictorials, there is no reason that pictorials cannot be designed from scratch and tailored to the target audience. As with reading comprehension, certain populations (e.g., the military) have specific knowledge that might not exist in the general population. With pre-testing, this knowledge could be discovered and used to create comprehensible pictorials for that particular population. For example, Eberhard and Green (1989) used focus groups composed of auto mechanics to create pictorials for use in service stations. After their creation, the pictorials were tested with a second set of mechanics for comprehension.

3.3.3 Target Populations

It was stated earlier that in order to construct a warning which can be comprehended by the target audience, one first must know something about the intended audience. Laughery and Brelsford (1991) implored warning designers to 'know thy user' with regard to (1) demographics (age, gender), (2) familiarity and experience with the product, (3) competence (technical knowledge, language, and reading ability), and (4) hazard perception. While this type of analysis may be too complex for some consumer products, it is a prudent recommendation to consider these issues.

3.3.4 Summary: Pictorial Comprehension

The conclusions which can be drawn about comprehension are not much different from those which were drawn regarding attention. Comprehension is a necessary but not sufficient condition for warning effectiveness. It is necessary because a warning must be understood if it is to have any potential impact on subsequent behavior. It is not sufficient because high levels of comprehension do not ensure high rates of compliance (see Strawbridge, 1986; Friedmann, 1988; Otsubo, 1988; Jaynes and Boles, 1990; Wogalter *et al.*, 1992). Thus, as with attention, comprehension should be considered a criterion of warning effectiveness which is largely independent of the behavioral criterion.

This section demonstrated different ways to test warnings on this criterion. It was shown that both verbal and pictorial comprehension are important in the overall scheme of information transmission. These two modes of communication should not be treated as independent components of a warning, but rather they should be thought of as a complete or whole unit (since they can convey both unique and redundant information). Comprehension of warnings is also dependent on the target audience—some audiences may have knowledge which is not readily available in the general population. The methods of Eberhard and Green (1989) are an example of holistic warning evaluation in the appropriate population.

3.4 MEMORY

Memory for warning information is important, but like the other stages in the model, it is a means to an end and not the end itself. It is not so important that people remember or recall warning information as much as it is important that they know the information that the warning contains. Memory can be an indicator of knowledge, but it is not necessarily so. People can recall information that they know little about (e.g., nonsense syllables). However, much of recall is an indicator of memory or knowledge. Memory for warning information is important also in situations where instructions or labels are not available, visible, or reviewable (e.g., Young and Wogalter, 1990). Where labels or instructions are not accessible, people must rely on their memory for the relevant warning information.

There are different ways to assess memory (recall, matching, recognition) and scoring it (strict or specific recall versus liberal or 'gist' recall). These different methods have been used to assess memory in the warnings literature, and each will be dealt with regarding warnings.

3.4.1 Open-ended Recall

One way to assess the degree of information that people get from a warning is to simply ask participants to recount all that they remember about it. This type of question can take two forms: non-cued recall and cued recall. Non-cued recall most often occurs when a study has few (or only one) warning. Under these conditions, the experimenter can request that the participant 'recall the content of the warning' (Godfrey and Laughery, 1984; Gill et al., 1987; Goldhaber and deTurck, 1988; Jaynes and Boles, 1990; Wogalter et al., 1991, 1992, 1993; Duffy et al., 1993; Wogalter and Young, 1994). This open-ended question does not give participants any information that would help them answer the question, except that there was a warning. With multiple warnings, one could ask the participants to recall all the warnings they remember seeing, but more likely one would ask about each warning by providing questions for each one (Young and Wogalter, 1988, 1990; Barlow and Wogalter, 1991; Kalsher et al., 1991; Kaskutas and Greenfield, 1991). An exception to the one–many rule is found in Otsubo (1988) and Strawbridge (1986), who asked participants about the individual components (i.e., the cause, the consequences, and the avoidance behavior described) of a single warning.

Numerous studies have used the open-ended technique with a single scoring criterion (see Gill et al., 1987; Otsubo, 1988; Jaynes and Boles, 1990; MacKinnon, Stacy, Nohre, and Geiselman, 1992; Duffy et al., 1993), but others have found that one scoring criterion is not sufficient for this type of research and have used 'strict' and 'lenient' scoring (Young and Wogalter, 1990; Wogalter et al., 1991, 1993). Strict scoring requires that the participants recall specific wording from the warning. Lenient scoring requires only that the participants recall the meaning or gist of the warning message. From the basic literature on the topic of memory, it is clear that much of what people remember of a verbal statement is a subset or kernel of the exact wording. That is, people remember the gist or meaning of a statement much more readily than they remember the wording verbatim. Since warning researchers are (in most cases) concerned that the meaning of a warning be conveyed, a lenient or gist scoring criterion is valid. As such, researchers often collect information about the specific wording users retain from a warning, as well as what general information or knowledge they receive.

The primary benefit associated with strict scoring is that it helps to uncover the specific material that participants gleaned from the warning. Even if participants knew about the hazard previously, they would have to obtain the specific wording from the warning itself. This method also virtually precludes the use of guessing on the part of participants. However, it is more difficult for participants to recall the exact wording of a warning even when they have read it. This is especially true when there are many different warnings to recall (i.e., Young and Wogalter, 1990).

The primary benefit associated with lenient or gist recall is that it is more ecologically valid. It is rare when the exact wording of a warning is so important that successful gist recall is not an indication of adequate recall. Gist recall is a more realistic test of how people process information and remember information in their everyday lives. However, lenient recall allows people to guess the warning information when they do not actually remember it. Also, it is more difficult to score consistently and reliably (inter-rater reliability is generally somewhat lower than with strict scoring). In many cases, both criteria are used, since both methods of evaluation have benefits and both can be used a posteriori.

3.4.2 Recognition

With a few exceptions (Loken and Howard-Pitney, 1988; Fischer *et al.*, 1989; MacKinnon *et al.*, 1992), recognition is used primarily in tests of pictorials or signs (Keller, 1972; Booher, 1975; Cahill, 1976; Easterby and Zwaga, 1976; Green and Pew, 1978; Cairney and Sless, 1982; Laux *et al.*, 1989; Young and Wogalter, 1990). In this context, recognition implies definition, since most studies require participants to provide a definition for (or to 'recognize') a pictorial. The prevalence of recognition tests in pictorial research is likely due to the visual or iconic nature of pictorials, which lend themselves to visual recognition tests more readily than to verbal descriptions. Lerner and Collins (1980) suggest that recognition tests for symbols (i.e., definition and multiple choice tests) are the 'most suited for large-scale testing of specific symbol sets.'

Recognition has also been used in the more classic sense—identification of or remembering having seen a warning, pictorial, or sign. For example, Main, Rhoades, and Frantz (1994) showed 100 Canadian residents a government-mandated symbol for adhesive products and asked them if they had ever seen it. Mori and Abdel-Halim (1981) measured exposure times and eye movements to determine how long it took people to recognize different road signs. In a similar vein, Shinar and Drory (1983) asked people to recognize (and recall) road signs they had just passed.

Recognition tests are often more realistic tests of memory than recall, since safety signs and warnings are supposed to cue people about potential hazards. If people see a warning and recognize it and its meaning, then independent recall or memory of safety-related information is a less important issue. Since people generally do not encounter potential hazards in the absence of external cues (as with recall tests), recognition tests determine how well people use such cues (signs, warnings, etc.) in the evaluation of hazardous situations. However, recognition tests prompt participants for information and therefore promote guessing.

3.4.3 Matching

Matching involves having participants pair two similar things. In the warning's literature, matching is used primarily for associating pictorials with verbal statements, but some exceptions exist. Collins et al. (1982) had participants match the meaning (or referent) of the pictorial. Like recognition, matching provides cues that aid in recall and may be more realistic than non-cued recall tests. Easterby and Zwaga (1976) stated that matching tests for pictorials were superior in some instances to open-ended recall tests. However, matching does lend itself to guessing (quite possibly even more so than with recognition).

3.4.4 Other Memory Tests

MacKinnon et al. (1992) gave participants an indirect recall test which in part consisted of word-stem completion. The word-stem completion task presented participants with the first two letters of each of four risks and the four distractors (e.g., 'consumption of alcoholic beverages impairs your ability to dr_____.', p. 155). Participants had to recall the missing word given the sentence cue. Barlow and Wogalter (1991), Gill et al. (1987), and Fischer et al. (1989) have assessed memory for the location of a warning, independent of the warning's content. In addition, Barlow and Wogalter (1991) assessed participants' memories for the configuration (plain rectangle, rectangle with a signal icon, and circle and arrow) of the warning. These different dependent measures can be used to determine how much participants remember about the warning, independent of its content.

3.4.5 Summary: Memory

All the tests described here can be used to assess memory for warning information. When using these tests, some general issues need to be considered. First, all tests except the open-ended recall with strict scoring are subject to guessing. In order to control for guessing, instruct participants on what their guessing strategies should be and, if guessing is not desired, instruct participants not to guess. Instructions are a very powerful instrument for guiding participants about the desirability of certain kinds of responses, and they should be used properly. In addition, analyze correct and incorrect responses. If a particular participant is found to be guessing wildly, his or her data may have to be discarded. Another issue concerns the order of multiple memory tests. For example, Young and Wogalter (1990) presented participants with eight warnings (with or without pictorials) in an owner's manual under incidental exposure conditions. Open-ended recall was used for the verbal content, while recognition and matching were used for the pictorials. In this case, the open-ended recall test was administered first, since it provided no cues or information which would bias the subsequent tests. Otherwise it is generally advised that the order of materials should be counterbalanced or at least randomized. Order should be considered carefully if multiple tests are to be administered; the appropriate method of ordering depends on the nature of the information that is to be collected.

3.5 ATTITUDES/BELIEFS

Attitudes and beliefs are, in a sense, the truth as we see it. They are interpretations of the world around us, and they are based on previous experience, familiarity, knowledge, etc. This section will address each of the components separately.

3.5.1 Attitudes

Attitudes encompass a mentality or outlook—a habitual or characteristic mental attitude. They also include inclinations, propensities, dispositions, tendencies, or leanings— attitudes of mind that favor one alternative over others. The most commonly researched attitude in the warning/safety literature is risk perception. In this context, risk is defined as the perceived 'chance of injury, damage, or loss' (*Webster's New Universal Unabridged Dictionary*, 1983). Perceived risk is considered important because it has the potential to influence people's intent to seek out warning information and comply with warnings (Wogalter, Desaulniers, and Brelsford, 1987; DeJoy, 1989, 1991; Dingus, Wreggit, and Hathaway, 1993).

Measuring and defining risk perceptions

Generally risk is conceived as a construct with multiple contributing variables: familiarity with the product, hazardousness of the product, likelihood of being injured, severity of potential injury, etc. The most common way to incorporate all of these notions is to ask people to evaluate products on each dimension using numbered, Likert-type scales (see Fischhoff, Slovic, Lichtenstein, Read, and Combs, 1978; Rethans, 1980; Godfrey, Allender, Laughery, and Smith, 1983; Karnes and Leonard, 1986; Wogalter *et al.*, 1987; Desaulniers, 1989). These scales usually range from zero to seven, but almost any scale over 3 or 4 points will do. Another method for evaluating perceived risk is to rank-order a list of products according to their riskiness (see Dunn, 1972; Dorris and Tabrizi, 1978; Desaulniers, 1989).

When using Likert-type data, generally there are two methods of evaluating the data: univariate or multivariate analysis. The most common univariate method is stepwise regression (see Wogalter *et al.*, 1987; Young, Wogalter, and Brelsford, 1992). However, this method is no longer considered very robust, since it allots variance shared between individual variables. A more appropriate analysis is multivariate, principal components analysis (see Fischhoff *et al.*, 1978; Vaubel and Young, 1992; Young, 1995). This method allows the researcher to examine how the different individual variables (which compose risk perceptions) interact with one another to form perceptions of risk. The information gained from this type of analysis is not only more generalizable, it is also more informative.

Evaluating the quality of risk perceptions

The analyses described above can be used to determine the nature or composition of risk perceptions. However, they are useful also to evaluate the risk perception itself—the level of risk that is attributed to a product as a result of the combination of variables which compose risk. This type of evaluation has been done by comparing people's risk perceptions for products or situations with (a) risk ratings from experts (see Martin and Heimstra, 1973; Baber and Wankling, 1992) and (b) risk estimates from objective sources, such as the National Electronic Injury Surveillance System or actuarial databases (Fischhoff, 1977; Lichtenstein, Slovic, Fischhoff, Layman, and Combs, 1978; Slovic, 1978; Slovic, Fischhoff, and Lichtenstein, 1978, 1979; Combs and Slovic, 1979; Kasper, 1980). This type of comparison demonstrates that perceived ('subjective') risk and quantifiable ('objective') risk often are not very closely related.

Using risk perceptions in research

Perceived risk has been used in behavioral research to determine its effects on compliance rates. Chy-Dejoras (1992) and Donner (1990) correlated risk perceptions with compliance in order to determine (a posteriori) if people's risk perceptions affected their behavior. On the other hand, Otsubo (1988) used risk level as an independent variable in a behavioral compliance study. Otsubo presented participants with a high or a low risk product and evaluated differences in compliance rates between the two.

Using risk perceptions to develop warnings

Risk perceptions have been used also in the development of warnings. Specifically, the perceived risk associated with different components of warning signs (e.g., color, shape, etc.) has been evaluated experimentally (see Kanouse and Hayes-Roth, 1980; Polzella *et al.*, 1992). 'Hazard association value' (the level of hazard associated with a particular item) has been used (Bresnahan and Bryk, 1975), as have other, more general measures (e.g., 'Would you quit your job if you had to work with a product which contained this warning?;' Farid and Lirtzman, 1991). The other major component of signs which has been evaluated with regard to perceived risk is signal words (see Leonard and Matthews, 1986; Leonard, Hill, and Karnes, 1989).

3.5.2 Beliefs

Beliefs include any cognitive content which is held as being true. They can also be impressions, notions, or ideas about what is true. Much of the research on beliefs has revolved around two issues: beliefs about the ability to control hazards, and the believability of warning messages.

Beliefs regarding the ability to control hazards

Risk perceptions are affected greatly by people's perception of their ability to control hazards (Laux and Brelsford, 1989). This, in turn, influences their willingness to seek out and comply with warning information. Locus of control has been examined with univariate and multivariate scales. The univariate scales revolve around the control that people have over the hazards or the extent to which exposure to the hazard is voluntary (e.g., 'How much control do you have over the hazards?,' from 0 'no control' to 7 'complete control'; 'Is exposure to the hazard voluntary or involuntary?,' from 0 'voluntary' to 7 'involuntary') (see Rethans and Albaum, 1980; Fischhoff *et al.*, 1978; Young, 1995). The multivariate scales have been more involved (e.g., the product safety locus of control scale; Laux and Brelsford, 1989), classifying people as having an internal or external locus of control. People who believe that they encounter most hazards voluntarily, and who have a high internal locus of control are people who are more likely to look for and comply with warnings. Therefore the nature of the experimental task in behavioral compliance studies becomes very important. If people do not believe that they need to take precautions (because of their belief system), then the design of warnings is less relevant to their behavior.

Believability of warning messages

People access and interpret the information they receive from warnings in light of what they already believe (DeJoy, 1991). Because beliefs are strongly held and resistant to change, warning information which contradicts previously held beliefs is likely to be discounted or ignored (Beltramini, 1988). Information can be contradictory on a content level (e.g., smokers do not believe information about the hazards of smoking; Loken and Howard-Pitney, 1988), on an irrational level (e.g., people do not evacuate their homes during tornadoes because they simply do not want to believe that they could lose everything; Perry, 1983), on a control level (e.g., people do not wear seat belts because of their belief that they have control over potential accidents; Robertson, O'Neill, and Wixom, 1972), or on an experiential level (e.g., people may not heed a warning because of their belief that it is a false alarm; Loomis and Porter, 1982; Mallett, Vaught, and Brnich, 1993). The believability of warning messages has been studied using univariate, Likert-type scales ('how believable is this message?,' from 0 'not at all believable' to 7 'extremely believable') (see Andrews, Netemeyer, and Durvasula, 1991), and questionnaires (Loken and Howard-Pitney, 1988). These studies demonstrate that people can be resistant to warning messages, but that this is not universally so. Demonstrations and feedback sometimes can enhance the believability of safety information (Zohar, Cohen, and Azar, 1980), but the potential for people to resist some safety information should be considered when designing behavioral experiments.

3.5.3 Summary: Attitudes/Beliefs

Mallet *et al.* (1993) demonstrated the effect of attitudes/beliefs on underground coal miners' interpretation of and propensity to comply with warnings. Of 21 miners who escaped a fire in a mine, the researchers found a tendency to discount information in the warnings (a) because of previous false alarms (at least until the hazard was upon them), and (b) because of their reliance upon other experienced miners instead of warning signs for relevant information. Attitudes and beliefs influence not only people's willingness to comply with warning information, but also their interpretation and processing of information at lower levels or stages of information processing (see DeJoy, 1991; Lehto and Papastavrou, 1993; Shinar and Drory, 1983). While a great deal of research has been conducted on how to make warnings more noticeable, comprehensible, and memorable, little research has been conducted to determine how warnings could be constructed to influence attitudes or beliefs. The importance of this gap in the literature will become apparent in the final section.

3.6 GENERAL CONCLUSIONS

The purpose of this chapter has been to review important methodological issues which affect the study of intermediate information-processing stages (attention, comprehension/memory, attitudes/beliefs). There is a myriad of issues that are involved in the proper design of experimental or field studies. It should be noted that the methodological considerations presented here are more than just experimental nit-picking. Proper design and manipulation of independent variables, as well as proper measurement and scoring of dependent variables, is critical to the validity of the conclusions drawn from any parcel of data. In an attempt to sum up the main points of this chapter, a concrete example may be helpful.

McCarthy, Finnegan, Krumm-Scott, and McCarthy (1984) concluded, from a review of about 400 papers, that warnings have no measurable impact on user behavior and safety. Disregarding the nature and quality of research on which this conclusion is based, it is clear from the present discussion that their conclusion is overly broad and unsupported. A seminal behavioral study (Dorris and Purswell, 1977) makes this point—of 100 participants using a hammer, none complied with the warning on it. While that might have been taken by some as evidence that warnings are not effective (in general), the authors asked the students whether or not they noticed the warning. None of the students noticed or saw the warning prior to using the hammer, making the behavioral question moot. The failure in processing occurred at the attentional level, suggesting only that the warning should be made more salient. No other inferences about warnings or warning effectiveness can be drawn from this study. We do not suggest that compliance would have been perfect had the students noticed the warning. Rather, we suggest that no indication about the behavioral effects of a warning can be drawn so long as there are serious questions regarding the flow of information through the intermediate stages of processing.

Dorris and Purswell (1977) is an extreme example of failure at an intermediate stage of processing. Other studies have shown differing levels of behavioral compliance when participants actually did notice and read warnings (Strawbridge, 1986; Friedmann, 1988; Otsubo, 1988; Jaynes and Boles, 1990; Wogalter et al., 1992). In general, these studies demonstrate that higher rates of information flow through the intermediate stages (noticing, reading, and recalling) produce higher rates of behavioral compliance. However, compliance rates often are significantly lower than the rates for the initial stages of processing, suggesting that, 'even if users notice, read and recall the warning, there is no guarantee that users will follow the warning' (Strawbridge, 1986, p. 720). That is because there has been a general failure in the literature to control for or manipulate higher level processes (attitudes/beliefs) and/or external, mediating factors (e.g., cost of compliance). In most behavioral studies, these issues are either ignored completely, or they are mentioned in the discussion section as potential reasons for the lack of observed compliance.

Disregarding the external factors, making participants believe that they are in a hazardous situation (requiring compliance to warnings), making them believe that compliance will in fact prevent the hazard, and making them believe that they cannot otherwise control exposure to the hazard is extremely difficult in controlled laboratory studies (for ethical reasons). Hatem (1993) demonstrated this concern in a study of behavioral compliance, in which participants were instructed to increase ventilation in a room in which they were using glue. While 87% of all participants noticed the odor from the glue, only 1 participant in 59 complied with the warning. Post-experiment interviews with 38 of the participants demonstrated that about half felt that compliance was not necessary because of the short exposure duration, 20% felt that the experimental nature of the task indicated that they were in no real danger, 15% thought that the room was sufficiently large to dissipate the odor, about 20% thought that they were not allowed to open the window or turn on the fan, and about 10% thought they did not have to increase ventilation for other, unspecified reasons. Otsubo (1988) demonstrated that, of the participants who did not comply but did recall the warning, 79% 'explained that they were normally careful, and felt no need for protection' (p. 539). Given the generally high rates of information flow through the initial processing stages (up to the attitudes/beliefs stage), it is possible that a major failure mode exists in the attitudes/beliefs stage.

This fact points to one major issue with respect to the design of research studies in the area of warnings: that failure to control for all the stages of processing, including attitudes and beliefs, limits the ability of researchers to make statements regarding the effect of lower-level manipulations (e.g., pictorials) on behavior. Take an example of a hypothetical study in which it is the desire of the researcher to demonstrate the effect of pictorials on behavioral compliance. The researcher produces two identical warnings, except for the presence of a pictorial, and tests them on a valid population of participants under ideal conditions. Let us then say that the researcher observes 0% compliance in both the pictorial and non-pictorial conditions. From that one could conclude that pictorials are ineffective in altering behavior, even if it were demonstrated that they produced higher rates of noticing and recall. However, what if the verbal content of the warning instructed participants to buy $10 000 worth of specialized safety equipment? Knowing this, we would expect compliance to be 0%. However, an implicit assumption in the study is that the only difference between conditions (and hence the only factor assumed to produce the observed compliance rates) is the pictorials. While this example is extreme, it points to the fact that research must account for all stages of processing when conducting research. The attentional, comprehension, and memory benefits afforded by lower level manipulations do not determine, in themselves, whether the entire warning will have an effect on behavior.

The information-processing stages presented in this chapter are serially linked—that is, information flows from one stage to the next. Because of this, it has been stressed that each stage is an independent criterion for warning effectiveness—one which is necessary but not sufficient for behavioral effectiveness. However, as was demonstrated in the section on attitudes and beliefs, the stages are not completely independent of one another. The processing of information at the early stages is not determined entirely by the stimulus itself (e.g., the warning) (see Lehto and Papastavrou, 1993; Shinar and Drory, 1983). According to DeJoy's (1991) value–expectancy theory, attitudes and beliefs influence how people perceive information: the way they attend to information, the way they interpret it, etc. People's attitudes and belief structures are firmly held and difficult to modify. People interpret messages in light of these concerns (Slovic *et al.*, 1979), making attitudes and beliefs important in the processing that occurs at the earlier stages of the C-HIP model (attention and comprehension) described in Chapter 2 (by Wogalter, DeJoy, and Laughery). The interactivity of processing stages should be considered when designing studies of warnings and safety communication.

In the end, we are all interested in making warnings more effective. The intermediate processing stages discussed in this chapter are but means to that end. However, they are important means, and they cannot be ignored. Researchers conducting behavioral studies should consider the effect that all the information presented to participants, whether manipulated experimentally or not, will have on eventual compliance. Moreover, many different dependent measures should be collected in behavioral studies. Failures of processing can occur at any stage in the chain—some participants will not comply because they failed to see the warning, others because they do not comprehend the message, etc. Collecting data on the intermediate stages of processing will allow one to determine the location of failure and how, possibly, to remedy it. Optimizing content and format will not, in itself, guarantee compliance to warnings, since warnings are not deterministic and since they depend on human volition. However, optimizing these components will allow the warning to do all that it is capable of doing, i.e., transmitting relevant safety information in a way that is noticeable, understandable, memorable, believable, and compelling.

REFERENCES

ADAMS, S.K. and TRUCKS, L.B. (1976) A procedure for evaluating auditory warning signals. In *Proceedings of Human Factors Society 20th Annual Meeting*. Santa Monica, CA: Human Factors Society, pp. 166–172.

ANDREWS, J.C., NETEMEYER, R.G., and DURVASULA, S. (1991) Effects of consumption frequency on believability and attitudes toward alcohol warning labels. *Journal of Consumer Affairs*, 25, 323–338.

ASPER, O. (1972) The detectability of two slow vehicle warning devices. *Journal of Safety Research*, 4, 85–89.

BABER, C. and WANKLING, J. (1992) An experimental comparison of text and symbols for in-car reconfigurable displays. *Applied Ergonomics*, 23, 255–262.

BARLOW, T. and WOGALTER, M.S. (1991) Alcohol beverage warnings in print advertisements. In *Proceedings of Human Factors Society 35th Annual Meeting*. Santa Monica, CA: Human Factors Society, pp. 451–455.

BELTRAMINI, R.F. (1988) Perceived believability of warning label information presented in cigarette advertising. *Journal of Advertising*, 17, 26–32.

BOOHER, H.R. (1975) Relative comprehensibility of pictorial information and printed words in proceduralized instructions. *Human Factors*, 17, 266–277.

BRESNAHAN, T. and BRYK, J. (1975) The hazard association values of accident-prevention signs. *Professional Safety*, 17–25.

CAHILL, M.C. (1975) Interpretability of graphic symbols as a function of context and experience factors. *Journal of Applied Psychology*, 60, 376–380.

CAHILL, M.C. (1976) Design features of graphic symbols varying in interpretability. *Perceptual and Motor Skills*, 42, 647–653.

CAIRNEY, P.T. and SLESS, D. (1982) Communication effectiveness of symbolic safety signs and different user groups. *Applied Ergonomics*, 13, 91–97.

CANADIAN INTER-MARK (1972a) *Magnascope National Awareness Survey: Warning Symbols—April, 1972*. Ottawa, Ontario, Canada: Department of Consumer and Corporate Affairs.

CANADIAN INTER-MARK (1972b) *Magnascope National Awareness Surveys: Warning Symbols—April and August, 1972*. Ottawa, Ontario, Canada: Department of Consumer and Corporate Affairs.

CHILDERS, T.L., HECKLER, S.E., and HOUSTON, M.J. (1986) Memory for the visual and verbal components of print advertisements. *Psychology and Marketing*, 3, 137–150.

CHILDERS, T.L. and HOUSTON, M.J. (1984) Conditions for a picture superiority effect on consumer memory. *Journal of Consumer Research*, 11, 643–654.

CHRIST, R.E. (1974) Color research for visual displays. In *Proceedings of Human Factors Society 18th Annual Meeting*. Santa Monica, CA: Human Factors Society, pp. 542–546.

CHY-DEJORAS, E.A. (1992) Effects of an aversive vicarious experience and modeling on perceived risk and self-protective behavior. In *Proceedings of Human Factors Society 36th Annual Meeting*. Santa Monica, CA: Human Factors Society, pp. 603–607.

COLEMAN, M. and LIAU, T.L. (1975) A computer readability formula designed for machine scoring. *Journal of Applied Psychology*, 60, 283–284.

COLLINS, B.L., LERNER, N.D., and PIERMAN, B.C. (1982) NBSIR 82-2485, *Symbols for Industrial Safety*. Washington, DC: National Bureau of Standards.

COMBS, B. and SLOVIC, P. (1979) Newspaper coverage of causes of death. *Journalism Quarterly*, Winter 1979, 873–843, 849.

DEJOY, D.M. (1989) Consumer product warnings: review and analysis of effectiveness research. In *Proceedings of Human Factors Society 33rd Annual Meeting*. Santa Monica, CA: Human Factors Society, pp. 936–940.

DEJOY, D.M. (1991) A revised model of the warnings process derived from value-expectancy theory. In *Proceedings of Human Factors Society 35th Annual Meeting*. Santa Monica, CA: Human Factors Society, pp. 1043–1047.

DESAULNIERS, D.R. (1989) Consumer products hazards: what will we think of next? In *Proceedings of INTERFACE '89, The Sixth Symposium on Human Factors and Industrial Design in Consumer Products*. Santa Monica, CA: Human Factors Society, pp. 115–120.

DINGUS, T.A., WREGGIT, S.S., and HATHAWAY, J.A. (1993) Warning variables affecting personal protective equipment use. *Safety Science*, 16, 655–674.

DONNER, K.A. (1990) *The effects of warning modality, warning formality, and product on safety behavior*. Masters thesis, Rice University, Houston, TX.

DORRIS, A.L. and PURSWELL, J.L. (1977) Warnings and human behavior: implications for the design of product warnings. *Journal of Product Liability*, 1, 255–263.

DORRIS, A.L. and TABRIZI, M.F. (1978) An empirical investigation of consumer perception of product safety. *Journal of Products Liability*, 2, 155–163.

DREYFUSS, H. (1970) SAE 700103, *Visual Communications: A Study of Symbols*. Society of Automotive Engineers.

DUFFY, R.D., KALSHER, M.J., and WOGALTER, M.S. (1993) The effectiveness of an interactive warning in a realistic product-use situation. In *Proceedings of Human Factors Society 37th Annual Meeting*. Santa Monica, CA: Human Factors Society, pp. 935–939.

DUFFY, T.M. (1985) Readability formulas: what's the use? In DUFFY, T.M. and WALLER, R. (eds), *Designing Usable Texts*. Orlando, FL: Academic Press, pp. 113–140.

DUFFY, T.M. and KABANCE, P. (1982) Testing a readable writing approach to text revision. *Journal of Educational Psychology*, 74, 733–748.

DUNN, J.G. (1972) Subjective and objective risk distribution. *Occupational Psychology*, 46, 183–187.

EASTERBY, R.S. and ZWAGA, H.J.G. (1976) Evaluation of Public Information Symbols, ISO Tests: 1975 Series. AP Report No. 60, March.

EBERHARD, J. and GREEN, P. (1989) UMTRI-89-26, *The Development and Testing of Warnings for Automotive Lifts*. Ann Arbor, MI: University of Michigan Transportation Research Institute.

FARID, M.I. and LIRTZMAN, S.I. (1991) Effects of hazard warning on workers' attitudes and risk-taking behavior. *Psychological Reports*, 68, 659–673.

FIDELL, S. (1978) Effectiveness of audible warning signals for emergency vehicles. *Human Factors*, 20, 19–26.

FISCHER, P.M., RICHARDS, J.W., BERMAN, E.J., and KRUGMAN, D.M. (1989) Recall and eye tracking study of adolescents viewing tobacco advertisements. *Journal of the American Medical Association*, 251, 84–89.

FISCHHOFF, B. (1977) Cognitive liabilities and product liability. *Journal of Product Liability*, 1, 207–219.

FISCHHOFF, B., SLOVIC, P., LICHTENSTEIN, S., READ, S., and COMBS, B. (1978) How safe is safe enough? A psychometric study of attitudes towards technological risks and benefits. *Policy Sciences*, 9, 127–152.

FLESCH, R.F. (1948) A new readability yardstick. *Journal of Applied Psychology*, 32, 384–390.

FRANTZ, J.P., MILLER, J.M., and LEHTO, M.R. (1991) Must the context be considered when applying generic safety symbols: a case study in flammable contact adhesives. *Journal of Safety Research*, 22, 147–161.

FRIEDMANN, K. (1988) The effect of adding symbols to written warning labels on user behavior and recall. *Human Factors*, 30, 507–515.

FUNKHOUSER, G.R. (1984) An empirical study of consumers' sensitivity to the wording of affirmative disclosure messages. *Journal of Public Policy and Marketing*, 3, 26–37.

GALLUP CANADA, INC. (1989) *Consumer Survey on the Labeling and Packaging of Hazardous Chemical Products*, Vol. 1, *Detailed Findings*. Hull, Quebec, Canada: Consumer and Corporate Affairs Canada.

GALLUSCIO, E.H. and FJELDE, K. (1993) Eye movement and reaction time measures of the effectiveness of caution signs. *Safety Science*, 16, 627–636.

GILL, R.T., BARBERA, C., and PRECHT, T. (1987) A comparative evaluation of warning label designs. In *Proceedings of Human Factors Society 31st Annual Meeting*. Santa Monica, CA: Human Factors Society, pp. 476–478.

GODFREY, S.S., ALLENDER, L., LAUGHERY, K.R., and SMITH, V.L. (1983) Warning messages: will the consumer bother to look? In *Proceedings of Human Factors Society 27th Annual Meeting*. Santa Monica, CA: Human Factors Society, pp. 950–954.

GODFREY, S.S. and LAUGHERY, K.R. (1984) The biasing effects of product familiarity on consumers' awareness of hazard. In *Proceedings of Human Factors Society 28th Annual Meeting*. Santa Monica, CA: Human Factors Society, pp. 483–486.

GODFREY, S.S., LAUGHERY, K.R., YOUNG, S.L., VAUBEL, K.P., BRELSFORD, J.W., LAUGHERY, K.A., and HORN, E. (1991) The new alcohol warning labels: how noticeable are they? In *Proceedings of Human Factors Society 35th Annual Meeting*. Santa Monica, CA: Human Factors Society, pp. 446–450.

GOLDHABER, G.M. and deTURCK, M.A. (1988) Effects of consumers' familiarity with products on attention to and compliance with warnings. *Journal of Products Liability*, 11, 29–37.

GRAY, W.B. (1975) *How to Measure Readability*. Philadelphia: Dorrance and Co.

GREEN, P. and PEW, R.W. (1978) Evaluating pictographic symbols: an automotive application. *Human Factors*, 20, 103–114.

HATEM, A.T. (1993) *The effect of performance level on warning compliance*. Masters thesis, Purdue University, Lafayette, IN.

HODGKINSON, R. and HUGHES, J. (1982) Developing wordless instructions: a case history. *IEEE Transactions on Professional Communication*, 25, 74–79.

JAYNES, L.S. and BOLES, D.B. (1990) The effect of symbols on warning compliance. In *Proceedings of Human Factors Society 34th Annual Meeting*. Santa Monica, CA: Human Factors Society, pp. 984–987.

KALSHER, M.J., CLARKE, S.W., and WOGALTER, M.S. (1991) Posted warning placard: effects on college students' knowledge of alcohol facts and hazards. In *Proceedings of Human Factors Society 35th Annual Meeting*. Santa Monica, CA: Human Factors Society, pp. 456–460.

KANOUSE, D.E. and HAYES-ROTH, B. (1980) *Product Labeling and Health Risks*. Banbury Report 6, Cold Spring Harbor Laboratory.

KARNES, E.W. and LEONARD, S.D. (1986) Consumer product warnings: reception and understanding of warning information by final users. In *Trends in Ergonomics/Human Factors III*, Part B, *Proceedings of the Annual International Industrial Ergonomics and Safety Conference*, pp. 995–1003.

KASKUTAS, L. and GREENFIELD, T. (1991) Knowledge of warning labels on alcoholic beverage containers. In *Proceedings of Human Factors Society 35th Annual Meeting*. Santa Monica, CA: Human Factors Society, pp. 441–445.

KASPER, R.G. (1980) Perceptions of risk and their effects on decision making. In SCHWING, R.C. and ALBERS, W.A. (eds), *Societal Risk Assessment*. New York: Plenum Press, pp. 71–84.

KELLER, A.D. (1972) Evaluation of graphic symbols for safety and warning signs. Masters thesis. University of South Dakota.

KLARE, G.R. (1974–1975) Assessing readability. *Reading Research Quarterly*, 10, 62–102.

KRAMPEN, M. (1965) Signs and symbols in graphic communication. *Design Quarterly*, 62, 3–31.

KREIFELDT, J.G. and RAO, K.V.N. (1986) Fuzzy sets: an application to warnings and instructions. In *Proceedings of Human Factors Society 30th Annual Meeting*. Santa Monica, CA: Human Factors Society, pp. 1192–1196.

LANGLOIS, J.A., WALLEN, B.A., TERET, S.P., BAILEY, L.A., HERSHEY, J.H., and PEELER, M.O. (1991) The impact of specific toy warning labels. *The Journal of the American Medical Association*, 265, 2848–2850.

LAUGHERY, K.R. and BRELSFORD, J.W. (1991) Receiver characteristics in safety communications. In *Proceedings of Human Factors Society 35th Annual Meeting*. Santa Monica, CA: Human Factors Society, pp. 1068–1072.

LAUGHERY, K.R., VAUBEL, K.P., YOUNG, S.L., BRELSFORD, J.W., and ROWE, A.L. (1993) Explicitness of consequence information in warnings. *Safety Science*, 16, 597–614.

LAUGHERY, K.R. and YOUNG, S.L. (1991) An eye-scan analysis of accessing product warning information. In *Proceedings of Human Factors Society 35th Annual Meeting*. Santa Monica, CA: Human Factors Society, pp. 585–589.

LAUX, L.F. and BRELSFORD, J.W. (1989) Locus of control, risk perception, and precautionary behavior. In *Proceedings of INTERFACE '89, The Sixth Symposium on Human Factors and Industrial Design in Consumer Products*. Santa Monica, CA: Human Factors Society, pp. 121–124.

LAUX, L.F., MAYER, D.L., and THOMPSON, N.B. (1989) Usefulness of symbols and pictorials to communicate hazard information. In *Proceedings of INTERFACE '89, The Sixth Symposium on Human Factors and Industrial Design in Consumer Products*. Santa Monica, CA: Human Factors Society, pp. 79–83.

LEHTO, M.R. and PAPASTAVROU, J.D. (1993) Models of the warning process: important implications towards effectiveness. *Safety Science*, 16, 569–596.

LEONARD, S.D., CREEL, E., and KARNES, E.W. (1990) On warning words. *Safety News*, Human Factors Society's Safety Technical Group Newsletter, December, 3–4.

LEONARD, S.D., CREEL, E., and KARNES, E.W. (1991a) Commonly used hazard descriptors are not well understood. In KARWOWSKI, W. and YATES, J.W. (eds), *Advances in Industrial Ergonomics and Safety*, Vol. 3. London: Taylor and Francis, pp. 731–738.

LEONARD, S.D., CREEL, E., and KARNES, E.W. (1991b) Adequacy of responses to warning terms. In *Proceedings of Human Factors Society 35th Annual Meeting*. Santa Monica, CA: Human Factors Society, pp. 1024–1028.

LEONARD, S.D., HILL, G.W., and KARNES, E.W. (1989) Risk perception and use of warnings. In *Proceedings of Human Factors Society 33rd Annual Meeting*. Santa Monica, CA: Human Factors Society, pp. 550–554.

LEONARD, S.D. and MATTHEWS, D. (1986) How does the population interpret warning signals? In *Proceedings of Human Factors Society 30th Annual Meeting*. Santa Monica, CA: Human Factors Society, pp. 116–120.

LERNER, N.D. and COLLINS, B.L. (1980) PB81-185647, *The Assessment of Safety Symbol Understandability by Different Testing Methods*. Washington, DC: National Bureau of Standards.

LICHTENSTEIN, S., SLOVIC, P., FISCHHOFF, B., LAYMAN, M., and COMBS, B. (1978) Judged frequency of lethal events. *Journal of Experimental Psychology: Human Learning and Memory*, 4, 551–578.

LOKEN, B. and HOWARD-PITNEY, B. (1988) Effectiveness of cigarette advertisements on women: an experimental study. *Journal of Applied Psychology*, 73, 378–382.

Loo, R. (1978) Individual differences and the perception of traffic signs. *Human Factors*, 20, 65–74.

LOOMIS, J.P. and PORTER, R.F. (1982) The performance of warning systems in avoiding controlled-flight-into-terrain (CFIT) accidents. *Aviation, Space, and Environmental Medicine*, 53, 1085–1090.

MCCARTHY, J.V. and HOFFMANN, E.R. (1977) *The Difficulty that Traffic Signs Present to Poor Readers*. Commonwealth of Australia, Department of Transport.

MCCARTHY, R.L., FINNEGAN, J.P., KRUMM-SCOTT, S., and MCCARTHY, G.E. (1984) Product information presentation, user behavior, and safety. In *Proceedings of Human Factors Society 28th Annual Meeting*. Santa Monica, CA: Human Factors Society, pp. 81–85.

MACGREGOR, D.G. (1989) Inferences about product risks: a mental modeling approach to evaluating warnings. *Journal of Products Liability*, 12, 75–91.

MACKETT-STOUT, J. and DEWAR, R.E. (1981) Evaluation of symbolic public information signs. *Human Factors*, 23, 139–151.

MACKINNON, D.P., STACY, A.W., NOHRE, L., and GEISELMAN, R.E. (1992) Effects of processing depth on memory for the alcohol warning label. In *Proceedings of Human Factors Society 36th Annual Meeting*. Santa Monica, CA: Human Factors Society, pp. 538–542.

MAIN, B.W., RHOADES, T.P., and FRANTZ, J.P. (1994) Conveying flammability hazards and precautions to lay users: an examination of current labeling systems. *National Fire Protection Association Journal*, January/February, 71–76.

MALLETT, L., VAUGHT, C., and BRNICH, M.J. (1993) Sociotechnical communication in an underground mine fire: a study of warning messages during an emergency evacuation. *Safety Science*, 16, 709–728.

MARTIN, G.L. and HEIMSTRA, N.W. (1973) The perception of hazard by children. *Journal of Safety Research*, 5, 238–246.

MORI, M. and ABDEL-HALIM, M.H. (1981) Road sign recognition and non-recognition. *Accident Analysis and Prevention*, 13, 101–115.

MORRIS, L.A. and KLIMBERG, R. (1986) A survey of aspirin use and Reye's syndrome awareness among parents. *American Journal of Public Health*, 76, 1422–1423.

OTSUBO, S.M. (1988) A behavioral study of warning labels for consumer products: perceived danger and use of pictographs. In *Proceedings of Human Factors Society 32nd Annual Meeting*. Santa Monica, CA: Human Factors Society, pp. 536–540.

PERRY, R.W. (1983) Population evacuation in volcanic eruptions, floods, and nuclear power plant accidents: some elementary comparisons. *Journal of Community Psychology*, 11, 36–47.

PLUMMER, R.W., MINARCH, J.J., and KING, L.E. (1974) Evaluation of driver comprehension of word versus symbol highway signs. In *Proceedings of Human Factors Society 18th Annual Meeting*. Santa Monica, CA: Human Factors Society, pp. 202–208.

POLZELLA, D.J., GRAVELLE, M.D., and KLAUER, K.M. (1992) Perceived effectiveness of danger signs: a multivariate analysis. In *Proceedings of Human Factors Society 36th Annual Meeting*. Santa Monica, CA: Human Factors Society, pp. 931–934.

POWELL, K.B. (1981) Readability formulas: used or abused? *IEEE Transactions on Professional Communication*, 24, 43–44.

PYRCZAK, F. and ROTH, D.H. (1976) The readability of directions on non-prescription drugs. *Journal of the American Pharmaceutical Association*, 16, 242–243, 267.

RETHANS, A.J. (1980) Consumer perceptions of hazards. In *PLP-80 Proceedings*, pp. 25–29.

RETHANS, A.J. and ALBAUM, G.S. (1980) Towards determinants of acceptable risk: the case of product risks. *Advances in Consumer Research*, 8, 506–510.

ROBERTSON, L.S., O'NEILL, B., and WIXOM, C.W. (1972) Factors associated with observed safety belt use. *Journal of Health and Social Behavior*, 13, 18–24.

SAMET, M.G., GEISELMAN, R.E., and LANDEE, B.M. (1982) A human performance evaluation of graphic symbol-design features. *Perceptual and Motor Skills*, 54, 1303–1310.

SHINAR, D. and DRORY, A. (1983) Sign registration in daytime and nighttime driving. *Human Factors*, 25, 117–122.

SIEGEL, A.I., LAMBERT, J.V., and BURKETT, J.R. (1974) NTIS No. AD 786 849, *Techniques for Making Written Material More Readable/Comprehensible*. Lowry Air Force Base: Air Force Human Resources Laboratory.

SILVER, N.C., LEONARD, D.C., PONSI, K.A., and WOGALTER, M.S. (1991) Warnings and purchase intentions for pest-control products. *Forensic Reports*, 4, 17–33.

SIMPSON, C.A. and WILLIAMS, D. (1980) Response time effects of alerting tone and semantic context for synthesized voice cockpit warnings. *Human Factors*, 22, 319–330.

SLOVIC, P. (1978) The psychology of protective behavior. *Journal of Safety Research*, 10, 58–68.

SLOVIC, P., FISCHHOFF, B., and LICHTENSTEIN, S. (1978) Accident probabilities and seat belt usage: a psychological perspective. *Accident Analysis and Prevention*, 10, 281–285.

SLOVIC, P., FISCHHOFF, B., and LICHTENSTEIN, S. (1979) Rating the risks. *Environment*, 21, 14–39.

SLOVIC, P., FISCHHOFF, B., and LICHTENSTEIN, S. (1980) Informing people about risk. In MORRIS, L.A., MAZIS, M.B., and BAROFSKY, I. (eds), *Product Labeling and Health Risks*. Banbury Report 6, Cold Spring Harbor Laboratory, pp. 165–180.

SNYDER, H.L. and TAYLOR, G.B. (1979) The sensitivity of response measures of alphanumeric legibility to variations in dot matrix display parameters. *Human Factors*, 21, 457–471.

STRAWBRIDGE, J.A. (1986) The influence of position, highlighting, and imbedding on warning effectiveness. In *Proceedings of Human Factors Society 30th Annual Meeting*. Santa Monica, CA: Human Factors Society, pp. 716–720.
TESTIN, F.J. and DEWAR, R.E. (1981) Divided attention in a reaction time index of traffic sign perception. *Ergonomics*, 24, 111–124.
TREISMAN, A. (1986) Features and objects in visual perception. *Scientific American*, 255, 114–124.
TREISMAN, A. (1989) Features and objects, The Fourteenth Bartlett Memorial Lecture. *Quarterly Journal of Experimental Psychology*, 19, 1–17.
VAUBEL, K.P. (1990) Effects of warning explicitness of consumer product purchase intentions. In *Proceedings of Human Factors Society 34th Annual Meeting*. Santa Monica, CA: Human Factors Society, pp. 513–517.
VAUBEL, K.P. and BRELSFORD, J.W. (1991) Product evaluations and injury assessments as related to preferences for explicitness in warnings. In *Proceedings of Human Factors Society 35th Annual Meeting*. Santa Monica, CA: Human Factors Society, pp. 1048–1052.
VAUBEL, K.P. and YOUNG, S.L. (1992) Components of perceived risk for consumer products. In *Proceedings of Human Factors Society 36th Annual Meeting*. Santa Monica, CA: Human Factors Society, pp. 494–498.
Webster's New Universal Unabridged Dictionary (1983) New York: Simon & Schuster, Inc.
WHEALE, J.L. (1983) Evaluation of an experimental central warning system with a synthesized voice component. *Aviation, Space, and Environmental Medicine*, June, 517–523.
WICKENS, C. (1984) *Engineering Psychology and Human Performance*. Columbus, OH: Merrill.
WOGALTER, M.S. and BARLOW, T. (1990) Injury severity and likelihood in warnings. In *Proceedings of Human Factors Society 34th Annual Meeting*. Santa Monica, CA: Human Factors Society, pp. 580–583.
WOGALTER, M.S., DESAULNIERS, D.R., and BRELSFORD, J.W. (1987) Consumer products: how are the hazards perceived? In *Proceedings of Human Factors Society 31st Annual Meeting*. Santa Monica, CA: Human Factors Society, pp. 615–619.
WOGALTER, M.S., KALSHER, M.J., and RACICOT, B.M. (1992) The influence of location and pictorials on behavioral compliance to warnings. In *Proceedings of Human Factors Society 36th Annual Meeting*. Santa Monica, CA: Human Factors Society, pp. 1029–1033.
WOGALTER, M.S., RASHID, R., CLARKE, S.W., and KALSHER, M.J. (1991) Evaluating the behavioral effectiveness of a multi-modal voice warning sign in a visually cluttered environment. In *Proceedings of Human Factors Society 35th Annual Meeting*. Santa Monica, CA: Human Factors Society, pp. 718–722.
WOGALTER, M.S. and YOUNG, S.L. (1994) Enhancing warning compliance through alternative product label designs. *Applied Ergonomics*, 25, 53–57.
WOGALTER, M.S., RACICOT, B.M., KALSHER, M.J., and SIMPSON, S.N. (1993) Behavioral compliance to personalized warning signs and the role of perceived relevance. In *Proceedings of Human Factors Society 37th Annual Meeting*. Santa Monica, CA: Human Factors Society, pp. 950–954.
WOGALTER, M.S., SOJOURNER, R.J., and BRELSFORD, J.W. (1997) Comprehension and retention of safety pictorials. *Ergonomics*, 40, 531–542.
WOLFF, J.S. and WOGALTER, M.S. (1998) Comprehension of pictorial symbols: effects of context and test method. *Human Factors*, 40, 173–186.
YOUNG, S.L. (1995) *Components of Perceived Risk for Consumer Products*. Doctoral dissertation, Rice University, Houston, TX.
YOUNG, S.L. (1991) Increasing the noticeability of warnings: effects of pictorial, color, signal icon, and border. In *Proceedings of Human Factors Society 35th Annual Meeting*. Santa Monica, CA: Human Factors Society, pp. 580–584.
YOUNG, S.L. and WOGALTER, M.S. (1990) Comprehension and memory of instruction manual warnings: conspicuous print and pictorial icons. *Human Factors*, 32, 637–649.

YOUNG, S.L. and WOGALTER, M.S. (1988) Memory of instruction manual warnings: effects of pictorial icons and conspicuous print. In *Proceedings of Human Factors Society 32nd Annual Meeting*. Santa Monica, CA: Human Factors Society, pp. 905–909.

YOUNG, S.L., WOGALTER, M.S., LAUGHERY, K.R., MAGURNO, A., and LOVVOLL, D. (1995) Relative order and space allocation of message components in hazard warning signs. In *Proceedings of Human Factors and Ergonomics Society 39th Annual Meeting*. Santa Monica, CA: Human Factors and Ergonomics Society, pp. 969–973.

YOUNG, S.L., WOGALTER, M.S., and BRELSFORD, J.W. (1992) Relative contribution of likelihood and severity of injury to risk perceptions. In *Proceedings of Human Factors Society 36th Annual Meeting*. Santa Monica, CA: Human Factors Society, pp. 1014–1018.

ZOHAR, D., COHEN, A., and AZAR, N. (1980) Promoting increased use of ear protectors in noise through information feedback. *Human Factors*, 22, 69–79.

CHAPTER FOUR

Methodological Techniques for Evaluating Behavioral Intentions and Compliance

MICHAEL S. WOGALTER
North Carolina State University

THOMAS A. DINGUS
Virginia Polytechnic Institute and State University

The ultimate criterion of warning effectiveness is actual behavioral compliance. Given its importance, there are surprisingly few behavioral studies in the warnings literature, probably because their implementation is difficult. Instead, many studies use questionnaires to measure behavioral intentions to warning-related variables. While a link between behavior and behavioral intentions has been established in the social psychology literature, the association has not been confirmed in the warning literature. Nevertheless, sometimes questionnaire type studies that include measures of behavioral intention are the best one can do given limited resources. The main purpose of this chapter is to describe the techniques for examining behavioral intentions and actual behavioral compliance. Self-report, observational, physical trace and epidemiological methods are described. It is hoped that researchers will incorporate (or adapt) some of the techniques in future studies.

4.1 INTRODUCTION

The ultimate goal of warnings is to reduce personal injury and property damage. For this goal to be met warnings need to influence people so that they do not behave in ways that lead to personal injury and property damage. Chapter 3 by Young and Lovvoll describes methods that permit measurement of warning effects at the intermediate or pre-behavior stages of the communication-human information processing (C-HIP) model. While there is no doubt that the processes of attention, memory, and the other stages are important for effective warnings, it is the last stage of the model, the behavior stage, that is the most important. The occurrence of safe behavior is the ultimate measure of whether the warning works. If a warning is effective at the behavioral stage, the warning is probably adequate at the earlier stages. Indeed, behavioral data are so important that if only one measure of warning effectiveness can be obtained, a compliance test is the best one to do. It is superior to all other tests. While other measures (e.g., of attention or memory) are capable of evaluating important aspects relevant to warning effectiveness, the effects may not always translate into behavior.

This chapter focuses on the methods of measuring behavioral intentions and actual behavioral compliance. Behavioral intentions usually are assessed by questionnaire, while behavioral compliance usually is assessed by observing whether warning-directed behavior occurs. The main goal of this chapter is to increase understanding of the techniques that have been used or could be used. By familiarizing researchers with the range of potential methods, we hope that this chapter will facilitate the conduct of future compliance research. Chapter 11 by Silver and Braun describes specific outcomes and conclusions from studies using these techniques.

The term 'behavior' can be interpreted very broadly. Frequently it is defined as being some observable, measurable overt response with respect to some internally or externally generated stimulus. This definition is quite general and could include just about everything. For example, this definition would permit questionnaire responses to be classified as behavior. In this chapter we take a more restricted view of behavior. We define behavior as whether people do what the warning asks them to do. Although behavioral intentions (e.g., judgments of how careful they would be, etc.) probably are related to warning pertinent behavior, we do not consider them the same as actual behavioral compliance.

Given that compliance is such an important outcome for warnings, one might expect that most research would measure it. In fact, there are relatively few behavioral compliance studies in the warning literature. Most studies on warnings use the kinds of measure and technique described in Chapter 3 by Young and Lovvoll, mainly rating scales and questionnaires. More surprisingly, very few of these questionnaire studies measure behavioral intentions, which is perhaps the closest indication of behavior that paper and pencil techniques are capable of measuring. Why do so few studies measure behavior when we know it is the ultimate criterion of warning effectiveness?

The main answer to this question is that behavioral compliance research and testing are difficult to do for various reasons. The most compelling reason is that it is unethical to expose research participants to hazards. For example, it would be improper to test warnings in true life threatening circumstances such as diving into shallow water. For example, one could not ethically remove a NO DIVING sign to make a comparison to warning-present conditions. More generally, it would be unethical to use any but the best possible warning in this and other hazardous situations. Most behavioral studies involve specially created 'hazardous situations' which appear realistic but where the safety of the research participants is always protected.

A second reason for the difficulty of conducting behavioral compliance tests is control. One may not have the opportunity to directly manipulate the conditions that prompt a hazard warning. Warnings for severe weather where the events are random and infrequent is an example. Also one can not usually measure compliance to products like drain cleaners and condoms in natural environments such as people's private homes.

Finally, beyond the above-mentioned problems, compliance tests frequently are prohibited because of limited resources and capabilities. Behavioral compliance research is time and labor intensive and expensive. Additionally, it is difficult to manipulate warnings printed on labels of products sold in the stores. Permissions and appropriate label making capabilities would be needed to conduct such studies. Limited resources is a particularly common plight for university researchers studying warning issues as there are virtually no federal grants available to fund the research. However, funding for warnings should not be a problem for large companies who may be selling hazardous products. US tort law says that product manufacturers are responsible for providing adequate warnings for associated hazards that can not be (practically) eliminated. It would seem then that companies—with their significant financial resources, superior knowledge, and legal responsibility—would take steps to assess the adequacy of the warnings for their products through behavioral compliance testing.

We recognize, however, that situations do occur where it is not feasible to conduct behavioral testing. In some instances, the warning may be needed before behavioral testing can be conducted. However, generally it is still possible (and advisable) to perform follow-up warning effectiveness tests after the initial warning has been placed in the stream of use. Some of these 'post' tests could be behavioral in nature, and could assist in determining whether the warning is adequate or should be replaced with a better warning. Thus, while behavioral testing can not always be performed, we want to encourage its use whenever possible.

In the remainder of this chapter, we describe and comment on methodological issues associated with assessing behavioral intentions, followed by a similar but more extensive review of the techniques involved in behavioral compliance research.

4.2 BEHAVIORAL INTENTIONS

In this section we describe some of the methods that could be and have been used to measure behavioral intentions. Given that sometimes behavioral testing cannot be done, an important issue is whether behavioral intentions predict behavior. Considerable research in social psychology (Ajzen and Fishbein, 1977; Eagly and Chaiken, 1993, 1996; Kim and Hunter, 1993; Eckes and Six, 1994; Kraus, 1995) as well as in other fields such as medicine and health (Taylor, 1991; Brannon and Feist; 1992) indicate that behavioral intentions do predict behavior. In other words, what people say they intend to do reflects to varying degrees what they actually do. Not all studies show a relationship between intentions and actual measured behavior, but in general the bulk of the research indicates that prediction of actual behavior from intention judgments depends on whether the context/ scenario/state-of-mind during which the intentions are taken are similar to the specific situation in which the behavior is to occur. The closer the match, the better the prediction. Nevertheless, the fact that some research shows no match between intentions and actual behavior should make interested parties somewhat uncomfortable when questionnaire/ interview results are not subsequently confirmed by behavioral data of some type. Prudence would recommend withholding judgment until the results are backed up by other studies— preferably each using a different research methodology. Multiple methods showing the same effect would allow stronger statements regarding the generalizability of the warnings-related phenomenon.

Additionally, we would be more confident about the utility of behavioral intention measures with respect to warnings if there were a lot of research showing a direct tie to behavior. However, specific data on the relationship between behavior and behavioral intentions in the warnings domain is virtually nonexistent, and much of the supporting evidence is indirect. The necessary research would determine people's perceptions, beliefs and intentions *before* behavioral compliance is assessed. That is, behavioral intentions measurement needs to be made at a point in the task sequence before compliance might take place. Behavioral intention data concordant with actual behavior would indicate useful prediction capability of the behavioral intentions measure. As we have said, no studies in the warnings to date have done this. Nevertheless, we believe that knowledge gained in other domains on this topic has at least some generalizability to the warning domain.

There is one side note to this issue. A number of warning compliance studies have had participants complete a questionnaire *after* the main behavioral compliance measurement phase is over. DeJoy (1989; see also Chapter 9) reviews several behavioral compliance studies that also include a questionnaire asking whether they noticed and read the warning. The data show an across-the-board drop in the percentages of participants who report noticing the warning, who report reading the warning, and who comply with the warning. Frequently these questionnaire measures show significant correlations with behavioral compliance (e.g., Friedmann, 1988; Jaynes and Boles, 1990; Otsubo, 1988; Wogalter, Godfrey, Fontenelle, Desaulniers, Rothstein, and Laughery, 1987; Wogalter, Kalsher, and Racicot, 1993b). One example of such data is by Otsubo (1988). For a task involving a circular saw, 74% of the participants noticed the warning, 52% said they read it, and 38% were observed to comply. On the other hand, for a task involving a jig saw, 54% of the participants noticed the warning, 25% read it, and 13% complied. Thus, we can see an indication that noticing and reading roughly concurs with compliance levels. However, besides the fact that it is problematic that these predictors are assessed after the event they are supposed to predict, there also is the possibility that compliance or noncompliance might influence how people answer the questions on the follow-up questionnaire. For example, participants who did not comply in the situation may subsequently respond that they did not see the warning when they actually did. Additionally, answers to the questionnaire items may be affected by social desirability and demand characteristics (e.g., they answer in the way that they think the experimenter wants them to). Thus, we can not say with certainty that the responses on the post-task questionnaire are accurate and useful because the noticing and reading measures are retrospective reports that are subject to various biases. More useful for predicting of behavioral compliance would be the recording of precursor events, such as looking behavior (whether people are observed looking in the direction of the warning and appearing to be reading it).

4.2.1 Methodology

Behavioral intentions go by different names in the literature: precautionary intent, cautionary intent, intended carefulness, likelihood of complying, and willingness to comply. In behavior intentions research, participants typically are asked one or several questions about whether they would comply with a warning for a particular product or environmental hazard. The questions sometimes request dichotomous 'yes' or 'no' answers, but most research employs Likert-type rating scales. Participants are asked for a judgment on the extent to which they would comply with a warning. The points on the scale range might range from 1 to 7 (or some other set of numbers) with the values labeled along some or all of the points of the scale with verbal anchors. For example, the end point anchors might be 'definitely would not comply' and 'definitely would comply' with the intermediate values labeled 'somewhat likely to comply,' 'likely to comply,' and so forth. The number of scale points can vary from study to study (and within studies). The lowest ratings on most scales is zero or one. In some studies, participants are asked to estimate the percentage of people who would likely comply with the warning (the number of people out of 100 who would comply) which is similar to asking for a judgment along a 101-point (0–100) scale.

In attempting to predict behavior from behavioral intentions, some variability in the scores is necessary. Suppose that the scores on a behavioral compliance measure are all very low (near 0%) or all very high (near 100%). With very low or very high scores, it is not possible to show a statistical relationship with another variable—in this case, compliance with something else. In other words, if no one complies then you can not predict when compliance will occur from another variable. Some variability in the scores is needed to allow prediction.

It is important also that researchers do not misrepresent behavioral intentions research as behavior-based. Researchers should tell readers that the measurement involves intended or self-reported behavior. For example, Staelin (1978) uses the term 'actual behavior' to describe what is really a behavioral intention. To be fair, the point of Staelin's paper was to compare people's normative behavior (what they are supposed to do) to what they would personally do (the self-reported 'actual' behavior). Nevertheless, it is important for warning researchers to be specific and to use unambiguous terms. Readers of research articles should pay close attention to the study's method to be sure that the 'behavior' discussed is actual compliance behavior.

4.2.2 Other Behavioral Intention Measures

In this section we describe several other kinds of behavioral intention measures used in published research. Farid and Lirtzman (1991) used a very interesting behavior intention measure while assessing Egyptian workers' perceptions of different warning labels for hazardous chemicals. They assessed the workers' intention to quit, and found that workers exposed to high-hazard labels were significantly more likely to say that they would quit their jobs than workers exposed to low-hazard labels.

Intention to purchase is another behavioral intentions measure (Ursic, 1984; Silver, Leonard, Ponsi, and Wogalter, 1991; Laughery, Vaubel, Young, Brelsford, and Rowe, 1993). Sometimes this can be a highly appropriate way to assess how effectively a warning conveys the message that a particular product (e.g., an over-the-counter medication) is or is not appropriate for certain people with certain health conditions or that it should not be used in certain tasks and environments. If people say that they will purchase it for the right condition and not purchase it for a contraindicated condition, then this provides evidence that the labeling is doing its job. Research has also asked how much people would pay for a product with different warnings labels (Barlow and Wogalter, 1991; Wogalter, Forbes, and Barlow, 1993a).

4.3 BEHAVIORAL COMPLIANCE

Since the mid 1980s there has been solid growth in the number of published articles using behavioral compliance as a measure of warning effectiveness. These studies have been conducted in various creative ways and a description of the methods used in these studies is the main focus of the remainder of this chapter.

Careful thought and planning are required to create a situation, either experimental or observational, that provides interpretable behavioral compliance data. Some of the main approaches considered in compliance testing are described in the following sections.

4.3.1 Value Added

An important concept related to the influence of warnings is the extent to which a warning or a component of a warning adds value. One of the main reasons for including a control group (in which no warning is present) in experimental research is to determine the extent to which people would perform the safety behavior anyway, without the warning being present; the no-warning control condition provides a 'base' rate for the target behavior. A warning condition that shows higher levels of behavior than a control's base rate (with everything else held constant) in essence shows the value added by the warning. For example, consider the comparison between a condition without a warning (control) with a condition that has a warning added. If the control condition had 20% compliance and the warning condition had 35% compliance, the added value of including a warning is the difference between the two, or 15% compliance. Lehto and Miller (1988) call this effect 'efficiency,' which considers prior incident rate of the desired response and the rate after the warning is presented. Thus, the value-added/efficiency measure expresses whether and how much the warning makes a difference.

The added value of particular warning features can also be determined by comparing two experimental conditions that systematically differ on some dimension, e.g., comparing two colors like yellow versus blue. If a yellow warning has higher compliance than a blue warning, it indicates that the yellow color adds value to the warning's effectiveness relative to the blue color. Additionally, experiments can be designed so that warnings differ in multiple systematic ways so that one can determine quantitatively the relative value of the different features and the interactions among them. For example, Adams and Edworthy (1995) showed that the effect of changing character size produces a greater impact on effectiveness ratings than a change in border thickness. Assuming that this finding is confirmed in a behavioral compliance study, this result would have the implication that warning designers should take greater effort in trying to increase the size of the print than the thickness of the warning border. By having more than one factor in a single experiment, it is possible to see whether and how they interact. For example, one might find that combining larger sized characters with a thick border adds value beyond that expected by the simple linear addition of each of the component effects. Unfortunately, except for a few studies (e.g., Braun and Silver, 1995; Wogalter et al., 1993a), researchers have not yet extensively employed experimental designs that can give the relative effect sizes between component factors (Cox, Wogalter, Stokes, and Murff, 1997). Our knowledge about warning design would be benefited greatly by research manipulating more than one factor in the same experiment. Such research would aid our understanding of the relative importance of certain warning features. See Edworthy and Adams (1996) for an extensive discussion of this point.

4.3.2 Incidental Exposure Paradigm

Many of the behavioral compliance studies in the literature use an incidental exposure experimental paradigm. In this paradigm, participants are not informed that the study deals with warnings. Participants are led to believe the purpose of the research is something other than warnings, i.e., a 'cover' story is given. The warning is presented to participants in the context of a set of tasks that they are trying to accomplish—it occurs incidentally, simulating how people are most often exposed to warnings in real life. For example, some behavioral compliance research studies have used a chemistry demonstration task. In this protocol, participants are led to believe that the study concerns how people perform a series of steps in a chemistry laboratory demonstration procedure. Across the many studies that have used this paradigm, the warning is exposed in various systematic ways (e.g., in task instructions, on a sign, from a digitized voice recording) without any explicit mention (or implicit suggestion). The warning occurs as part of the situation in which the participants' goal is to measure and mix various chemical substances and solutions. Before and during these procedures, nothing is mentioned about warnings until the study is over when participants are debriefed about the true purpose of the research. In short, the incidental paradigm makes the experimental situation realistic in that people are trying to perform a set of tasks where a warning could be present.

The incidental paradigm is similar to work conducted in the human memory and social psychology literature. In the human memory literature, the incidental paradigm often is contrasted with an intentional memory paradigm. With intentional conditions, participants are told that they need to learn the material (e.g., explicitly told to memorize a list of words) and the point of the study is readily apparent to the participant. With incidental conditions (e.g., Craik and Lockhart, 1972), participants are led to believe that they are being exposed to the material for some other reason (e.g., to get their subjective, qualitative judgments of the material), and later they are given a surprise memory test of the material. Most questionnaire research on warnings could be categorized as being intentional, given that participants are asked explicitly to evaluate a set of warnings. Most behavioral compliance studies are incidental in that participants are led to believe initially that the research has some other (non-warning) purpose.

The incidental exposure paradigm does involve some level of deception. Ethics committees (Institutional Review Boards or IRBs) at universities and other organizations are sometimes concerned that deception is contradictory to the notion of 'informed consent,' a hallmark procedure enabling participants to play a role in choosing their own exposure to risk. Nevertheless, IRB committees will grant permission to conduct incidental exposure studies if (a) the situation is carefully planned so that there is virtually no risk of harm (e.g., the participant stopped prior to any hazardous action); (b) the study's rationale shows that the potential benefits of the research outweigh the costs of not immediately informing participants of the ruse (the benefit, of course, is that the results can help to produce better warnings, and ultimately, reduce injuries); and (c) a full/complete debriefing takes place immediately following the study's completion that describes the true nature of the study and what specific factors were being examined. Compared to certain areas of social psychology, the deception in warning studies is quite mild. IRB committees also like to see that participants' names and their performance are held in a confidential manner and that the consent form makes it clear that participants can discontinue their participation in the study at any time without penalty. With these safeguards in place, most oversight committees will approve the procedures.

Behavioral compliance studies can also make valid use of the intentional exposure paradigm. Consider the following hypothetical evaluation of warning effectiveness that is similar to that used in the work of Geller and associates (Geller, Casali, and Johnson, 1980; Johnson and Geller, 1984; Roberts and Geller, 1994). Suppose people are explicitly warned that compliance to a seat belt regulation will be recorded by hidden cameras and that failure to comply may lead to substantial fines (let us say $50 or so). This blatant intentional warning will no doubt be highly effective in getting people to wear their seat belts. One way to measure the warning's effectiveness in this scenario is to collect base line seat belt use prior to a warning announcement and then compare that wearing rate to a similar period after the warning is presented. More complex variations of this procedure also could be employed. The point is that it may be appropriate to use an intentional exposure protocol in a behavioral compliance study in certain circumstances.

4.3.3 Participant Populations

Participants in warning research should be representative of the intended target population. However, in practice most studies fall short of this goal. One reason for this limitation is that it is difficult to bring nonstudents to university laboratory locations for a variety of cost and logistical reasons. Most laboratory-based behavioral compliance studies employ participants from a pool of undergraduate students taking introductory psychology courses for which there is usually a research participation requirement. It is important to take steps to ensure that the details of an incidental-type study do not get communicated to participants before they arrive at the laboratory. One can reduce contamination by asking participants during debriefing not to tell anyone about the study until after the study is completed. Also it helps to tell them why they need to withhold telling others—that it is important that future participants know very little about the study beforehand because otherwise it would affect the results adversely. With this information, most participants are willing to adhere to the request. Also, if one is using good experimental procedures like randomly assigning participants to conditions, then any 'compromised' participants will be equally distributed across conditions and should not bias the final outcome seriously.

4.3.4 Demand Characteristics

The concept of demand characteristics is important to behavioral compliance research because an otherwise well executed experiment sometimes can provide incorrect or misleading conclusions about the effects. Specifically, an experiment has demand characteristics when participants are forced to behave in a certain way because of the circumstances of the particular experimental situation. For example, suppose that research participants are told to use a hammer to accomplish some task. Suppose further that there is a warning on the hammer that says 'Do not use if handle is cracked,' and the only hammer provided has a cracked handle, then essentially the research participants are being encouraged to ignore the warning. Participants may believe that failing to do the task may jeopardize the receipt of course credit or promised monetary compensation associated with their participation in the study. Now let us consider a slightly different situation. Here the situation is identical to the one described above except that another hammer is available that does not have a cracked handle. Assuming the crack in one hammer is apparent and the warning on both hammers is prominent and conspicuous, the results would surely be different than the earlier-described single hammer study. Generally people will comply with warnings when it does not require much effort to do so. If there is no alternative hammer, then the only correct solution would be to discontinue participation in the experiment, which as we have said, might be perceived as 'costly'. In other words, certain characteristics of the situation can sometimes 'demand' participants behave in a certain way.

4.3.5 Data Collection Techniques in Behavioral Compliance Studies

Behavioral compliance research can be classified in a number of ways. We have already discussed one distinction: incidental versus intentional. Another categorization involves the extent to which the research is laboratory-based, field-based or something in between. Laboratory studies tend to be (a) more highly controlled, and thereby have greater internal validity, (b) more sensitive to manipulations between conditions in the study, and (c) frequently involve the use of undergraduate students as participants in the research. Field studies tend to be (a) less well controlled, (b) less sensitive in detecting effects between manipulated variables except when many participants are involved, (c) more externally valid (i.e., concurs with real-life situations), and (d) tend to involve nonstudents (i.e., a wider range of participant demographics). Field studies tend to be more similar to real life situations than laboratory studies.

Some behavioral compliance studies are not so easy to categorize as 'laboratory' or 'field.' Thus, a study conducted at a shopping center mall involving dish washing cleaning solutions is an example in this gray area. Because the types of behavior are being studied outside of the normal place that dishes are washed, i.e., the home kitchen sink, we would not classify this research as a true field study or a true laboratory study. This type of study might be termed a quasi-field study (see also Cox et al., 1997).

There are other ways to classify behavioral compliance research. In large part, most of the studies that have been performed to date have been experimental; that is, they have involved some explicit manipulation of variables by the researchers. This kind of study has the potential for giving the most solid cause–effect conclusions. Another kind of study is an *ex post facto* (after the fact) study. Several examples were mentioned briefly earlier in this chapter. Naturally occurring severe weather conditions are extreme circumstances where one has very little control over the conditions that prompt a warning. To study the effectiveness of such warnings, one might have to examine the effects after they were used. Different locales receiving different warnings would need to be matched with regard to relevant criteria so that a valid comparison can be made. Consider another example of *ex post facto* research, in this case the implementation of a law that requires a warning on a hazardous product. If no warning effectiveness measures were collected before the mandated warning is placed on the product, then the best a researcher can do to investigate the effectiveness of the warning is to find another population (matched on multiple demographic characteristics) that does not have the law and then make a comparison between the two groups. The basic weakness of an *ex post facto*-type study is the lack of researcher-controlled manipulation of conditions and absence of baseline data, and therefore, one can not draw strong conclusions about a factor's influence (despite extraordinary efforts to match conditions for everything except the variable under consideration) as would be the case with a tightly controlled experiment. Nevertheless, *ex post facto* studies sometimes make extremely valuable contributions to our knowledge that otherwise might not be obtained (see e.g., Greenfield and Kaskutas, 1993; MacKinnon, 1995; Mayer, Smith, and Scammon, 1991).

4.3.6 Method-oriented Taxonomy

Besides the above-mentioned categorizations of behavioral research, a method-oriented taxonomy is perhaps the most informative in the context of this chapter, because our primary objective is to delineate useful techniques for data collection and analysis. 'Method' in this context refers to both data collection and data analysis techniques. While most studies to date have used classical experimental methods, the actual implementation of these methods varies greatly with respect to design complexity, venue selected, measures collected and data quality. Though most studies have used classical experimental techniques, some of the most valuable contributions have come from innovative, non-experimental, observations. The methodological categories include the following.

Self-reports of behavioral compliance. The data generated by this method are subjective, gathered via questionnaire or interview. Unlike behavioral intention data, however, they are collected *ex post facto* and therefore might be contaminated, as we pointed out earlier. Such reports are different and perhaps more reliable than intention data since the participant is not predicting future behavior, but is instead reporting previous behavior. Self-report data are especially useful in field research where often the act of compliance cannot be practically observed.

Observation of behavioral compliance. Many studies simply observe participants either complying with, or not complying with, a warning. Such observation can be made directly by human observers or indirectly by other means such as a video or still camera.

Physical traces of behavioral compliance. A number of innovative techniques have been employed which allow the objective measurement of compliance without direct observation of the behavior itself. This methodology has the advantage of unobtrusively measuring compliance, thus increasing the internal validity of the data generated.

Epidemiological analysis of compliance. This technique involves the use of prospective or retrospective analysis of objective data generated either from archival or observational sources. An advantage of the use of this method is that data from a very large sample, or an entire population, can be utilized.

We describe each of these categories in more detail in the sections below.

4.4 SELF-REPORTS OF BEHAVIORAL COMPLIANCE

In many cases, it is either impossible or detrimental to observe behavioral compliance directly. In some circumstances, the behavior simply cannot be observed directly. Such is the case when a study must be performed in a location to which the experimenter cannot gain access to observe the compliance behavior. For example, Dershewitz (1979) studied two groups to determine if mothers would use safety devices to 'safety-proof' their homes. The experimental group consisted of 101 families receiving health information on home safety-proofing. The control group consisted of 104 families. Each of the 205 families was given Kindergards (plastic locking devices for cupboards and cabinets) and electric outlet covers. Once the mothers were given the free devices, two methods were available to determine whether or not they were installed: direct observation and self-report. To effectively employ direct observation, participants must be willing to let the experimenter enter their home, so that he or she can observe the outlets and cabinets in question. Direct observation in this instance is also expensive and time consuming, because it involves scheduling visits to each of a large number of locations.

Another method used by Dershewitz (1979) is self-reports. Participants were asked via telephone whether they had installed the devices. The problems associated with self-reports are similar to those discussed previously for behavioral intentions. The data are not as accurate as directly observing compliance where, barring any observation error, scoring reliability is generally 100%. While pertinent data are limited, self-reports have been shown to match well with actual behavior. For example, a study by Hunn and Dingus (1992) compared self-report data to actual physical evidence of compliance in a consumer product warning scenario. The authors found evidence that self-reported compliance and physical evidence of compliance differed by less than 5%.

Self-report compliance measures have been used by researchers in a number of domains. Planek, Schupack, and Fowler (1972) studied the impact of the National Safety Council's defensive driving course (DDC) on over 8000 drivers from 26 states. These drivers self-reported their accidents and violations for the previous year by completing a questionnaire prior to participating in the course. One year after the program, graduates of the program responded to a similar questionnaire. State records were analyzed to assess self-report accuracy. The results showed substantial agreement in what people said and what was known to have happened from driving violation records.

Self-reports of compliance are used quite often in the health domain (Taylor, 1991; Brannon and Feist, 1992). Compliance, or adherence to medication or other treatments is often critical to health maintenance or recovery. Self-reported compliance has been shown to be useful when supplemented by other measures.

4.5 OBSERVATION OF BEHAVIORAL COMPLIANCE

The primary way to measure behavioral compliance is to see whether people follow the warning-directed behavior or types of behavior while engaging in some task. A critical feature is that the observation be unobtrusive; that is, the experimental circumstances, including the experimenter's act of observing or the presence of a camera, do not influence compliance.

4.5.1 Measuring Observed Compliance

In most behavioral studies, direct observation provides information on whether individuals performed the appropriate safe behavior. Usually whether the person complied or not is completely clear and easily observed and recorded. However, sometimes, the question of whether compliance has occurred is less clear, and additional methods must be employed to handle the ambiguity. These include (a) enhancing or tightening up the classification of what constitutes acceptable compliance versus noncompliance behavior, (b) training of the experimenters or judges so that they know specifically which behavior types are recordable, and/or (c) using two experimenters or judges to record the observations concurrently as a reliability check.

Sometimes only a single measure of compliance is recorded. For example, Wogalter, Racicot, Kalsher, and Simpson (1994) used an electronic LED sign that directed the participants to behave in a single safety-related manner: to put on gloves to protect against chemical irritants involved in the task that they were asked to perform. When the participants complied, they were given a score of '1,' and when they did not, they were assigned a score of '0.' This kind of scoring can be transformed readily to usable descriptive summary statistics: a mean of these numbers gives the proportion that complied (and multiplied by 100 gives percentage complied).

Some warnings have multiple directives. For example, the warning in Jaynes and Boles (1990) requested participants to wear (a) a mask, (b) gloves, and (c) goggles; and the warnings in Wogalter, Barlow, and Murphy (1995) described the proper connection of an external disk drive to a computer and directed participants (a) to turn off the computer, (b) to eject a protective transport disk from the external drive, and (c) to physically touch the metal connection on the back of the computer to discharge any static electricity present.

When there are multiple directives, the data may be scored for analysis in several ways. One very basic method is to determine whether participants did *everything* that the warning requested. Thus, for the two example studies cited above, the answer to the question depends on whether they behaved in all three ways. If they did, participants were given a score of '1,' but if only two, one or none of the three warning-directed types of behavior occurred, they are assigned a score of '0.'

Some studies employing warnings with multiple directives analyze each of the compliance behavior types separately. So for the examples given above, three separate sets of scores for each participant would be collected (one for each type of warning-directed behavior) and analyzed. In the Wogalter et al. (1995) study separate analyses were performed for the behavior pattern of turning off the computer, another analysis for the behavior pattern of ejecting the transport disk, etc. A potential benefit of analyzing separate compliance behavior patterns is that a richer and more complete picture of compliance-related behavior may be obtained.

Statistical analysis of '1' and '0' scores generally requires chi-square or nonparametric statistical tests to determine whether there are significant differences between conditions. This is the conventional method of evaluating categorical or nominal data. However, Cochran (1950) argues that analysis of variance (ANOVA) is a valid, reasonably robust test of binomial (dichotomous) data. There are two important advantages of using ANOVA techniques: (a) we can use conventional follow-up tests (e.g., simple effects, Tukey HSD) to compare the mean (proportions) for significant ANOVA effects, and (b) we can more easily detect interactions among simultaneously manipulated independent variables. Nevertheless, it should be noted that some conservative statisticians might object to using ANOVA for data of this type because of the very small added chance of error.

Dichotomous scores (e.g., yes/no, 0/1, etc.) are by their very nature limited. Large numbers of participants are sometimes needed to have sufficient statistical power to detect small differences between conditions. This can increase the study's costs dramatically. Another way of scoring compliance to multiple directives is to sum each participant's scores across the types of warning-directed behavior. Thus, in the example of the 3-behavior-type studies given above, if a participant complies with two out of three types of behavior then that person's score is a 2, if he/she complies with all or none of the directives, then these individuals are given scores of 3 or 0, respectively. So rather than having three sets of dichotomous data, we now have scores on a 4-point scale which can be analyzed more readily using more statistically powerful (more sensitive) analyses such as ANOVA.

When simply summing the scores for the individual warning-related types of behavior, the component scores are given equal weight. However, some types of behavior may be more serious than others. For example, breathing a chemical may be worse than touching it, and thus the failure to use respiratory protective equipment is worse than failure to wear gloves. In such instances, one could differentially weight each of the multiple directives according to their importance (which in this case is related to injury severity). The weightings are multiplied against the 1 and 0 scores before summing the values. Although differentially weighting the component scores is both logical and reasonable, we do not know of any warning compliance studies that have done this.

Sometimes compliance levels produce floor or ceiling effects. A floor effect occurs when scores across all conditions are very low (at or near the lower limit, e.g., 0% or 10% compliance rates). A ceiling effect is the opposite situation, when scores are all very high (e.g., from 80% to 100%). In either case it is possible to miss a true effect because the scores cannot move or differ by much between conditions. For example, if the base rate behavior is at 90% without a warning being present, then it will be very difficult to show a statistically significant increase in compliance because the maximum increase that can be produced is 10%. We suggest that a pilot study with a limited set of participants be conducted to get an idea whether the compliance levels are near floor or ceiling levels, and if so, to make some adjustments to achieve a more moderate base rate of compliance, say 35–65%. Some nonresearchers have misinterpreted the results of studies showing moderate levels of compliance. Citing these levels, they suggest that warning experts are unable to design effective warnings because the best compliance levels in some experiments are not at or near 100%. However, they fail to understand that the research is designed and intended to delineate factors that make a difference. The absolute levels of compliance in research should not be assumed to be the maximum levels that can be achieved.

Another consideration is statistical versus practical significance. For example, suppose some research study shows that some factor produces a statistically significant effect. Suppose further that the factor is font type where the warning in one font produced greater compliance than another font. The statistical results, however, describe only whether the difference between the two conditions is not likely to be zero and that such a difference is repeatable (with a small margin of error). It does not give information on the variable's importance. With large samples, the size of the difference can be very small and still be statistically significant. Therefore the effect's practical importance can be minuscule relative to the effects of other influential factors. Nevertheless, it is difficult to determine how small a beneficial effect must be to lack practical significance.

In trying to determine the effects of an independent variable, we recommend that researchers use several measures of compliance. As we have already noted, some measures are more sensitive than others. For example, consider the study mentioned earlier (Wogalter *et al.*, 1995) in which the warning directed participants to turn off a computer, eject the transport disk and touch a connector plug to release static electricity while connecting one external disk drive. The particular compliance measure, involving whether participants do or do not turn off the computer, might be relatively insensitive to the warning manipulation. A possible reason why this measure might not show an effect is that people installing the disk drives might tend to turn the computer off regardless of *whether or not they see a warning* directing them to do so. The base rate of turning off the computer is high (near ceiling level) without the warning being present. Or in other words, the warning has little or no value to add in this case. However, other compliance measures (such as ejecting a transport disk and touching the connectors to limit static electricity) might show effects of the warning manipulation because their base rates are relatively low. Thus, if an experiment uses only one dependent variable and that measure turns out to be relatively insensitive to the warning feature being manipulated (e.g., its placement), the potential benefit of that feature might go undetected. The likelihood of such situations occurring and their potential repercussions on warning design criteria should not be underestimated. This illustration also points out why it is important to be cautious in interpreting null (nonsignificant) effects in research studies. With null findings, it is difficult to know whether the feature actually has no effect or whether the experiment was not sufficiently powerful to detect the feature's effect. Given that most behavioral experiments are costly to conduct, it is worthwhile to make the experiment as sensitive as possible.

Behavioral compliance studies usually are much less sensitive at detecting small differences between conditions as compared to studies using rating scales (e.g., those assessing behavioral intentions). Commonly, rating studies show differences that are not found in compliance studies (and, interestingly, it is hardly ever the other way around). This noncorrespondence has been mistakenly interpreted by some nonresearchers as indicating ratings are of limited value with respect to measuring warning effectiveness. The greater sensitivity of ratings to subtle warning design differences is attributable partially to heightened statistical power derived from the measures essentially being composed of scores having a wider range of values (e.g., a 9-point rating scale) as compared to the dichotomous measures (e.g., yes/no scores) used in compliance studies. Furthermore, the sensitivity difference is attributable also to the heightened attention to small differences between warning designs in an intentional exposure rating study versus the much more subtle manipulation in an incidental exposure compliance study. To be as sensitive as ratings, compliance research would need to employ substantially larger sample sizes than is commonly used in studies of this type, making them even more expensive in terms of labor and time commitment. Rating studies often produce results similar to compliance studies with respect to the basic patterns of scores exhibited between conditions. The problem is when the rating and compliance results exhibit very different patterns of results between similar warning conditions. As we have said before, you should probably give greater credence to the results of a well designed behavioral compliance study compared to those of a rating study.

4.5.2 Other Observational Measures of Compliance

In this section we describe several other kinds of observational technique to measure behavioral compliance. One potential measure that has been mentioned in the warning literature (e.g., Wogalter *et al.*, 1987) is how many people decide to discontinue their participation in the study. This is an interesting measure because it might indicate how risky the situation appears to be. Different interpretations can be inferred from the different points at which participants decide to quit. For example, if some participants quit early in the experimental procedures (such as during the consent form phase when they are given preliminary instructions about the study) then this might indicate that participants believe that the situation has some believable level of risk. Presumably, something in the situation is causing people to decide that it is 'just not worth taking a chance of getting hurt.' In our experience the number of individuals who decide to discontinue participation is extremely low, making it difficult to show statistically significant differences. However, as we discussed earlier, the costs of quitting may be too high. Participants may worry that to do so would jeopardize their receiving course credit for their participation or some other incentive offered to them earlier. Additionally, they might hold the belief that scientists and their employer would not let injurious events occur to volunteer participants.

There are two other kinds of potentially useful behavioral indicants that have received relatively little attention in the research literature: (a) task sequencing, and (b) latency or speed of compliance. In some situations, it is important to perform certain actions in a particular order and to do them quickly to protect against injury or property damage. The logic is that a person who puts on protective equipment before getting involved with a potential hazard is acting more safely than a person who puts on the protective equipment while they are actually at risk. Most compliance studies count compliance as adequate only if it is performed *before* the performance of particular acts. Conversely, a person who dons a piece of protective equipment *after* initiating a risky act would not be counted as having complied. While responding in due haste is important in some situations, in others a more deliberate approach may be more appropriate. In these cases, longer latencies before engaging a potential hazard could indicate greater safety. A lock-out tag-out warning that is located to protect a disengaged power switch from being improperly engaged (e.g., while maintenance or repair work is being performed) is an example where the appropriate response is to wait until there is assurance that there is no potential danger to anyone (or to the equipment) before the equipment is serviced and restarted.

4.5.3 Laboratory studies

In this section, we will describe some of the methods used to measure behavioral compliance in the laboratory. Most of this work has been done at university-based laboratories under highly controlled conditions.

Chemical hazard

The chemistry laboratory paradigm has been employed in numerous studies since the mid-1980s. The basic methodology has proven to be successful in demonstrating the influence of numerous factors on warning compliance, including the effect of location of the warning in a set of instructions (Wogalter et al., 1987), cost of compliance and social influence (Wogalter, Allison, and McKenna, 1989), video modeling (Racicot and Wogalter, 1995), message personalization (Wogalter et al., 1994), voice (Wogalter et al., 1993b; Wogalter and Young, 1991), clutter (Wogalter et al., 1993a), pictorials (Jaynes and Boles, 1990), color (Braun and Silver, 1995; Rodriguez, 1991), shape (Jaynes and Boles, 1990; Rodriguez, 1991), container label design (Wogalter and Young, 1994), and time stress (Wogalter, Magurno, Rashid, and Klein, 1998). Many of these effects using the chemistry paradigm have been supported using other methodologies, giving at least some indication that results from experiments using the chemistry paradigm can be generalized to other situations. Because of its appreciable use in investigating various warning-related factors, the basic methodology of this technique will be presented in more detail than other techniques that we review.

At the outset of the chemistry task procedure, participants are told that the research is an engineering psychology study designed to determine how people perform a set of steps involving the measuring and mixing of chemicals. The opening description is actually accurate (i.e., not really deceptive), but it does not refer to warnings being the real purpose of the study. In other words, participants are incidentally exposed to the warning as part of the overall task of using the chemicals.

In the initial overview, participants are told: (1) that they will be mixing and weighing a set of chemical substances and solutions, (2) that they should complete the laboratory task as quickly and as accurately as possible, (3) that they will have a limited amount of time to complete the task, and (4) that the final product will be evaluated for accuracy. All participants are then shown how to use a triple-beam balance on a nearby desk top.

A variety of chemistry equipment including: beakers, flasks, graduated cylinders, stirring rods, measuring spoons, disposable vinyl gloves and paper surgical masks are located on a laboratory table in an adjacent room. The substances and solutions are disguised to make them appear somewhat novel and potentially hazardous. For example, food coloring is combined with water to make solutions of different colors. Other containers hold substances of different colors and graininess, e.g., pink table sugar, corn meal, and yellow powdered sugar. Some studies have added a small amount of 'chemical'-type odor (e.g., ammonia) to help make the situation more believable to participants that they were mixing potentially hazardous chemicals. Figure 4.1 shows a typical chemistry laboratory set up.

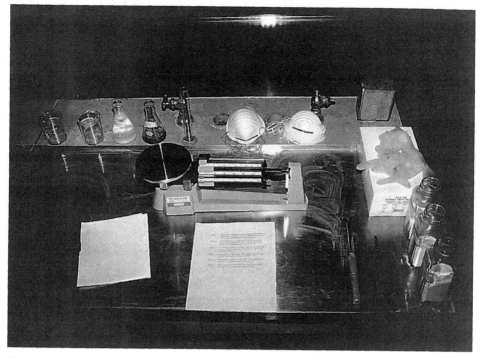

Figure 4.1 Typical chemistry laboratory set up (from Wogalter et al., 1994).

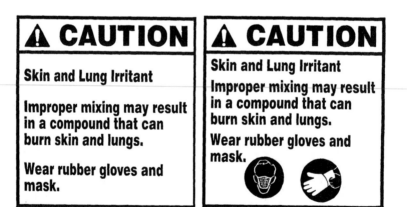

Figure 4.2 Example warning sign used in a chemistry laboratory experiment (from Wogalter et al., 1993a).

At some point in the procedure participants are exposed to a warning (e.g., in the instructions, as a separate posted sign, from an audio tape player, or in a video, etc.) that might say the following: 'WARNING: Wear gloves and masks while performing the task to avoid irritating fumes and possible irritation of skin.' A depiction of an example sign is shown in Figure 4.2. The instruction sheet contains several steps describing how to measure and mix certain quantities of substances and solutions. The primary behavioral compliance measure is whether participants put on the mask and gloves before beginning to handle the chemicals. Some studies have also recorded the use of goggles and a lab coat.

Chemical products are particularly good 'hazards' to use in warnings research because participants cannot easily tell, just by looking at them, the true extent of risk. The difficulty of discriminating the potential hazardousness of chemical products can enhance participants' belief that they may indeed be working with a product that could be dangerous. Several studies have had participants use different kinds of chemically based consumer products in incidental exposure tasks. One popular item in warning research is glue. Glue is a good product because its fundamental purpose can be used as a foundation of the subterfuge task that participants are asked to perform. Strawbridge (1985) looked at the effect of several warning variables, including the embedded placement of warnings on labels of a glue product. The warning label stated that the glue contained acid and had to be shaken to avoid severe burns. The behavioral compliance measure was whether participants shook the container before using it. Hatem and Lehto (1995), while exploring the possible use of odor as a potential hazard cue, used a glue container that had the warning 'Danger: Toxic fumes may cause respiratory problems. Open all windows and doors before using. Turn on a fan if available.' The behavioral compliance measure was whether participants attempted to open a nearby window or turn on an accessible fan.

Other chemical products have also been employed. The chemicals used by Friedmann (1988) were a drain cleaner and wood cleaner. Participants' safety was protected by stopping the experiment after participants removed the lids (so they never actually worked with the substances). The primary measure was whether participants put on the provided safety equipment before removing the lid. Frantz (1993, 1994) measured participants' compliance while using a drain cleaner and water repellent sealer when examining the positioning of warnings with respect to the directions for use (integrated versus separated) and the explicitness of the procedures described. Compliance to the labeled warnings was measured and compared between conditions. Dingus, Wreggit, and Hathaway (1993) measured the use of gloves and mask for a 'newly formulated' household cleaner that participants were asked to try out at their home. Participants returned later with the materials, and compliance was measured according to whether there was a disturbance in the condition of the gloves (whether they were stretched) and the mask (whether a loose knot was untied).

An experiment by Chy-Dejoras (1992) involved a floor tile adhesive remover. The study was described as a marketing survey and participants watched a video of a person using adhesive remover. The behavior of the actor(s) in the video was manipulated. In one video, there was a single actor who used the product without gloves, and in another video, there were two actors, one of whom wore gloves and one who did not. Also, the aversive effects of the product were manipulated. In one video, the effects were benign. In a second condition, the depicted event was slightly aversive showing a person spilling the adhesive remover and vocally expressing pain. In the third condition, the video depicted a highly aversive experience showing the spilling of adhesive remover and pictures of a burned hand. After viewing the tape, participants prepared to perform a floor tiling task that required the use of the adhesive remover. The behavioral measure was whether or not they outfitted themselves with gloves before starting the tiling task.

Mechanical hazard

Several behavioral compliance studies have involved participants in tasks that require the use of tools and devices having mechanical injury risks. One tool that has been used is the power saw. Obviously such mechanical implements are dangerous, but also they are somewhat familiar (at least to some participants). Setting up the situation so that participants are at very little or no risk requires careful planning.

Otsubo (1988) assigned participants to one of eight experimental conditions which corresponded to a factorial combination of two levels of product danger (high danger represented by a circular saw and low danger represented by a jigsaw), and four warning label formats (words only, pictograph only, words plus pictograph, and no warning). The study measured whether users donned gloves as directed by the warning label. Participants were stopped just before they used the saws as a precaution against possible injury.

Zeitlin (1994) arranged for four groups of college students differing in tool-using experience and exposure to safety training to use an electric chain saw to perform several tasks. Compliance with safety warnings contained in the chain saw operating instructions was measured. Participants actually used the saw in this study; but the author describes that adequate safety precautions were taken.

Research has involved other products that have a risk of mechanical injury. Frantz and Rhoades (1993) asked participants to unpack and arrange office furniture and supplies in a room. The office furniture and supplies included a file cabinet that displayed a label warning of a tipping hazard. Warning placement was manipulated: it was printed on the shipping carton, placed inside the bottom surface of the top drawer, placed on the front of the file cabinet as well as on the bottom of the top drawer, and placed in the top drawer on a piece of cardboard as well as on the surface of the drawer. The compliance measure was whether participants first placed materials in the bottom drawer to prevent the cabinet from tipping over.

Two other studies have examined the effectiveness of warnings for mechanical hazards. Dorris and Purswell (1977) employed a hammer with a cracked handle with or without a warning present that said the hammer should not be used if the handle is cracked. Another study, by Kalsher, Wogalter, and Silver (1998), had participants use a drill to construct parts of a bird house. The presence or absence of a tactile (raised border) warning telling users to wear gloves, a mask, and goggles was manipulated.

Electrical hazard

Several studies have involved electrical hazards. Duffy, Kalsher, and Wogalter (1995) led participants to believe that they were helping the experimenter set up some video and recording equipment for the 'real' study. In this context, participants connected the electrical cords to power outlets in which they were incidentally exposed to one of three warnings on the available extension cords. The resulting connection can be used to determine compliance, eliminating the need for direct observation. In this study, the effectiveness of an interactive label that required physical manipulation before it could be used was compared to a standard label. Compliance was based on whether the electrical cords were properly connected.

Gill, Barbera, and Precht (1987) had participants perform a series of tasks, one of which required them to use an extension cord to connect to an electric heater. The warning attached to the heater was the standard warning printed on the back of the unit, a color 'ski pass' label attached to the cord or a color-coded interactive label attached to the plug. The warning directed users not to use anything but a heavy-duty extension cord, whereas the only extension cord available in the room was light-duty. The number of people who used the inappropriate extension cord was recorded.

Wogalter et al. (1995) asked participants to connect an external disk drive to a computer. In this study, a short safety directive label was placed at various locations (on the cover page of the manual, on the shipping box, in an accompanying leaflet, on the disk drive cable, and on the front of the drive). The label requested that installers first read the second page of an accompanying owner's manual which instructed them to take three precautions: (a) to turn off the computer, (b) to eject a shipping disk, and (c) to touch metallic plugs to discharge any static electricity. A similar study by Conzola and Wogalter (1999) using the same task manipulated the presentation of voice versus print warning/ directives. Unlike most warning compliance research, the risk in this study was product damage, not personal injury.

4.5.4 Field and Quasi-field Studies

Several studies have observed compliance to warnings in field settings. As previously discussed, field studies have the advantage of increased external validity, but often at the expense of experimental control.

A common field research technique is to collect data from large numbers of consumers in a shopping mall. For example, Venema (1989) studied 330 participants visiting a home exhibition. Participants were asked to perform tasks involving methylated spirits (methyl alcohol). In one task, participants were asked to assume they were having a fondue and needed to refill the burner. In a second task, they were to assume they needed to remove paint from a table in their house using the spirits. Three versions of labels were studied: a neutral label with no safety information, the current label used for each product, and an improved layout constructed in accordance with recommendations found in labeling standards. The degree to which participants read and followed precautions stated in the label was observed.

Wogalter and Young (1991) observed 531 shoppers as they approached a simulated slippery-floor hazard near the entrance of the shopping center. Placed in the area were a set of orange cones and a mop inside of a bucket. There were four warning conditions: (a) none, (b) voice only, (c) print only, and (d) voice and print combined. Both the print and voice warnings stated 'Warning! Wet Floor. May be Slippery.' The voice (when present) emanated from a tape recorder inside a nearby mop bucket. Compliance was based on the proportion of individuals who walked through the area avoiding a specific section of the floor near the cones.

Other studies have employed similar, unobtrusive observational techniques in public areas of buildings. Wogalter et al. (1989) examined the use of stairs in a college dormitory where a warning was posted indicating that the elevator was broken. In another elevator-use study, Wogalter, Begley, Scancorelli, and Brelsford (1997) measured compliance to various signs directing individuals to use the stairs if they were only going up one floor or down two floors so that elevator users traveling between more distant floors would have better service. The researchers rode the elevators of six multi-storey buildings for specified intervals and recorded the numbers of persons who failed to comply with the signs.

Wogalter et al. (1987) describe several field studies utilizing unsuspecting 'participants' in public buildings. These studies measured: (a) the use of telephones and copy machines when warnings stating that the machines were broken were present or absent; (b) the use of a water fountain having an enhanced versus unenhanced contaminated-water warning; and (c) the use of exit doors in the presence or absence of broken-door warnings. In all of these, the researchers recorded the numbers of people who complied or did not comply.

Reisinger and Williams (1978) conducted a hospital study testing three educational/persuasive programs designed to increase the crash protection of infants in cars by increasing the use of infant car seats. The behavior of the women participating in the programs (all of whom were new mothers) was then compared to new mothers who received no crash protection education. A total of 1200 babies were observed during the study; the three program groups, as well as the control group, had about 300 subjects in each condition. Compliance was measured in terms of whether the infant car seats were positioned properly and securely in the car.

Field studies have been conducted also in work environments to test the effectiveness of a variety of safety programs. Zohar, Cohen, and Azar (1980) administered hearing tests to selected workers in a noisy metal fabrication plant and gave some of them feedback that they had incurred noise-induced shifts in hearing sensitivity during their work shifts. Over a period of five months, the use of hearing protectors by workers receiving this feedback was compared to a matched control group who did not receive the feedback.

Gomer (1986) conducted a field study in the context of litigation that was directed toward measuring the effectiveness of a label which warned about the risk of delayed lung disease. The study attempted to reconstruct the conditions and labeling requirements corresponding to the state-of-the-art in the mid-1960s. Seventeen employees handled bags of limestone in a dusty environment over a period of two days. On the second day, strong warnings of the hazard of limestone dust were placed on the bags that recommended respirators be worn. The number of workers who saw the warning and who requested respiratory protection was recorded.

Summala and Pihlman (1993) describe a safety campaign in which all 30 000 truck drivers in Sweden were sent a music tape that provided information about driving in work zones. The tape emphasized the concerns of workers in work zones regarding large vehicles that pass by too closely at excessive speeds. The study was conducted over a period of four months. Drivers were unobtrusively observed by camera, and vehicle speed and lane position in the work zone were measured. Figure 4.3 shows one of the scenes at a work zone.

Field research has been conducted successfully also in recreational settings. Hathaway and Dingus (1992) conducted a study investigating the effects of cost of compliance and warning information content in a racquetball venue. Cost of compliance consisted of two levels: in the high cost condition, no eye protection was provided, whereas in the low cost condition, eye protection was provided in a salient location just outside the court area. The warning information factor was comprised of three levels: (a) no warning provided, (b) an ANSI standard warning, and (c) an ANSI standard warning plus specific consequence information. The proportion of the 420 racquetball players who wore eye protection was observed unobtrusively.

Lehto and Foley (1991) conducted a field study of ATV operator behavior in six states that did or did not have helmet laws in 1988 and 1989. The use of helmets and other personal protective equipment was observed. Also recorded was: (a) the presence of warning labels; (b) the presence and enforcement of state regulations governing ATV use; (c) whether operator training courses had been taken; (d) self-reported reading of owner's manuals; and (e) operator attitudes.

Another study observed participants at a university automotive repair garage. Wogalter, Glover, Magurno, and Kalsher (1999) measured the effectiveness of warnings on battery booster cables to convey the proper procedure of connecting them to jump-start an automobile with a dead battery. In the context of several car-related service tasks, participants were asked to perform the jump-start procedure (both cars had realistic-appearing

Figure 4.3 Swedish truckers were sent a cassette audio tape that included safety information about hazardous driving at work zone areas. Compliance was assessed by measuring the truck speed and distance of the vehicle to the side of the work zone (from Summala and Pihlman, 1993).

fake batteries). The warning (when present on the cables) contained verbal and pictorial information that described the hazards associated with car batteries as well as a pictorial diagram showing the sequence of steps that should be performed in the jump-start procedure. The number of people who properly connected the cables in the warning present and absent conditions was assessed.

4.6 PHYSICAL TRACE MEASURES OF COMPLIANCE

In most behavioral compliance studies, participants are observed more or less unobtrusively by the experimenter, because the presence of observers can influence compliance levels (Wogalter *et al.*, 1989). The best type of measurement would have no apparent observer and take place in a natural environment (i.e., where the product or equipment usually is used, such as in people's own homes).

One way to measure natural compliance behavior is to measure physical trace data. Physical trace data refer to any change or 'signature' in the environment associated with the compliance situation. An early example of this approach was to use 'glue-sealed' pages in magazines to assess advertising exposure (Politz, 1958 as cited by Ramond, 1976). Between each pair of pages in a magazine, a small glue spot was placed inconspicuously near the binding. The glue was configured such that it would not re-adhere once broken. Advertising exposure was then measured by counting the percentage of pages with a broken seal.

A primary advantage of physical trace measurement is that it can be used in field settings where direct observation is not feasible. A study by Hunn and Dingus (1992) illustrates this value. Compliance involved the use of protective gloves while using a cleaning product in a common household spray bottle. Participants were told that the study dealt with the comparison of products of different strengths and qualities to their normal brand. The participants were told that they were testing a new cleaning formulation and could use as much of it as they wished. This study tested several factors including: two levels of information type and two levels of compliance cost (high cost: no gloves provided; low cost: gloves provided). The participants were instructed to take the packaged product home (in one half of the cases gloves were included in the package) and to use it for a week, at which time they were to return and complete a questionnaire asking about their experiences with the product. When they returned the package, they were given a questionnaire which contained a variety of product quality and marketing distracter questions, as well as label memory and compliance questions. In addition to questionnaire responses, physical trace measures were taken to verify whether the participants had worn the gloves and to ensure that they used the product on at least one occasion. Glove use was apparent from deformities at the finger-tips that occur after a very short period of use. Physical trace data of product use was assessed by the placement of a small paint dot on the threads of the spray bottle. If the paint dot was intact, it meant that the participant had not turned the bottle to the 'on' position, and therefore could not have used the product as intended. In all but a very few instances, the questionnaire responses indicating glove use were in agreement with the physical trace data.

In a follow-up study by Dingus et al. (1993), a different product was used and both gloves and respirator masks were provided as part of the packaging. The gloves and masks were prominently displayed in the consumer package so that participants would know of their availability. As with the Hunn and Dingus study, the consumers were informed that they would have to bring the contents of the package back to the same location after approximately one week had passed. A convenient time for the participant was noted and a majority of the participants did return to the mall. If a participant could not return to the mall at the specified time an alternative time was scheduled or the researcher arranged a time to pick up materials at the person's place of residence. In addition to the glove physical trace data, the mask elastic straps were tied in such a way that it was necessary for the participants to untie a simple knot to use the mask, thus giving a physical trace measure of use.

Trace measurements are in common use in health-related compliance (or adherence) research. Taylor (1991) describes the use of pill counts, that is, the amount of medication left in a bottle when the course of medication is supposed to be completed, to measure nonadherence. However, Taylor states that, despite their objective nature, pill counts are subject to several forms of bias. Patients may remove some pills from the bottle, or they may have pills left over from a previous treatment that they take instead. In addition, pill counts only estimate how many pills were removed from the dispenser, and not whether they took them at the correct times (Meichenbaum and Turk, 1987).

Brannon and Feist (1992) describe a number of automated devices to facilitate pill counting and to determine whether or not medication is taken at the prescribed time. Cramer, Mattson, Prevey, Scheyer, and Ouellette (1989) describe the use of a microprocessor in the pill caps to record every bottle opening and closing. The microprocessor yields information concerning the time of day that the bottle is opened, but does not detect the number of pills removed with each opening. Thus, this procedure provides more data than the pill-count technique, but still it cannot ascertain the exact rate of adherence. Brannon and Feist (1992) also examined biochemical evidence as a physical trace measure of compliance. Biochemical indices are detected through blood or urine samples. However, problems exist with the technique, including individual differences in absorption and metabolism, and the reliability, accuracy, and cost of the assays.

4.7 EPIDEMIOLOGICAL COMPLIANCE DATA

Epidemiology involves the distribution and determinants of disease or injury in a population. Epidemiological techniques are particularly valuable for assessing the value of injury control interventions (like warnings) in a population. Commonly, very large samples (or even entire populations) are analyzed in conjunction with naturally occurring changes or formal interventions. The data used for this type of analysis can be either archival (e.g., sales or accident records) or observational (e.g., the number of people seeking medical advice after a public service announcement). Epidemiological studies can be prospective or retrospective.

There have been a number of epidemiological studies which have provided valuable insight into the effectiveness of warning interventions by taking advantage of a change in legislative mandate. For example, Schucker, Stokes, Stewart, and Henderson (1983) evaluated the impact of the Saccharin Study and Labeling Act, passed by Congress in 1977, which required, among other things, that manufacturers place a warning on labels of products containing saccharin stating that: 'Use of this product may be hazardous to your health. This product contains saccharin which has been determined to cause cancer in laboratory animals.' With the enactment of the saccharin labeling requirement, an objective means of testing label effectiveness was created. That is, by monitoring the sales of soft drinks containing saccharin, the rate of warning compliance could be calculated for the entire population of diet soft drink customers. However, as is typical in this type of research, there is no opportunity for tight experimental control. For example, in addition to the warning labels, there were concurrent news reports and other information sources that were providing saccharin information. Thus, it became very difficult to assess the exact causal factors associated with any change in soft drink sales.

To evaluate the impact of the warning labels, Schucker and his colleagues developed a model that specified soft drink sales as a dependent variable and the presence of the warning, price of the product, news reporting and diet-drink advertising as independent variables. Seasonal sales trends were taken into account also. Orwin, Schucker, and Stokes (1984) used an auto-regressive moving-average modeling approach to evaluate the effect of the saccharin warning label on the sales of diet soft drinks. This technique allowed the authors to attribute any change in sales to specific causes.

A similar type of study was conducted in response to a national anti-smoking campaign. Warner (1977) evaluated the effects of the campaign on annual US per capita cigarette consumption. The anti-smoking campaign was a collection of mostly uncoordinated, educational activities by a variety of organizations including the government, private voluntary agencies, and for-profit business firms. To evaluate the effects, current cigarette consumption was compared to projections based on cigarette consumption prior to enactment of the anti-smoking campaign.

Several studies using large samples of the population have measured the impact of the alcohol warning label that is required on all containers in the USA since November, 1989. The measures have included behavioral intentions, awareness and memory of the label and its contents and changes in attitudes and beliefs (Greenfield and Kaskutas, 1993; Hankin, Sloan, Firestone, Ager, Goodman, Sokol, and Martier, 1996; Greenfield, Graves, and Kaskutas, 1999; Nohre, MacKinnon, Stacy, and Pentz, 1999).

Epidemiological methods have been applied to the motor vehicle domain in a number of instances. For example, Edwards and Ellis (1976) evaluated the effects of a driver improvement training program implemented by the Texas Department of Public Safety. The research studied the effect of the program on driving records and developed a method for predicting the frequencies of violations and accidents for the 12 months following training. Robertson (1975) investigated the effectiveness of interlock and buzzer-light systems on the use of safety belts. The study was conducted at 138 sites in the cities of Baltimore, Houston, and Los Angeles, as well as suburbs of New York City, Richmond, VA, and Washington, DC. Use or non-use of safety belts by drivers was observed at each of these sites. To reduce any potential bias during the data collection phase, observers were 'blind' to (unaware of) the fact that buzzer-light and interlock systems were being compared.

Voevodsky (1974) studied the effectiveness of a center-mounted brake light as a means for preventing collisions under normal driving conditions. A total of 343 taxis operating with deceleration warning lights were compared to a control group of 160 taxis operating without the lights. After the light-equipped taxis had traveled a total of 12.3 million miles, rear-end collision rates were assessed.

Preusser, Ulmer and Adams (1976) studied compliance of drivers convicted of drinking and driving. Compliance in this case was the lack of a repeat offense. A program called the Nassau County Alcohol Safety Action Project Driver Rehabilitation Countermeasure ran from February, 1971 through June, 1973. The program's objective was to reduce the recidivism rate of drivers convicted of alcohol-related offenses. Random assignment of drivers to treatment and control groups was permitted by legislation. The experimental group consisted of approximately 3200 drivers who completed the rehabilitation program. The control group consisted of approximately 2600 drivers. The number of repeat offenses was measured and compared between these two groups.

4.8 SUMMARY AND IMPLICATIONS

In this chapter we reviewed studies that have measured behavioral intentions and compliance to warnings. In behavioral intention studies, participants make judgments of whether they would comply in a particular situation (or how careful or cautious they would be, etc.). Sometimes behavior intention studies are the best kind of warning assessment that can be obtained given the fact researchers cannot ethically expose participants to any type of substantial risk, and given practical considerations such as cost and time pressures. But when feasible, the best method of assessment is actual behavioral compliance. This chapter described various methods which have been used to conduct behavioral compliance research in laboratory and field settings. Most of these studies use some level of deception and use an incidental exposure paradigm where participants perform one or more tasks without being told that the study concerns warnings. Also described are techniques that examine physical trace indicators of compliance in naturalistic settings where direct observation is difficult or impossible. Finally, studies that have used epidemiological techniques are described. Because behavioral compliance is the ultimate measure of warning effectiveness, we hope that researchers will employ this measure more frequently in future research, and that this review will assist them in setting up future investigation, whether they make use of existing methods or create new ones.

Throughout this chapter, we have made recommendations with respect to the collection and analysis of warning compliance data. Several of the more general recommendations are worth summarizing here. We recommend that researchers collect data on several response measures in their studies whenever practical. Other kinds of measure, including subjective opinions, are valuable additions to the research literature because they can aid interpretation of compliance data.

Although direct behavioral measurement provides the most valid measure of warning compliance, collection of behavioral measures does not ensure that a study will provide valid and meaningful results. Critical aspects of the research design must be addressed to ensure successful evaluation of warning circumstances. These include unobtrusive measurement, an environment free of demand characteristics, and use of a scenario that does not contain inherent floor or ceiling effects. Generally a pilot study is recommended to ensure that critical features of the study are in order prior to actual data collection. Often even the most seasoned researchers are surprised by a particular study outcome while utilizing a new method or exploring a new content domain. In addition to the issues of measurement, the study must be carefully designed and must allow the determination of causes of compliance to be attained. Aspects of concern include: the use of proper baseline or control conditions and fair manipulation of selected variables, among others.

We reviewed some of the reasons why compliance measurement cannot be employed under certain circumstances. Behavioral intentions data can be substituted for measures of compliance, but specific research on the predictive significance of intentions in warnings applications is needed. Nevertheless, research in social psychology and other domains strongly suggests prediction is greater when the behavioral intentions data are assessed in situations that are similar to the actual compliance situation.

There are several important areas in the behavioral compliance research area that are likely to unfold in the next decade or so. One is the prediction of behavioral compliance, and the others are related to the rapid transition to powerful computers and people's interaction with them.

We expect that research will move towards more powerful models that predict behavioral compliance. With the variables that are discussed in this book, we already are able to predict and enhance compliance better than we were some 15 years ago, and we expect this trend to continue. Part of this will come from research started by Purswell, Schlegal, and Kejriwal (1986). They developed a questionnaire that was intended to measure risk-taking propensity which included items such as the percentage of time individuals used seat belts, whether they would use lifejackets when boating, and their reported tendency to cross a street against a light. In a set of tasks, participants were observed using a chemical drain opener, electric carving knife, sabre saw, and router. The researchers found that the questionnaire had significant value in predicting safe or unsafe behavior.

A second group of trends for research in this area will involve computer-based situations in which persons make risk decisions. These situations can be very lifelike considering that computers comprise a substantial portion of many people's lives. Thus, warnings during actual computer use procedures can be extremely real in the situation that they present to users. A substantial amount of work may be involved if a wrong decision is made (e.g., see Cox, 1995).

Perhaps the most exciting trend will be in the use of multimedia simulations of actual life events (other than computers) that can put participants in a 3D-like environment using a 2D computer or television screen (see Glover and Wogalter, 1997). In these simulations, individuals participate in a virtual environment (like those shown in sophisticated adventure games or in architectural design programs). Such programs, and even more sophisticated lifelike virtual reality environments, can put individuals into seemingly real risk environments without actual exposure to hazards (although it may appear that way). Using these programs, researchers will be able to place people into realistic hazard situations where warnings are present (in various conditions) and measurement can be made on whether they comply with them.

REFERENCES

ADAMS, A.S. and EDWORTHY, J. (1995) Quantifying and predicting the effects of basic text display variables on the perceived urgency of warning labels: tradeoffs involving font size, border weight and colour. *Ergonomics*, 38, 2221–2237.

AJZEN, I. and FISHBEIN, M. (1977) Attitude–behavior relations: a theoretical analysis and review of empirical research. *Psychological Bulletin*, 84, 888–918.

BARLOW, T. and WOGALTER, M.S. (1991) Increasing the surface area on small product containers to facilitate communication of label information and warnings. In *Proceedings of Interface 91*. Santa Monica, CA: Human Factors Society, pp. 88–93.

BRANNON, L. and FEIST, J. (1992) *Health Psychology*, 2nd Edn. Belmont, CA: Wadsworth.

BRAUN, C.C. and SILVER, N.C. (1995) Interaction of signal word and color on warning labels: difference in perceived hazard and behavioral compliance. *Ergonomics*, 38, 2207–2220.

CHY-DEJORAS, E.A. (1992) Effects of an aversive vicarious experience and modeling on perceived risk and self-protective behavior. In *Proceedings of the Human Factors Society 36th Annual Meeting*. Santa Monica, CA: Human Factors Society, pp. 603–607.

COCHRAN, W.G. (1950) The comparison of percentages in matched samples. *Biometrika*, 37, 256–266.

CONZOLA, V.C. and WOGALTER, M.S. (1999) Using voice and print directives and warnings to supplement product manual instructions. *International Journal of Industrial Ergonomics*, in press.

COX III, E.P. (1995) Compliance to product warnings. In *Proceedings of the Marketing and Public Policy Conference*. Atlanta: Department of Marketing, Georgia State University.

COX III, E.P., WOGALTER, M.S., STOKES, S.L., and MURFF, E.J.T. (1997) Do product warnings increase safe behavior? A meta-analysis. *Journal of Public Policy and Marketing*, 16, 195–204.

CRAIK, F.I.M. and LOCKHART, R.S. (1972) Levels of processing: a framework for memory research. *Journal of Verbal Learning and Verbal Behavior*, 11, 671–684.

CRAMER, J.A., MATTSON, R.H., PREVEY, M.L., SCHEYER, R.D., and OUELLETTE, V.L. (1989) How often is medication taken as prescribed? *Journal of the American Medical Association*, 261, 3273–3277.

DEJOY, D.M. (1989) Consumer product warnings: review and analysis of effectiveness research. In *Proceedings of the Human Factors Society 33rd Annual Meeting*. Santa Monica, CA: Human Factors Society, pp. 936–940.

DERSHEWITZ, R.A. (1979) Will mothers see free household safety devices? *American Journal of Diseases of Children*, 133, 61–64.

DINGUS, T.A., WREGGIT, S.S., and HATHAWAY, J.A. (1993) An investigation of warning variables affecting personal protective equipment use. *Safety Science*, 16, 655–673.

DORRIS, A.L. and PURSWELL, J.L. (1977) Warnings and human behavior: implications for the design of product warnings. *Journal of Products Liability*, 1, 255–263.

DUFFY, R.R., KALSHER, M.J., and WOGALTER, M.S. (1995) Interactive warning: an experimental examination of effectiveness. *International Journal of Industrial Ergonomics*, 15, 159–166.

EAGLY, A.H. and CHAIKEN, S. (1993) *The Psychology of Attitudes*. Fort Worth, TX: Harcourt Brace Jovanovich.

EAGLY, A.H. and CHAIKEN, S. (1996) Attitude structure and function. In GILBERT, D., FISKE, S., and LINDZEY, G. (eds), *The Handbook of Social Psychology*, 4th Edn. New York: McGraw-Hill.

ECKES, T. and SIX, B. (1994) Fakten und Fiktionen in der Einstellungs-Verhaltens-Forschung: Eine Meta-Analyse [Fact and fiction in attitude-behavior research: a meta-analysis]. *Zeitschrift fuer Sozialpsychologie*, 253–271.

EDWARDS, M.L. and ELLIS, N.C. (1976) An evaluation of the Texas driver improvement training program. *Human Factors*, 18, 327–334.

EDWORTHY, J. and ADAMS, A. (1996) *Warning Design: A Research Prospective*. London: Taylor & Francis.

FARID, M.I. and LIRTZMAN, S.I. (1991) Effects of hazard warning on workers' attitudes and risk-taking behavior. *Psychological Reports*, 68, 659–673.

FRANTZ, J.P. (1993) Effect of location and presentation format on user processing of and compliance with product warnings and instructions. *Journal of Safety Research*, 24, 131–154.

FRANTZ, J.P. (1994) Effect of location and procedural explicitness on user processing of and compliance with product warnings. *Human Factors*, 36, 532–546.

FRANTZ, J.P. and RHOADES, T.P. (1993) A task analytic approach to the temporal placement of product warnings. *Human Factors*, 35, 719–730.

FRIEDMANN, K. (1988) The effect of adding symbols to written warning labels on user behavior and recall. *Human Factors*, 30, 507–515.

GELLER, E.S., CASALI, J.G., and JOHNSON, R.P. (1980) Seat belt usage: a potential target for applied behavior analysis. *Journal of Applied Behavior Analysis*, 13, 669–675.

GILL, R.T., BARBERA, C., and PRECHT, T. (1987) A comparative evaluation of warning label designs. In *Proceedings of the Human Factors Society 31st Annual Meeting*. Santa Monica, CA: Human Factors Society, pp. 476–478.

GLOVER, B.L. and WOGALTER, M.S. (1997) Using a computer simulated world to study behavioral compliance with warnings: effects of salience and gender. In *Proceedings of the Human Factors and Ergonomics Society 41st Annual Meeting*. Santa Monica, CA: Human Factors and Ergonomics Society, pp. 1283–1287.

GOMER, F.E. (1986) Evaluating the effectiveness of warnings under prevailing working conditions. In *Proceedings of the Human Factors Society 30th Annual Meeting*. Santa Monica, CA: Human Factors Society, pp. 712–715.

GREENFIELD, T.K., GRAVES, K.L., and KASKUTAS, L.A. (1999) Long-term effects of alcohol warning labels: findings from a comparison of the United States and Ontario, Canada. *Psychology & Marketing*, 16, 261–282.

GREENFIELD, T.K. and KASKUTAS, L.A. (1993) Early impacts of alcoholic beverage warning labels: national study findings relevant to drinking and driving behavior. *Safety Science*, 16, 689–708.

HANKIN, J.R., SLOAN, J.J., FIRESTONE, I.J., AGER, J.W., GOODMAN, A.C., SOKOL, R.J., and MARTIER, S.S. (1996) Has awareness of the alcohol warning label reached its upper limit? *Alcoholism: Clinical and Experimental Research*, 20, 440–444.

HATEM, A.T. and LEHTO, M. (1995) Effectiveness of glue odor as a warning signal. *Ergonomics*, 38, 2250–2261.

HATHAWAY, J.A. and DINGUS, T.A. (1992) The effects of compliance cost and specific consequence information on the use of safety equipment. *Accident Analysis and Prevention*, 24, 577–584.

HUNN, B.P. and DINGUS, T.A. (1992) Interactivity, information and compliance cost in a consumer product warning scenario. *Accident Analysis and Prevention*, 24, 497–505.

JAYNES, L.S. and BOLES, D.B. (1990) The effect of symbols on warning compliance. In *Proceedings of the Human Factors Society 34th Annual Meeting*. Santa Monica, CA: Human Factors Society, pp. 984–987.

JOHNSON, R.P. and GELLER, E.S. (1984) Contingent versus noncontingent rewards for promoting seat belt usage. *Journal of Community Psychology*, 12, 113–122.

KALSHER, M.J., WOGALTER, M.S., and SILVER, N.C. (1998) Behavioral compliance to tactile warnings using a bird house construction project scenario. Unpublished manuscript, Rensselaer Polytechnic Institute, Troy, NY.

KIM, M. and HUNTER, J.E. (1993) Attitude–behavior relations: a meta-analysis of attitudinal relevance and topic. *Journal of Communication*, 43, 101–142.

KRAUS, S.J. (1995) Attitudes and the prediction of behavior: a meta-analysis of the empirical literature. *Personality and Social Psychology Bulletin*, 21, 58–75.

LAUGHERY, K.R., VAUBEL, K.P., YOUNG, S.L., BRELSFORD, J.W., and ROWE, A.L. (1993) Explicitness of consequence information in warnings. *Safety Science*, 16, 597–614.

LEHTO, M.R. and FOLEY, J. (1991) Risk taking, warning labels, training, and regulation: are they associated with the use of helmets by all-terrain vehicle riders? *Journal of Safety Research*, 22, 191–200.

LEHTO, M.R. and MILLER, J.M. (1988) The effectiveness of warning labels. *Journal of Products Liability*, 11, 225–270.

MACKINNON, D.P. (1995) Review of the effect of the alcohol warning label. In WATSON, R.R. (ed.), *Drug and Alcohol Abuse Reviews*, Vol. 7, *Alcohol, Cocaine, and Accidents*. Totowa, NJ: Humana Press, pp. 131–161.

MAYER, R.N., SMITH, K.R., and SCAMMON, D.L. (1991) Evaluating the impact of alcohol warning labels. *Advances in Alcohol Research*, 18, 706–714.

MEICHENBAUM, D. and TURK, D.C. (1987) *Facilitating Treatment Adherence: A Practitioners Guidebook*. New York: Plenum.

NOHRE, L., MACKINNON, D.P., STACY, A.W., and PENTZ, M.A. (1999) The association between adolescents' receiver characteristics and exposure to the alcohol warning label. *Psychology & Marketing*, 16, 245–259.

ORWIN, R.G., SCHUCKER, R.E., and STOKES, R.C. (1984) Evaluating the life cycle of a product warning: saccharin and diet soft drinks, *Evaluation Review*, 8, 801–822.

OTSUBO, S.M. (1988) A behavioral study of warning labels for consumer products: perceived danger and use of pictographs. In *Proceedings of the Human Factors Society 32nd Annual Meeting*. Santa Monica, CA: Human Factors Society, pp. 536–540.

PLANEK, T.W., SCHUPACK, S.A., and FOWLER, R.C. (1972) *An Evaluation of the National Safety Council's Defensive Driving Course in Selected States*. Chicago: National Safety Council Research Department.

POLITZ, A. (1958) *The Readers of the Saturday Evening Post*. New York: Curtis Publishing.

PREUSSER, D.F., ULMER, R.G., and ADAMS, J.R. (1976) Driver record evaluation of a drinking driver rehabilitation program. *Journal of Safety Research*, 8, 98–105.

PURSWELL, J.L., SCHLEGAL, R.E., and KEJRIWAL, S.K. (1986) A prediction model for consumer behavior regarding product safety. In *Proceedings of the Human Factors Society 30th Annual Meeting*. Santa Monica, CA: Human Factors Society, pp. 1202–1205.

RACICOT, B.M. and WOGALTER, M.S. (1995) Effects of a video warning sign and social modeling on behavioral compliance. *Accident Analysis and Prevention*, 27, 57–64.

RAMOND, C. (1976) *Advertising Research: The State of the Art*. New York: Association of National Advertisers.

REISINGER, K.S. and WILLIAMS, A.F. (1978) Evaluation of programs designed to increase the protection of infants in cars. *Pediatrics*, 62, 280–297.

ROBERTS, D.S. and GELLER, E.S. (1994) A statewide intervention to increase safety belt use: adding to the impact of a belt use law. *American Journal of Health Promotion*, 8, 172–174.

ROBERTSON, L.S. (1975) Safety belt use in automobiles with starter-interlock and buzzer-light reminder systems. *American Journal of Public Health*, 65, 1319–1325.

RODRIGUEZ, M.A. (1991) What makes a warning label salient? In *Proceedings of the Human Factors Society 35th Annual Meeting*. Santa Monica, CA: Human Factors Society, pp. 1029–1033.

SCHUCKER, R.E., STOKES, R.C., STEWART, M.L., and HENDERSON, D.P. (1983) The impact of the saccharin warning label on sales of diet soft drinks in supermarkets. *Journal of Public Policy and Marketing*, 2, 46–56.

SILVER, N.C., LEONARD, D.C., PONSI, K.A., and WOGALTER, M.S. (1991) Warnings and purchase intentions for pest-control products. *Forensic Reports*, 4, 17–33.

STAELIN, R. (1978) The effects of consumer education on consumer product safety behavior. *Journal of Consumer Research*, 5, 30–40.

STRAWBRIDGE, J.A. (1985) The influence of position, highlighting, and imbedding on warning effectiveness. In *Proceedings of the Human Factors Society 30th Annual Meeting*. Santa Monica, CA: Human Factors Society, pp. 716–720.

SUMMALA, H. and PIHLMAN, M. (1993) Activating a safety message from truck drivers' memory: an experiment in a work zone. *Safety Science*, 16, 675–687.

TAYLOR, S.E. (1991) *Health Psychology*, 2nd Edn. New York: McGraw-Hill.

URSIC, M. (1984) The impact of safety warnings on perception and memory. *Human Factors*, 26, 677–684.

VENEMA, A. (1989) Research Report No.69. *Product Information for the Prevention of Accidents in the Home and During Leisure Activities: Hazard and Safety Information on Non-durable Products.* The Netherlands: Institute for Consumer Research, SWOKA.

VOEVODSKY, J. (1974) Evaluation of a deceleration warning light for reducing rear-end automobile collisions. *Journal of Applied Psychology,* 59, 270–273.

WARNER, K.E. (1977) The effects of the anti-smoking campaign on cigarette consumption. *American Journal of Public Health,* 67, 645–650.

WOGALTER, M.S., ALLISON, S.T. and MCKENNA, N.A. (1989) Effects of cost and social influence on warning compliance. *Human Factors,* 31, 133–140.

WOGALTER, M.S., BARLOW, T., and MURPHY, S. (1995) Compliance to owner's manual warnings: influence of familiarity and the task-relevant placement of a supplemental directive. *Ergonomics,* 38, 1081–1091.

WOGALTER, M.S., BEGLEY, P.B., SCANCORELLI, L.F., and BRELSFORD, J.W. (1997) Effectiveness of elevator service signs: measurement of perceived understandability, willingness to comply, and behaviour. *Applied Ergonomics,* 28, 181–187.

WOGALTER, M.S., FORBES, R.M., and BARLOW, T. (1993a) Alternative product label designs: increasing the surface area and print size. In *Proceedings of Interface 93.* Santa Monica, CA: Human Factors and Ergonomics Society, pp. 181–186.

WOGALTER, M.S., GLOVER, B.L., MAGURNO, A.B., and KALSHER, M.J. (1999) Safer jump-starting procedures with an instructional tag warning. In ZWAGA, H.J.G., BOERSEMA, T., and HOONHOUT, H.C.M. (eds), *Visual Information for Everyday Use: Design and Research Perspectives.* London: Taylor & Francis.

WOGALTER, M.S., GODFREY, S.S., FONTENELLE, G.A., DESAULNIERS, D.R., ROTHSTEIN, P.R., and LAUGHERY, K.R. (1987) Effectiveness of warnings. *Human Factors,* 29, 599–612.

WOGALTER, M.S., KALSHER, M.J., and RACICOT, B.M. (1993b) Behavioral compliance with warnings: effects of voice, context, and location. *Safety Science,* 16, 637–654.

WOGALTER, M.S., MAGURNO, A.B., RASHID, R., and KLEIN, K.W. (1998) The influence of time stress and location on behavioral warning compliance. *Safety Science,* 29, 143–158.

WOGALTER, M.S., RACICOT, B.M., KALSHER, M.J., and SIMPSON, S.N. (1994) The role of perceived relevance in behavioral compliance in personalized warning signs. *International Journal of Industrial Ergonomics,* 14, 233–242.

WOGALTER, M.S., and YOUNG, S.L. (1991) Behavioral compliance to voice and print warnings. *Ergonomics,* 34, 79–89.

WOGALTER, M.S., YOUNG, S.L. (1994) Enhancing warning compliance through alternative product label designs. *Applied Ergonomics,* 25, 53–57.

ZEITLIN, L.R. (1994) Failure to follow safety instructions: faulty communication or risky decisions? *Human Factors,* 36, 172–181.

ZOHAR, D., COHEN, A., and AZAR, N. (1980) Promoting increased use of ear protectors in noise through information feedback. *Human Factors,* 22, 69–79.

PART THREE

Research on Warnings: Stages of the Model

This section reviews and summarizes research on warnings. The research is organized around the stages of the communication-information processing (C-HIP) framework. In addition to providing an organizing framework, the model has utility in explaining research findings as well as why warnings may succeed or fail in application.

CHAPTER FIVE

Source

ELI P. COX III
University of Texas at Austin

This chapter discusses the importance of source characteristics in determining the effectiveness of a communication. The nature of the source is discussed and the research evaluating the importance of each of its dimensions is reviewed. Contemporary information processing models are introduced to help understand the interactions among the major elements of the communication-persuasion model. Finally, product warnings are discussed within the content of the communication-persuasion model.

5.1 INTRODUCTION

The communication-human information processing model (C-HIP) discussed in Chapter 2 by Wogalter, DeJoy, and Laughery presents a theoretical framework for understanding the process by which warning information is communicated. This chapter focuses on the warning communication *source*—the individual or entity responsible for initiating the communication intended to protect product users. McGuire (1980) estimated that more than 1000 empirical studies on persuasive communication are published annually. Certainly, the numerous scholarly papers published each year concerning the effectiveness of product warnings are included in this estimate. Unfortunately, the characteristics of the communication source and their impact on warning effectiveness constitute a neglected subject.

This chapter consists of four sections. First, the nature of the communication source is discussed briefly. Second, the major categories of source characteristics that have been found to increase communication effectiveness are reviewed. Third, two contemporary theories of information processing are introduced as a means of explaining some of the apparent inconsistencies found within communications research. Finally, the significance of source characteristics in warnings communications is discussed.

5.2 WHO IS THE COMMUNICATION SOURCE?

The source initiates communication by *encoding* the desired information in a message that is transmitted to the intended receiver. Typically, the source also selects the communication channel. It is very important that source characteristics be considered in attempting to design an effective communication because the receiver combines the available information about the source (and the channel as well) with that obtained directly from the message in the process of *decoding* the message's meaning. Thus, the words 'this car runs like a Swiss watch' have an entirely different meaning depending upon whether they are uttered by a used car salesman or the mechanic you have trusted for years.

Identifying the source is straightforward where interpersonal communication is involved. It is the nurse who speaks to a patient about therapy after surgery or the voter who writes to the President concerning gun control. Identifying the source is somewhat more complicated in organizational communications.

When the American Heart Association attempts to raise funds from potential donors through the use of volunteers calling door-to-door, it has chosen oral communication as the channel but has delegated the job of encoding and transmitting the final message to the individual volunteers. Alternatively, an automobile manufacturer employs an advertising agency to develop an advertising campaign using a celebrity spokesperson, such as Michael Jordan. In this case, the agency may select television advertising as the channel and write the message to be delivered by the celebrity.

In these instances, the volunteer and the celebrity spokesperson typically would be viewed as the communication source. However, the source is dual faceted to the extent that the potential donor incorporates existing knowledge about the Heart Association with the newly acquired knowledge about the previously unknown volunteer, or the car buyer combines existing knowledge about the manufacturer with the existing knowledge about the celebrity spokesperson.

5.3 CHARACTERISTICS OF AN EFFECTIVE SOURCE

The role of source characteristics in communicating safety information is illustrated in McGuire's (1980) communication-persuasion model, which combines a 10-step information processing model with a 5-element communication model. Table 5.1 presents this

Table 5.1 McGuire's (1980) communication-persuasion matrix.

Output: dependent variables (response steps mediating persuasion)	Input: independent variables (communication aspects)				
	Source	Message	Channel	Receiver	Destination
1 Exposure to communication					
2 Attended to it					
3 Reacted effectively to it					
4 Comprehended it					
5 Yielded to its arguments					
6 Stored and retained it					
7 Searched and retrieved it					
8 Decided to use it or not					
9 Behaved according to it					
10 Postbehavioral consolidating					

model as an input–output matrix where the rows represent the steps in the information processing model and the columns represent the elements of the communication model. Additional columns may be added within each of the five communication model elements to represent the characteristics found to influence its effectiveness. By considering the characteristics of each element in the communication model as independent variables and the receivers' responses at the sequential stages of the information processing model as dependent variables, the interaction between the component models becomes clear.

Lipstein and McGuire (1978) employ the communication-persuasion model to organize the 7000 articles included in their bibliography on advertising effectiveness. In that bibliography, they categorize the literature on source characteristics as follows, with the numbers in parentheses indicating the number of citations found in each category: credibility (282), likeability (257), power (73), quantitative aspects (52), and demographics (93). These source characteristics may be added as columns which comprise the larger column representing the communication source within McGuire's matrix, as mentioned in the previous paragraph. They are used here to review the source literature.

5.4 CREDIBILITY

Hovland and Weiss (1951) initiated the long tradition of evaluating source credibility. They found that individuals were more likely to believe in the feasibility of an atomic submarine when the attributed source was the noted scientist, J. Robert Oppenheimer, rather than the Soviet paper *Pravda*.

Highly credible sources have been found to: (a) produce more positive attitude changes than less credible ones; (b) elicit more behavioral change; (c) enhance fear appeals; and (d) inhibit counterargument to a message as people tend to lower their defenses and not think of as many cognitive responses (Mowen, 1995).

Hovland, Janis, and Kelly (1953) stated that credibility is determined by the perceived expertise and trustworthiness of the source. In a meta-analysis of studies examining 745 independent variables employed in 114 empirical studies, Wilson and Sherrell (1993) found that source effects accounted for 9% of the total variation in effectiveness of the messages being studied. Approximately 16% of that variation was due to the expert versus non-expert manipulation.

Sources have *expertise* if they have special levels of knowledge not generally available. This knowledge can be acquired through education, training or experience. Soumerai, Ross-Degnan, and Spira Kahn (1992) indicate that the authoritative medical sources concerning Reye's syndrome spread concern about the illness rapidly throughout the medical community despite equivocal information appearing at the same time in the popular press. Mallet, Vaught, and Brnich (1993) found that miners who had just survived a fire had trusted a person with a good understanding of the mine more than any kind of written sign or symbol found in the mine.

The second dimension of credibility is *trustworthiness*. Sources are viewed as trustworthy if their communication on a subject appears legitimate and there is no apparent conflict of interest. The magazine *Consumer Reports* accepts no advertising and purchases the products it tests through the same retailers used by consumers in order to maintain its independence and trustworthiness. In interpersonal communication, receivers find conversations are more believable if they are overheard because there is obviously no attempt to influence the receiver.

Craig and McCann (1978) found that a message sent to heavy electricity users from the Chairman of the New York State Public Service Commission was more effective than one from the Manager of Consumer Affairs at Consolidated Edison in increasing requests for additional information (18% versus 10%) and in decreasing the amount of electricity consumed (by about $4.50). In contrast, a celebrity endorser in a television commercial may not be credible to the critical viewer because it is clear to all that he or she has been paid a large sum of money by the manufacturer.

Locander and Hermann (1979) found that individuals were more likely to rely on independent sources of information (e.g., *Consumer Reports*, friends and neighbors) than sources advocating a position (e.g., advertisements, point-of-purchase displays and sales clerks) for products with a high perceived risk associated with them (i.e., stereo and lawn mower as opposed to paper towels and aftershave/cologne). (While 'perceived risk' in the warnings literature refers to the danger associated with the use of a product and is a function of both the severity and the likelihood of injury, 'perceived risk' appears in the marketing literature as the 'danger' of making a wrong decision and is a function of both the severity and the likelihood of that occurring; see Bauer, 1967.) In other words, as the importance of the decision increased so did the reliance on credible information sources.

5.5 LIKEABILITY

Sources who are physically attractive and likeable generally have been found to be more successful communicators than unattractive ones. Dion, Berscheid, and Walster (1972) found that college men and women rated physically attractive people to be more sensitive, warm and happy. Chaiken (1979) found that physically attractive communicators are more effective than unattractive ones and that they possessed characteristics other than physical attractiveness that could enable them to communicate a message more effectively (e.g., higher grade point averages and SAT scores). The dimension of likeability helps to explain why some public figures such as Bill Cosby and Michael Jordan do extensive product endorsements and other equally well known individuals such as Carl Lewis and Mike Tyson do not.

5.6 POWER

If the purpose of communication is to influence others and power is one's ability to influence others, then power is exerted through communication. Deutsch and Gerrard (1955) indicate that social influence is achieved through *information influence* and *normative influence*. Information influence is brought about through message content alone, in contrast to normative influence which is determined by the characteristics of the source and the source's relationship with the receiver.

French and Raven (1959) delineate six types of social power: reward, coercive, legitimate, referent, expert and informational power. All but the last of these are source characteristics providing a basis for normative influence. *Reward power* and *coercive power* refer to the ability of the source to provide positive and negative reinforcement, and involve the use of power in its conventional sense. Parents' instruction of their children is closely intertwined with the use of rewards and punishments. Traffic lights and stop signs indicate the appropriate behavior explicitly and implicitly point to the consequences of disobedience (a ticket or even arrest by a police officer). Sources are seen as having *legitimate power* if they have the right to prescribe behavior. Thus a grandparent may instruct a child even though no rewards or punishments are imminent.

Referent power stems from the receiver's ability to identify with the source and use him or her as a model for attitudes and behavior. The importance of referent power is evident in the forthcoming discussion of the demographic characteristics where it is seen that a source may be a more effective communicator if he or she is similar to the receiver. *Expert power* was discussed previously as a foundation for source credibility.

5.7 QUANTITATIVE ASPECTS

Quantitative aspects involve instances where the source is a group rather than an individual and refer to the group's size and degree of its unanimity. Early research on social influence emphasized the power of a group over an individual. For example, Asch's (1956) study indicated that individuals yielded to group pressure in giving estimates of the lengths of lines which clearly were incorrect. More recently, researchers have distinguished between informational influence and normative influence, and have suggested that a minority can influence the majority. Burnkrant and Cousineau (1975) found that individuals' ratings of an instant coffee were influenced when they were informed of the previous ratings of others. However, a consensus among the ratings of others did not produce a greater rating increase in conformity than did the experimental condition where there was much disagreement among the previous ratings. Additionally, Moscovici, Lage, and Naffrechoux (1969) found that a minority in a group could influence the responses of the majority in judging colors.

5.8 DEMOGRAPHICS

Demographic characteristics also have been examined, with the focus primarily on the sex, age, and race of the source and receiver. Rosen and Jerdee (1973) presented students and bank supervisors with one of six versions of a written description of a supervisory problem and asked them to evaluate the effectiveness of the supervisory styles. The gender of the supervisor and of the subordinates was varied in the descriptions. A 'helping style' was evaluated uniformly as the most effective style and the 'threatening style' was evaluated uniformly as the least effective. However, a 'reward style' was evaluated as more effective for male supervisors and a 'friendly dependent style' was more effective when the sex of the supervisor and subordinate differed.

Underlying demographics is the similarity of the source and the receiver, and researchers have found receivers may prefer a similar source over an expert when they are considering the appropriateness to them of the information they are receiving. Brock (1965) found that purchases made by retail paint customers were influenced more by sales clerks who had experience similar to the customers than by clerks with a greater level of experience. Additionally, peer interventions were found to be more effective than the recommendations and guidelines of national experts in changing the practices of physicians (Greco and Eisenberg, 1993).

5.9 SOURCE AND CONTEMPORARY INFORMATION PROCESSING MODELS

Research findings inconsistent with the conventional communication-persuasion model have been found since the model began to be studied empirically, especially where interactions among the elements of the communication model exist. McGuire (1969) cites research which found that receivers exposed to a high credibility source experienced more attitude change than those exposed to a low credibility source, but they had not learned the message content better than the group exposed to the low credibility source. Further, receivers exposed to a source with intermediate credibility experienced an intermediate level of attitude change but had learned more of the message content than either the high or low credibility group. The conventional model would have predicted that a source with higher credibility would have led to greater message comprehension which, in turn, would have resulted in a greater attitude change. However, the intermediate step of comprehension was not necessary for attitude change to occur in this case.

The relationship between source credibility and attitude change has been found to interact with characteristics of the message. Sternthal, Phillips, and Dholakia (1978) found that a moderate credibility source was more persuasive than a high credibility source when the receivers were favorably disposed to the message. A similar result was found by Mausner and Mausner (1955). Sternthal et al. (1978) speculated that the less credible source motivated receivers to acquire more supporting arguments to maintain their previous attitudes.

Additionally, the relationship between source credibility and attitude change has interacted with characteristics of the receiver. Petty and Cacioppo (1986) have indicated that high credibility sources are effective when the motivation and/or ability to process a message are low, but source characteristics are relatively unimportant when the ability and motivation to process the message are high. Further, individuals have been found to vary in their *need for cognition*—the degree to which they are inclined to engage in deliberative decision making. Haugvedt, Petty and Cacioppo (1992) found that receivers who have a low need for cognition are more likely to be influenced by source characteristics.

Researchers for more than a decade have been employing the elaboration likelihood model (ELM) to investigate the complex interactions among the input elements of the communication-persuasion model (Petty and Cacioppo, 1986). The ELM posits that there is a *central route* of information processing where individuals actively gather information, combine it with the information in memory and process it. Alternatively, there is a *peripheral route* of information processing where individuals passively gather and process information. Individuals employing the central route are said to be involved in *high involvement processing* while those employing the peripheral route are involved in *low involvement processing*.

A message sent to an individual consists of a variety of pieces of information or cues that include: the denotative or literal meaning of the message; its connotative meaning (word choice as well as accompanying sights, sounds, and even smells); characteristics of the source (as described earlier in this chapter) and characteristics of the channel (as described in Chapter 6 by Mazis and Morris). Cues that are actively deliberated when high involvement processing is involved are referred to as *central cues* and come primarily from the denotative meaning of the message. Cues that influence a decision when low involvement processing is involved are referred to as *peripheral cues* and include source characteristics and the other factors listed above.

The ELM helps to resolve many of the inconsistencies found in previous research with the dual routes of information processing. Typically, researchers create low involvement and high involvement subjects by manipulating the relevance of the judgments they are about to make (as the first independent variable). Also typically, message characteristics and source characteristics are manipulated (as second and third independent variables) and measures of agreement with the message content are used as dependent variables.

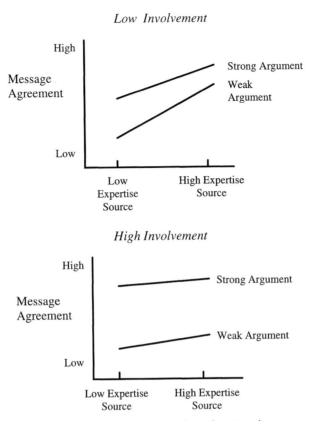

Figure 5.1 Argument agreement as a function of product involvement, argument quality and source expertise (adapted from Petty and Cacioppo, 1986).

Results of a typical ELM experiment are shown in Figure 5.1. In this 2 × 2 × 2 design, the manipulations are high and low involvement, strong and weak message arguments and high and low source expertise. While the figure shows positive main effects for both source expertise and message strength, the most notable finding is the different result for low and high involvement subjects. For low involvement subjects represented in the upper panel of the graph, the expertise of the source (as indicated by the slope of the two lines) is more important than argument strength (as indicated by the distance between the lines). By contrast, the lower panel illustrates that the impact of argument strength is substantial, while the impact of source expertise is minimal. Similar findings are predicted for other source characteristics, as they are considered peripheral cues which have greater impact on low involvement subjects (see Petty and Cacioppo, 1986, Chapter 6).

An alternative approach to information processing is the heuristic-systematic model (HSM) proposed by Chaiken (see Chaiken, 1980; Eagly and Chaiken, 1993). In this model, *systematic processing* corresponds to the ELM's central route where high involvement processing takes place. *Heuristic processing* is a more limited mode of information processing analogous to the ELM's peripheral route processing. Heuristic processing is not based upon message comprehension (Rathneswar and Chaiken, 1991). Rather, it employs special heuristics, or decision rules triggered by peripheral cues (to employ ELM terminology). One such heuristic might be to invoke the belief that 'statements by experts can be trusted' when the communication source is recognized as an expert. Other heuristics might include 'consensus implies correctness' or 'a long argument implies a strong argument.'

Generally the ELM is considered to treat the two processes as alternatives. By contrast, HSM suggests that the two systems are parallel processes for all decisions, but the effect of heuristic cues is attenuated as the amount of message processing and comprehension increases. After employing the HSM model to design a study evaluating written descriptions of a telephone answering system, Chaiken and Maheswaran (1994) concluded that source credibility influenced attitudes toward the product in three ways. First, source credibility served as an heuristic for subjects with little motivation, as discussed previously. Additionally, source credibility influenced attitudes of highly motivated subjects who read an ambiguous message. Second, the credibility of the source had a direct impact on the positive judgment of the telephone answering system. Third, source credibility had an indirect impact on product judgments by altering the degree to which subjects considered and took seriously the arguments contained in the message.

An additional advantage claimed for the HSM is that it provides a more complete understanding of low involvement processing (Rathneswar and Chaiken, 1991, p. 53). However, both the ELM and HSM help explain the interactions among the basic elements of the communication model. Both models also help us to understand how individuals with differing levels of motivation and ability process the same message differently. Wilson and Sherrell (1993, Table 5) review research employing these models in the examination of source effects.

Although the ELM and the HSM have not been applied in a warnings context, they provide a theoretical foundation for integrating past research findings, and are suggestive for future research. Consistent with past warnings research, these models would predict that product users are more likely to find, read and follow warning information for products with a high perceived hazardousness because central route processing tends to be involved. The models would predict also that heuristics associated with the color and shape of the warning and the signal word it contains may be employed if the product involves low perceived risk or if the user does not have the knowledge or skill to evaluate the message content. Considerations such as the perceived hazardousness of a product and the expertise of a product user have received much attention within the warnings literature. However, the ELM and the HSM provide an opportunity for researchers to study the complex interactions among source, message, and receiver.

5.10 SOURCE AND WARNING EFFECTIVENESS

Almost 50 years of research have documented that the effectiveness of a communication can be enhanced by varying source characteristics. The literature on health risk communications, including diverse topics such as food ingredient labeling and anti-smoking campaigns, is extensive and has examined source effects (see Morris, Mazis, and Barofsky, 1980; and in particular, McGuire, 1980). Unfortunately, the possibility that source characteristics can enhance the effectiveness of product warnings has been neglected by researchers and practitioners alike. A review of the product warnings literature for this chapter uncovered approximately a dozen publications discussing source effects in warning communications (e.g., Driver, 1987; Laughery and Laux, 1988) and only one that actually studied source effects (Lirtzman and Shuv-Ami, 1986).

Probably there are two reasons why the potential for source effects to enhance communications effectiveness has not been studied or utilized in the warnings area. First, the vast majority of product warnings are written in the form of on-product warnings, instruction booklets and signs. These warnings are the most impersonal of communications and cannot harness the power of source effects that are characteristic of other communication media. These media are in contrast to the medium of face-to-face communication, where the full range of source characteristics can be employed. It is also in contrast to the diverse media, including television, radio, newspapers, magazines, and direct mail, where advertisers have employed source effects so effectively. Thus, our purchase decisions are influenced because Michael Jordan wears a particular shoe or because the American Dental Association endorses a specific toothpaste.

The second reason why source effects have not been studied or utilized is that product warnings rarely identify the source explicitly. Exceptions are the fairly recent result of federal legislation concerning cigarette and alcohol warnings, which refer to the 'Surgeon General,' with alcohol warnings containing the additional statement: 'Government warning.' Consumers may assume that the source of warning information on, or accompanying, a product originated with the manufacturer but that is not actually the case for the Reye's syndrome warning for aspirin, the toxic shock syndrome (TSS) warning for tampons and the many other government-mandated warnings appearing on products such as pesticides.

The significance of not explicitly identifying the source of an on-product warning is not clear. McGuire (1980, p. 105) states that: 'Even though the current nonattributional practice can be defended, the possibility remains that leaving most warning labels without explicit source attribution may be neglecting an input component which could add to the label's impact.' He (McGuire, 1980, p. 104) also speculates that warning information is credible because consumers recognize that manufacturers are pointing out serious limitations of their products by printing a warning.

Beltramini (1988) evaluated the believability of five versions of the mandated cigarette warning but did not examine the significance of listing the Surgeon General as the source. Lirtzman and Shuv-Ami (1986) employed three surveys to assess the trustworthiness of consumer test labs, university researchers, a federal agency, the product's manufacturer and a labor union as a source of information concerning a dangerous or risky product. They found that the test labs were the preferred source. Unfortunately, they did not test to see if these differences in trustworthiness translate into greater acceptance of the same warning message.

Opportunities for utilizing strong source effects in product warnings are limited because the conventional product warning is small, impersonal and is devoid of motion and sound. However, research is needed to explore the full potential of source effects in enhancing the effectiveness of warnings even if that potential is limited. Research is needed to discover what inferences are made by product users when the warning source is not identified. Additionally, researchers need to discover whether the effectiveness of warnings can be enhanced by identifying the source and whether the most effective source is the product's manufacturer or some independent organization. Previous research would lead to the hypothesis that the most credible, independent sources are the most effective, but this is a matter of empirical testing. Further research opportunities will be discussed in the concluding section of the chapter.

5.11 OTHER MEDIA

Greater opportunity exists for employing source effects when other communication media exist. While the significance of the communication channel in warnings communication will be discussed in Chapter 6 by Mazis and Morris, following is a brief discussion of the interaction of source and channel effects.

Safety training in an industrial setting allows for the full range of source characteristics to be employed, including power, as the employer has the ability to observe worker performance and reinforce it. Training films used in corporate and educational settings and videos that are now accompanying new cars can employ most of the range of source characteristics.

Organizations such as the National Fire Protection Association and the National Propane Gas Association print and make available to their members various safety materials. While the NFPA lists its name on the front and/or back cover, this is not always the case with the NPGA. What difference this makes to the customers of propane dealers who distribute these safety materials to them is a matter of speculation.

Wright (1979) studied the effectiveness of a 5-second 'warning message' included as part of a 30-second television commercial for antacids. A version of the warning message containing the concrete recommendation that the package warning be read in the store before buying the product and portraying a model following that recommendation resulted in short term increases in package inspections and in the tendency to read in-store warning signs. Thus, the use of television as a warning medium increases the opportunity for employing source effects as a means of increasing warning effectiveness.

Loken and Howard-Pitney (1988) studied the effect of the presence of a healthy, attractive woman smoker and a written warning on the evaluation of print advertisements. They found that the presence of the model increased the rated attractiveness, persuasiveness, and credibility of the ad, and that a specific rather than a general warning decreased the ad's attractiveness and persuasiveness. Smokers rated the ads with specific warnings as being lower in attractiveness and credibility. All nine versions of the tested warning cited the Surgeon General and there was no ad shown without a warning.

Racicot and Wogalter (1995) examined the effectiveness of a warning video sign in a simulated chemistry laboratory experiment. In one version of the warning, a written statement was shown for 30 seconds. In the second version, the statement was shown for 10 seconds, safety equipment was shown for 8 seconds and a 12-second clip was shown of an individual putting the equipment on. The third version was the same as the second except that a male voice read the written statement while it was shown during the first 10 seconds. The researchers found that the addition of the display of the safety equipment and the modeling of safe behavior increased warning compliance substantially (50% to 92%). The addition of the audio increased compliance (to 100%) but the contrast between this condition and the two experimental conditions without audio was not statistically significant.

5.12 MULTIPLE WARNING SOURCES

As a practical matter, every individual communication must be viewed in the context of a history of communications on the same topic from a variety of personal and impersonal sources. An individual's attitudes prior to receiving a particular safety communication are in part a composite of the central and peripheral cues of messages received in the past.

At any one time, an individual may also be contending with more than one source of warning information. For example, a product user may evaluate the expertise of other individuals performing the same task and assess the consequences they experience from following or disregarding warning instructions. In an experiment by Chy-Dejoras (1992) significantly more subjects wore gloves (87% compared with 50%) if they saw the film demonstrating their use in addition to being exposed to a warning on a container of adhesive remover. Additionally, more subjects wore gloves (87% as opposed to 57%) when a model in the film wore gloves. Wogalter, Allison, and McKenna (1989) found that students were more likely to follow written chemistry laboratory instructions and wear a mask and gloves when a confederate wore them (100%) than when the confederate did not comply (33%).

It would be a serious mistake for a warnings designer to ignore these other sources of warning information, especially where the messages are not all in agreement. Hence, an individual warning must be designed as if it were in competition with rival sources of information. Additionally, the warnings designer should consider a mix of warning messages utilizing media which have a greater opportunity for capitalizing on source effects. Perhaps for this reason, a gas company may use a mix of appliance labels, bill stuffers, brochures, public service announcements, and oral communications from repair personnel to ensure the safe use of gas appliances.

Health risk warnings lend themselves to a multimedia approach. For example, Warner (1977) estimates that the anti-smoking campaign (the collective but uncoordinated activities of government agencies, private voluntary agencies, and for-profit firms) reduced per capita cigarette consumption by 20–30%.

Soumerai *et al.* (1992) report that the combined effects of medical journals, the FDA and CDC, consumer advocacy organizations, and the Aspirin Foundation resulted in the lowest level of Reye's syndrome since its monitoring began in the mid-1970s. The initial decline in aspirin use resulted from changes in physicians' knowledge and behavior rather than those of the parent because of industry resistance to acknowledgment of the association between Reye's syndrome and aspirin use.

5.13 SUMMARY AND CONCLUSION

The immense body of communications research provides much insight into the process of conveying warning information. The warnings literature demonstrates clearly that the use of source characteristics can add significantly to the effectiveness of communications. Although source characteristics should add to the effectiveness of product warnings, this is little more than speculation because of the dearth of research on the topic. Wilson and Sherrell (1993) found that 9% of the total variation in message effectiveness was due to source effects, and certainly this is an area warranting exploration.

Recent developments in information processing models have clarified the means by which the array of cues from source, message, and media are decoded in the process of receiving a message. Of particular significance is the finding that the cues processed and the manner in which they are processed vary by individual and circumstance. This literature would lead to the hypotheses that systematic or central route processing would be employed and source characteristics are less important when (i) the product users have a high *need for cognition*; (ii) the perceived hazardousness is high because of product danger or unfamiliarity; or (iii) the individual is capable of comprehending the warning information.

By contrast, peripheral route or heuristic processing would be employed and source characteristics are more important when (a) the product users have a low *need for cognition*; (b) the perceived hazardousness is low; or (c) the individual is incapable of comprehending the warning because the information is too complex or ambiguous, or there are time pressures or distractions.

Warnings research investigating these groups and situations could contribute significantly to the effectiveness of warnings.

It is hoped that the rich tradition of communication research and the recent theoretical developments discussed previously will result in hypotheses applicable to product warnings that can be tested empirically. It is also hoped that warning practitioners will be able to capitalize on the understanding derived through communication-persuasion research, and produce warnings capable of protecting product users from personal injury or damage to their property.

REFERENCES

ASCH, S.E. (1956) Studies of independence and conformity. I. A minority of one against a unanimous majority. *Psychological Reports*, 70 (whole No. 416).

BAUER, R.A. (1967) Consumer behavior and risk taking. In COX, D.F. (ed.), *Risk Taking and Information Handling in Consumer Behavior*. Boston: Division of Research, Harvard Business School.

BELTRAMINI, R.F. (1988) Perceived believability of warning label information presented in cigarette advertising. *Journal of Advertising*, 17, 26–33.

BROCK, T.C. (1965) Communicator–recipient similarity and decision choice. *Journal of Personality and Social Psychology*, 1, 650–654.

BURNKRANT, R.E. and A. COUSINEAU (1975) Informational and normative social influence in buyer behavior. *Journal of Consumer Research*, 2, 206–214.

CHAIKEN, S. (1979) Communicator physical attractiveness and persuasion. *Journal of Personality and Social Psychology*, 37, 1387–1397.

CHAIKEN, S. (1980) Heuristic versus systematic information processing and the use of source versus message cues in persuasion. *Journal of Personality and Social Psychology*, 39, 752–766.

CHAIKEN, S. and MAHESWARAN, D. (1994) Heuristic processing can bias systematic processing: effects of source credibility, argument ambiguity, and task importance on attitude judgment. *Journal of Personality and Social Psychology*, 66, 460–473.

CHY-DEJORAS, E.A. (1992) Effects of an aversive vicarious experience and modelling on perceived risk and self-protective behavior. In *Proceedings of the Human Factors Society 36th Annual Meeting*. Santa Monica, CA: The Human Factors Society, pp. 603–606.

CRAIG, C.S. and MCCANN, J.M. (1978) Assessing communication effects on energy consumption. *Journal of Consumer Research*, 5, 82–88.

DEUTSCH, M. and GERARD, H.B. (1955) A study of normative and informational social influences upon individual judgment. *Journal of Abnormal and Social Psychology*, 51, 629–636.

DION, K., BERSCHEID, E., and WALSTER, E. (1972) What is beautiful is good. *Journal of Personality and Social Psychology*, 24, 285–290.

DRIVER, R.W. (1987) A communication model for determining the appropriateness of on-product warnings. *IEEE Transactions on Professional Communication*, PC 30, 157–163.

EAGLY, A.H. and CHAIKEN, S. (1993) *The Psychology of Attitudes*. Fort Worth, TX: Harcourt Brace Jovanovich.

FRENCH JR., J.R.P. and RAVEN, B. (1959) The bases of social power. In CARTWRIGHT, D. (ed.), *Studies in Social Power*. Ann Arbor, MI: University of Michigan, pp. 150–167.

GRECO, P.J. and EISENBERG, J.M. (1993) Changing physicians' practices. *New England Journal of Medicine*, 329, 1271–1273.

HAUGVEDT, C.P., PETTY, R.E., and CACIOPPO, J.T. (1992) Need for cognition and advertising: understanding the role of personality variables in consumer research. *Journal of Consumer Psychology*, 1, 239–260.

HOVLAND, C. and WEISS, W. (1951) The influence of source credibility on communication effectiveness. *Public Opinion Quarterly*, 15, 635–650.

HOVLAND, C.I.I., JANIS, L., and KELLEY, H.H. (1953) *Communication and Persuasion: Psychological Studies of Opinion Change*. New Haven, CT: Yale University Press.

LAUGHERY, K. and LAUX, L.F. (1988) Communicating with the worker through the MSDS. In BRANSFORD JR, J.S. (ed.), ASTM STP 1035 *Hazard Communication*, II, *Material Safety Data Sheets Are A Communication Tool*. Philadelphia: American Society for Testing and Materials, pp. 71–83.

LIPSTEIN, B. and MCGUIRE, W.J. (1978) *Evaluating Advertising: A Bibliography*. New York: Advertising Research Foundation.

LIRTZMAN, S.I. and SHUV-AMI, A. (1986) Credibility of sources of communication on product safety hazards. *Psychological Reports*, 58, 707–718.

LOCANDER, W.B. and HERMANN, W. (1979) The effect of self-confidence and anxiety on information seeking in consumer risk reduction. *Journal of Marketing Research*, 16, 268–274.

LOKEN, B. and HOWARD-PITNEY, B. (1988) Effectiveness of cigarette advertising on women: an experimental study. *Journal of Applied Psychology*, 73, 378–382.

MCGUIRE, W.J. (1969) The nature of attitudes and attitude change. In LINDZEY, G. and ARONSON, E. (eds), *Handbook of Social Psychology*, 2nd Edn. Reading, MA: Addison-Wesley.

MCGUIRE, W.J. (1980) The communication–persuasion model and health-risk labeling. In MORRIS, L.A., MAZIS, M.B., and BAROFSKY, I. (eds), *Product Labeling and Health Risks*, Banbury Report No. 6. Cold Spring Harbor Laboratory, pp. 99–122.

MALLET, L., VAUGHT, C., and BRNICH JR., M.J. (1993) Sociotechnical communication in an underground mine fire: a study of warning messages during an emergency evacuation. *Safety Science*, 16, 709–728.

MAUSNER, B. and MAUSNER, J. (1955) A study of the antiscientific attitude. *Scientific American*, 192, 35–39.

MORRIS, L.A., MAZIS, M.B., and BAROFSKY, I. (1980) *Product Labeling and Health Risks*, Banbury Report No. 6. Cold Spring Harbor Laboratory.

MOSCOVICI, S., LAGE, E., and NAFFRECHOUX, M. (1969) Influence of a consistent minority on the responses of a majority in a color perception task. *Sociometry*, 32, 365–380.

MOWEN, J.C. (1995) *Consumer Behavior*, 4th Edn. Englewood Cliffs, NJ: Prentice-Hall.

PETTY, R. and CACIOPPO, J.T. (1986) *Communication and Persuasion: Central and Peripheral Routes to Attitude Change*. New York: Springer-Verlag.

RACICOT, B.M. and WOGALTER, M.S. (1995) Effects of a video warning sign and social modeling on behavioral compliance. *Accident Analysis and Prevention*, 27, 57–64.

RATHNESWAR, S. and CHAIKEN, S. (1991) Comprehension's role in persuasion: the case of its moderating effect on the persuasive impact of source cues. *Journal of Consumer Research*, 18, 52–56.

ROSEN, B. and JERDEE, T.H. (1973) The influence of sex-role stereotypes on evaluations of male and female supervisory behavior, *Journal of Applied Research*, 57, 44–48.

SOUMERAI, S.B., ROSS-DEGNAN, D. and SPIRA KAHN, J. (1992) Effects of professional and media warnings about the association between aspirin use in children and reyes syndrome. *The Milbank Quarterly*, 70, 155–182.

STERNTHAL, B., PHILLIPS, L., and DHOLAKIA, R. (1978) The persuasive effect of source credibility: a situational analysis. *Public Opinion Quarterly*, 42, 285–314.

WARNER, K.E. (1977) The effects of the anti-smoking campaign on cigarette consumption. *American Journal of Public Health*, 67, 645–650.

WILSON, E.J. and SHERRELL, D.L. (1993) Source effects in communication and persuasion research: a meta-analysis of effect size. *Journal of the Academy of Marketing Science*, 21, 101–112.

WOGALTER, M.S., ALLISON, S.T., and MCKENNA, N.A. (1989) Effects cost and social influence on warning compliance. *Human Factors*, 31, 133–140.

WRIGHT, P. (1979) Concrete action plans in TV messages to increase reading of drug warnings. *Journal of Consumer Research*, 6, 256–269.

CHAPTER SIX

Channel

MICHAEL B. MAZIS
American University

LOUIS A. MORRIS
PRR Research Division

Understanding the most effective channel for communicating risk information requires knowledge of several factors. One important factor is whether the goal of the warning is to deliver information prior to product purchase, at the purchase stage, at the use stage, or subsequent to product use. Another factor is knowledge about how information is processed across the five communications channels: advertising, product labels, signs and signals, leaflets and owners' manuals, and face-to-face communication. Broadcast advertising warnings have the advantage of being relatively intrusive thereby generating high levels of attention to safety disclosures; however, since typically commercials are brief, there is little time to communicate detailed information. Often warnings in print advertising and on product labels are to communicate more detailed information, but gaining attention and motivating consumers to read the information is a challenge. Signs and signals are important vehicles for communicating appropriate use and handling of products and materials, but these messages must be conveyed at the appropriate time to be useful. Typically, leaflets and owners' manuals are used to communicate complex risk information; thus, collateral information must be organized effectively to be comprehended. Face-to-face communication is two-directional and must be adapted to meet the receiver's needs. Finally, technology, such as through CD-ROM and the Internet, is expected in the future to affect dramatically consumers' access to risk information. Consumers will be able to interact directly with information providers and will receive customized information that is consistent with their needs.

6.1 INTRODUCTION

Warnings and risk information have become commonplace in modern society. Manufacturers use warnings to protect consumers from product misuse; consumers seek risk information to avoid potential injury; the legal system, through products liability lawsuits, compels manufacturers to provide adequate warnings to consumers; and government agencies mandate disclosure of health hazards to protect society.

Warnings are conveyed through a number of channels. The purpose of this chapter is to discuss the factors that must be taken into account when attempting to communicate information through a particular channel. Channels have strengths and weaknesses, and knowledge of the benefits and limitations of channels is essential in the design of effective warning messages.

This chapter has four sections. First, an overview of communication channels is presented. Second, information processing issues related to various modes of communicating warnings are discussed. Third, five major information channels, including product labels, signs collateral product information (such as leaflets and owners' manuals), advertising, and face-to-face communication are examined in detail. Finally, some observations about future directions are offered.

6.2 COMMUNICATIONS MEDIA CHANNELS

There are a number of ways to view channel variables. The two most common approaches are by sensory modality, such as auditory, visual, olfactory, and gustatory senses, and by communications medium. The former approach is explored in greater detail in Chapter 7 by Wogalter and Leonard; the latter approach will be utilized in the current chapter.

A variety of communications media are used to convey product hazards. The traditional modes are product labels, leaflets, signs, owners' manuals, and advertisements. However, other approaches such as face-to-face communication are used, for example, to inform patients about medical risks. Also, the mass media, through newscasts and televised magazine shows, are becoming an increasingly important source of risk information. Finally, technology has made available additional channels, such as video tapes, which are more frequently accompanying new products, and toll-free telephone numbers and the Internet to communicate specific information to interested parties.

One approach to better understand the functions of various communications media in conveying risk information is to relate channel variables to the product purchase and use process. Product purchase and use may be viewed as a four-step process. The first step is the *pre-purchase* stage, which involves the information gathering undertaken by consumers prior to product purchase. For example, consumers may actively acquire information prior to purchase by noticing advertisements, by reading magazine articles, by consulting consumer testing magazines or by seeking information from friends or experts. In addition, consumers may acquire information passively by absorbing information from the environment without actively seeking specific data about a product. For example, consumers may observe a friend using a product, overhear a conversation about a product, or watch a newscast discussing a product. The information that is acquired is stored in memory and is retrieved later (Beales, Mazis, Salop, and Staelin, 1981).

Acquisition of risk information in the pre-purchase stage often takes place through advertising, mass media, and interpersonal sources. However, interpersonal transmission is affected strongly by advertisements and by media portrayals. Pre-purchase information acquisition plays an important role in the development of schemas about product safety. For example, initial advertising for all-terrain vehicles (ATVs) portrayed ATVs as stable over most terrains. Therefore, many prospective purchasers viewed ATVs as relatively safe and suitable for use by children. This (mis)perception led to substantial use of ATVs by children and adolescents and to subsequent high injury and death rates (Ford and Mazis, 1996). Product safety schemas affect subsequent information acquisition. If consumers are persuaded through advertising and the mass media that a product is safe, they may choose to ignore warnings on product labels.

At the pre-purchase stage, consumers often do not search for detailed hazard information. They may actively seek general information about the product's features, or they may passively acquire information from the environment. Therefore warnings presented in the pre-purchase stage must gain the consumer's attention and must be relatively simple. Warnings in advertisements must be sufficiently conspicuous to 'interrupt' the consumer's usual activities, such as reading a magazine article or watching a television program. In addition, relatively simple warnings are needed because the consumer is unlikely to make the effort to read and to remember complex disclosures.

The second step is the *purchase* stage, which concerns information acquisition at the point-of-sale or just prior to the actual decision. At this stage, consumers may search actively for information because a commitment is about to be made. Consumers tend to have at least a moderate degree of involvement or interest in risk information at the purchase stage, although information receptivity is greater for first-time decisions than for repetitive decisions. Prior knowledge has an effect on the amount and type of information search at the purchase stage (Brucks, 1985; Sujan, 1985).

Thus, warnings are more likely to be noticed in the purchase stage than in the pre-purchase stage. However, the hazard information is competing against other information confronting consumers, and consumers have only a limited amount of processing capacity available. Therefore, the warnings must be well organized in order to be comprehended. Often consumers are interested in learning about product risks, but the information presented must be easily processed; the benefits of information acquisition must outweigh the costs. The more effort it requires to process the information the less likely it is that consumers will use the information.

At the purchase or decision stage, a variety of sources, including product labels, signs, and face-to-face communication, may be available to provide risk information. This information may be used to assist in the decision about whether to accept or reject the alternative presented or to select one option among competing alternatives.

For example, when a pregnant woman is considering the purchase of an alcoholic beverage product, she must weigh the perceived benefits against potential risks. She might read the warning label, and this might affect her decision. However, since the same warning appears on all alcoholic beverage brands, she must make a buy/not buy decision for the entire category of alcoholic beverage products.

On the other hand, if consumers are confronted with insect infestation, they may wish to purchase an insecticide. Consumers may desire to purchase the product that offers the greatest benefits and the fewest safety hazards. Thus, comparative information among brands in the product category is required to make such a decision (Bettman, Payne, and Staelin, 1986). However, if a great deal of effort is required to make comparative judgments, consumers are unlikely to use the risk information (Russo and Leclerc, 1991).

One approach to reducing the costs of processing risk information is to use a common format and a common set of concepts in labeling hazardous products. A standard label facilitates a consumer's ability to compare similar brands and to encode hazard information about a new product once the format has been learned through prior experience with other labels (Bettman *et al.*, 1986). There is a trade-off, however, because standardization may reduce consumers' attention to the warning message.

The third step is the 'use' stage, which involves operation or consumption of the product. Activities at the use stage include pilots operating aircraft, workers handling chemicals, and consumers ingesting prescription drugs. Frequently at the use stage there is a high level of consumer involvement and a heightened interest in risk information. However, as with the purchase stage, the propensity to read risk information and to adhere to warnings is affected by the perceived severity of the risks and by previous product use experience. When risks are perceived to be minimal and when the product has been used previously, consumers may believe that the product is safe, and they may fail to notice and to read warnings (Wright, Creighton and Threlsall, 1982). In such situations, efforts are needed to ensure that consumers' attention is attracted to the risk information.

However, first-time users often are receptive to avoiding safety risks. Thus, typically risk information aimed at product use is focused on persuading consumers to adopt safe behavior. Use information might warn prescription drug users about potentially lethal drug interactions, riders of all-terrain vehicles about riding on paved roads, and automobile drivers about an impending steep grade. In contrast, risk information directed to the pre-purchase and purchase stages often seeks to inform the audience of a hazard, remind consumers about potential risks, or change attitudes about the product. In the pre-purchase and purchase stages, danger is not imminent.

A key factor in the effectiveness of warning messages at the use stage is alerting people to hazard at critical times within a task (Dorris and Purswell, 1977). That is, if possible, warning messages should be integrated into tasks. For example, warnings may be placed on a car jack for a person changing a tire or on a door to an electrical panel for someone performing electrical work (Lehto and Miller, 1988). However, often such placement is not feasible.

At the use stage, product labels and signs or auditory signals are the most common means of providing risk information. However, leaflets, video tapes, owners' manuals or other materials accompanying the product frequently are used. Product labels, signs, and auditory signals have the greatest potential to be integrated within typical product use; thus, they are most likely to inform users of the hazard at a propitious time. However, collateral materials may be more effective at communicating more complex messages, but they may not be read or if the information is read it may be forgotten.

The final step is the 'post-use' stage, which involves possible product disposal or health problems occurring subsequent to product use. For example, instructions are provided to consumers on product containers about disposal of motor oil and pesticides and to workers in written materials and on signs about disposal of hazardous chemicals and nuclear waste. Also, information is provided for medicines about the importance of side effects, such as a rash, stomach distress, or fever, and the need to consult a physician.

At the post-use stage often there is little concern about appropriate disposal because risks tend to be borne by others. Since consumers or workers may be asked to engage in time-consuming behavior that may have a high personal cost and little personal benefit, risk information often must be particularly persuasive. Messages must seek to portray the extensive benefits of proper disposal. Written materials and videotapes, for example, may be particularly effective; labels and owners' manuals also are used for this purpose. In addition, efficient collection is necessary to reduce the costs of disposing of materials.

On the other hand, often there is considerable concern about health problems occurring subsequent to product use. However, product users may not have read or may not remember health hazard information. Thus, written materials must be available and well organized for users to read about possible health effects of product use, and to take appropriate action.

6.3 AN INFORMATION-PROCESSING ANALYSIS OF COMMUNICATIONS MEDIA

As discussed in the previous section, there are a variety of modes available for communicating risk information. This section provides a brief information-processing analysis of five communications channels: advertising, product labels, signs, collateral information, and face-to-face communications. The advantage and limitations of each medium are examined from the perspective of recipients' information processing capabilities.

6.3.1 Advertising

Communicating safety hazards through advertising usually takes place at the pre-purchase stage. Risk disclosures in television or radio advertisements have the distinct advantage of being noticed by viewers or listeners. Television advertising, in particular, is intrusive, thereby generating potentially high levels of attention to the safety disclosures.

However, since broadcast commercials are relatively brief, there is little time to present detailed risk information within 30-second advertisements. Also, the sequence and rate of presentation of the risk information is not under the recipient's control (Bettman, 1979). (Of course, two-way cable systems may present the opportunity for consumers to receive on request detailed risk information.) While broadcast disclosures may be noticed, complex information is unlikely to be remembered and to be retrieved during the purchase or use stages. Thus, risk disclosures in broadcast ads should aim to alert the recipient about the existence of the hazard, to provide a brief overview of the hazard, or to direct the recipient to seek additional information (Bettman et al., 1986).

By contrast, disclosure of safety hazards within print advertisements may be more detailed because readers control the sequence and the speed at which the information is received (Bettman et al., 1986). However, readers process information in print ads more selectively. Thus, consumers may choose to ignore the risk information because it is not sufficiently conspicuous or because it is deemed less important than other elements of the print ad. Readers are also primarily interested in magazine or newspaper articles, and they may have insufficient time or desire to devote to reading the 'fine print' in advertisements.

6.3.2 Product Labels

Information on packages is a key source of risk information at the time of purchase. By using product labels, consumers do not have to rely on internal memory to access hazard information; an external memory source is readily available. However, frequently there are distractions in the point-of-purchase environment. Consumers may explore features of several brands and interact with sales personnel. Therefore, time pressure may limit the amount of processing capacity devoted to warnings. In addition, warnings on product labels may be relatively inconspicuous, such as in small print or on the backs of packages. As a result, typically risk information is less obtrusive on product packages than in broadcast ads, and is more easily ignored (Bettman, 1979). Also, if hazard information varies across brands within a product category, considerable effort may be required to compare relative risk levels for various brands.

Product labels also are an important information source at the use and post-use stages. Risk information that is targeted toward product use may be more complex than hazard disclosures that are relevant to the purchase stage. There are likely to be fewer distractions at the use stage, and often users are motivated to acquire safety-related information. However, the propensity to acquire such information and to adhere to instructions is affected by perceived product hazards, previous experience with the product, and individual propensity to assume risk. Thus, product labels must be sufficiently conspicuous to attract attention, and they must offer a compelling message to affect behavior, particularly for experienced users.

6.3.3 Signs and Signals

Warnings on signs and through auditory signals and verbal messages frequently are used to communicate risk information. Signs have been used at the point-of-sale to warn prospective purchasers about possible product hazards, such as warnings about the potential cancer risk from consuming certain products. Such messages often are designed to inform potential consumers rather to change behavior. Since signs, like product labels, frequently are relatively unobtrusive, conspicuous placement is essential in order for consumers to process the warning messages. One difficulty with posting signs at the point-of-sale is resistance from retailers, who may be concerned about the potentially reduced sales and about the effect on the aesthetics of the retail environment.

Signs, auditory signals, and verbal messages are most common at the use stage, especially in industrial settings where dangerous materials are being handled by workers and in transportation vehicles, such as airplanes and trucks. Typically, signs and other messages are designed to be noticed at the 'right' time when products are being used. Thus, attention to the warning may be governed even more by location and by timing than by conspicuity. However, the sign must still be sufficiently large and the auditory signal must be sufficiently loud to attract attention.

6.3.4 Leaflets

Collateral materials, such as leaflets, owners' manuals, and video tapes, frequently accompany products to inform purchasers about proper assembly, appropriate use, potential hazards, and necessary actions in the event the product causes harm. Such materials are designed to affect behavior at the use and post-use stages. However, frequently purchasers rely on these materials primarily to assemble or to use the product appropriately. Thus, warnings related to safe product use are most likely to be read if this information is integrated with other information about how to obtain the benefits from using the product.

However, while some consumers may read the use-related warnings immediately after purchase, other risk information may not be read or, if read, it may not be remembered. Also, the collateral materials may be lost or may be difficult to locate. Since there may be a significant amount of information accompanying the product, proper design of this information is crucial. Risk information must be organized so that it may be comprehended easily and so that the pertinent information may be located if the need arises in the post-use stage.

6.3.5 Face-to-Face Communication

Face-to-face communication is used frequently by health professionals to warn patients about medical risks and by training personnel to inform product users about appropriate use and handling of equipment and of chemicals. Such communications have the advantage of being flexible, permitting interaction between the communicator and the recipient of the information. Follow-up questions may be asked, and the presentation may be tailored to the needs and the concerns of the receiver. However, miscomprehension is common, and the recipient may feel rushed or intimated about asking questions. Also, warnings may not be remembered unless written information accompanies the oral presentation.

6.4 EXAMINATION OF INFORMATION CHANNELS

This section presents a detailed examination of the five channels previously discussed. Key issues related to effective use of a particular channel are raised, and an examination of research literature is presented.

6.4.1 Advertising

Information disclosures commonly appear in *television commercials*. Estimates of television advertising containing 'footnotes' (also referred to as 'supers') have ranged from 26% (Hoy and Stankey, 1993) to 66% (Kolbe and Muehling, 1992). The variations in estimates reflect differences in the definition of a disclosure. Those providing greater estimates include corporate trademarks ('Jeep is a Registered Trademark of the Chrysler Corporation') and citations ('J.D. Power and Associates 1995 Initial Quality Survey'), which are excluded in other studies. Among these disclosures are commonly used advisories or warnings: 'Check with your doctor,' 'Use only as directed,' 'Avoid use of product when operating a motor vehicle or equipment,' and 'Wear protective eyewear when using this product'. Approximately 14% of network ads were found to contain such statements (Kolbe and Muehling, 1992). Hazard warnings are most common for over-the-counter drug ads.

There tends to be considerable conflict over the use of disclosures in television advertising. Advertisers tend to avoid the use of hazard disclosures, if possible, because they fear that warnings will reduce ad effectiveness, drawing attention away from the advertiser's selling message. However, a study of warning messages in antacid commercials found that recall of Alka Seltzer ads was not affected adversely by the presence of relatively brief hazard information (Houston and Rothschild, 1980). Also, including a short 'corrective' message ('does not cure sore throats or lessen their severity') in television ads for Listerine mouthwash was found to have no effect on recall of key selling messages (Mazis, McNeill, and Bernhardt 1983).

However, another study reported that including two or four risk messages in a televised prescription-drug ad reduced recall of product benefits (Morris, Mazis, and Brinberg, 1989). Thus, relatively brief warning messages are likely to have little or no impact on recall and knowledge of key sales messages. However, when significant amounts of risk information are included in television commercials, consumers must make 'trade-offs' because of information processing limitations. They are unable to remember all of the product benefits and the risks in a complex commercial message.

On the other hand, some government officials and consumer-group representatives have sought lengthy and detailed information disclosures in television advertising. Some of these efforts have been denounced as being counterproductive. In 1974, the US Federal Trade Commission (FTC) proposed a trade regulation rule that would have mandated a six-second disclosure of several nutrients and their percentages of the US Recommended Dietary Allowances. Bettman (1975, p. 173) commented that '... it seems highly improbable that the consumer can take in and remember the information presented, because of limitations in processing ability.'

In 1988, the National Association of Attorneys General proposed including relatively lengthy disclosures in rental car ads presenting price information. However, research provides some support for the industry's assertion that a simple disclosure would be equally effective (Murphy and Richards, 1992).

Information specificity is another important issue in designing television advertising disclosures. General advisories such as 'read the label,' 'avoid excessive use,' and 'consult your physician' are common in television advertising. Are such general disclaimers an effective means of conveying to consumers that using the advertised product may be hazardous? In a laboratory study of warning messages embedded in Alka Seltzer television commercials, none of 135 subjects recalled seeing the general warning message 'Read the label. Use only as directed,' which was presented visually, without being read by an announcer (Houston and Rothschild, 1980).

In addition, studies have found that specific warnings in television ads are more effective than are general warnings. For example, a specific disclosure for an antacid product ('Some antacids may not be safe if you are on a low salt diet') was recalled more often than was a general disclosure ('Some antacids may not be safe for you') (Houston and Rothschild, 1980). Other studies for televised prescription drug ads and for alcoholic beverage ads also found that specific disclosures were recalled more often than general disclosures (Morris et al., 1989; Smith, 1990).

The corrective advertising literature also provides some useful insights into the effectiveness of televised information disclosures. Field evaluations of the corrective message for Listerine mouthwash found that only about 5% of the commercial's viewers mentioned the presence of the corrective message and about 20% of those remembering the commercial were able to identify the content of the disclosure after some prompting (Mazis et al., 1983). In addition, telephone surveys conducted during the one-year period the corrective message was aired report a reduction of only 10–20% in overall 'incorrect' beliefs about Listerine's effectiveness (Armstrong, Gurol, and Russ, 1979; Wilkie, McNeill, and Mazis, 1984). Thus, while televised advertising disclosures can be informative, the short term impact may not be large. A 7-year longitudinal study of a corrective advertising message for Hawaiian Punch ('has 10% fruit juice') found that beliefs changed slowly (Kinnear, Taylor, and Gur-Arie, 1983).

Conspicuity of the information disclosure in television commercials can have an impact on whether and on the rate consumers learn about a hazard. Unfortunately, the written words in many television commercials are displayed too quickly for viewers to read them (Best, 1989; Smith, 1990; Hoy and Stankey, 1993. The FTC recommends disclosures that would require reading rates of 108–180 words per minute (wpm). However, one study reported that 35% of disclosures were presented at 180 wpm or more and that 56% of disclosures were presented at rates faster than 132 wpm (Best, 1989).

In addition, the size of the print and the number of words in televised disclosures have an impact on viewers' comprehension of risk messages. Shorter disclosures and disclosures that are displayed in larger print are comprehended more readily than are longer disclosures and disclosures that appear in smaller print (Murray, Manrai, and Manrai, 1993). Also, dual-modality disclosures in which warning messages are superimposed on a television screen at the same time an announcer reads the message have been found to achieve much higher levels of message recall than single-modality disclosures, either print only (Houston and Rothschild, 1980; Smith 1990; Barlow and Wogalter, 1993; Murray et al., 1993) or audio only (Morris et al., 1989). However, a content analysis of 246 network broadcast commercials aired in 1990 found that 245 (99%) were video-only and one disclosure was audio-only. No disclosures combined voice-over with print, i.e., dual-modality (Hoy and Stankey, 1993).

The structure of the risk information when there is more than one hazard disclosure in a television commercial also has been found to have an impact on warning message effectiveness. Morris et al. (1989) found that warnings that were dispersed throughout the commercial were recalled more often than were warning messages that were grouped in a single section near the end of the commercial.

Finally, novel approaches are available for presenting warnings through television advertising. Wright (1979) incorporated into an antacid commercial a five-second visual demonstration of a woman reading the warning message on the back of the package. This message caused shoppers to devote more time attending to warnings when shopping for antacid products. Of course, television commercials warning the public about hazards such as smoking cigarettes, contracting AIDS, and drunk driving often appear on television. Such messages, if carefully designed, may be effective in alerting the public and in changing behavior. For example, a 6-month paid anti-drunk-driving advertising campaign targeted at 18–24-year-old males reduced reported drinking-and-driving episodes and traffic accidents (Stam, Murry, and Lastovicka, 1993). However, most public health mass media campaigns have not been successful. Frequently these campaigns are aired in donated TV time during undesirable time slots that the media could not sell. To achieve success, it is important to be able to control the number of placements and the target audience reached.

Information disclosures also appear in *print advertising*, such as magazine or newspaper ads. Typically, such disclosures appear in fine print at the bottom of advertisements. While there is only a limited amount of research on the effectiveness of 'fine print' disclosures, studies of physicians who read prescription drug ads (Baum, Schaeffer, Wideman, Reddy, and Yellin, 1983) and of college students who were exposed in a laboratory experiment to a 'mock' camera ad have found that very few people read this footnoted information (Foxman, Muehling, and Moore, 1988).

On the other hand, studies have reported that conspicuity has an impact on the effectiveness of warnings in print ads (Barlow and Wogalter, 1993; Foxman *et al.*, 1988) and in owners' manuals (Wogalter, Barlow, and Murphy, 1995). One study, however, found that the conspicuousness of chewing tobacco warnings had no effect on warning message recall (Popper and Murray, 1989). In general, conspicuous warnings are more likely to be noticed by readers because they do not blend in with the surrounding text and pictures.

Cigarette advertising health warnings have been the subject of some important research studies. While cigarette ad warnings may not be judged as conspicuous, they are more noticeable than the 'fine print' disclosures that appear at the bottom of many print advertisements.

Fischer, Richards, Berman, and Krugman (1989) conducted an eye-tracking study to measure adolescents' attention to warnings in cigarette print advertising. The findings show that there was a relatively low level of attention to the hazard disclosures. Subjects also performed only slightly better than random guessing in a recognition test. In addition, a tachistoscope study found that while subjects frequently were aware of a warning appearing in a cigarette ad, few were able to recall the warning's content (Fischer, Krugman, Fletcher, Fox and Rojas, 1993). Other studies have explored the impact of increased warning conspicuity. Studies exploring the impact of message shape on attention have produced mixed results. One study found that changing the shape of a warning message from a rectangle to an irregular shape had a small impact on recall (Bhalla and Lastovicka, 1984). Another study also found that the shape of the warning affected attention levels (Riley, Cochran, and Ballard, 1982). However, other research has found no effect of shape on recall (Barlow and Wogalter, 1993) and on compliance with warnings (Jaynes and Boles, 1990).

In addition, one study found that moving the warning from a peripheral to a central location in a print ad had little impact on brand attitudes (Clark and Brock, 1994).

Finally, message familiarity was found to have an effect on attention to risk information. New cigarette warnings in print ads attracted greater readership and quicker attention than 'standard' warnings (Krugman, Fox, Fletcher, Fischer and Rojas, 1994).

These studies provide a portrait of consumer information processing of warning messages in print advertising. First, if the print is quite small, the risk information will not be noticed. This is a problem for outdoor (billboard) advertising (Cullingford, Da Cruz, Webb, Shean, and Jamrozik, 1988; Davis and Kendrick, 1989) as well as for magazine and newspaper ads. Second, readers typically follow a 'top-down' processing approach. They scan the ad briefly and frequently notice the warning. However, they often fail to read the warning fully and to process the risk information. In many cases, they are quite familiar with the risk information, and the pictures and text of the ad are more interesting. Third, new information and a dramatically conspicuous presentation could increase attention to hazard information. Novel ideas include integrating risk information into the text of the ad, including a visual portrayal of the risk along with the verbal risk information, and developing separate or 'counter' ads containing risk information. However, while such approaches would result in greater recall of risk information, advertisers might be concerned about potentially lower levels of attention to selling messages.

6.4.2 Product Labels

Risk information on product labels may serve a number of purposes: it may warn potential buyers prior to purchase; it may inform prospective users before product use; and it may provide information to users about disposal hazards. Product labels also contain a variety of risk information. For example, food labels contain information about ingredients, such as sodium and fat, and about the presence of phenylketonurics, which serve to warn some users about potential health hazards. Pesticide labels often warn about possible adverse reactions and recommend treatment. Toilet bowl cleaner labels provide instructions about safe disposal.

For product labels, as well as for advertising, a key issue is getting the risk information noticed. Often warnings appear in small print and on the back of the container. As a result, many consumers fail to read this important information. However, attention levels vary considerably as a result of warning message design factors and of receiver characteristics.

A large body of research evidence has found that *message design factors*, such as the size and location of the risk information, affect warning message noticeability. For example, noticeability is improved by placing the message on the front label in a horizontal position (Funkhouser, 1984; Godfrey *et al.*, 1991; Laughery *et al.*, 1993). Also, reducing the clutter (e.g., promotional material) surrounding the warning message has been found to enhance attention to health hazard disclosures (Swasy, Mazis, and Morris, 1992; Laughery, Young, Vaubel, and Brelsford, 1993; Wogalter, Kalsher, and Racicot, 1993).

Studies have found also that pictorials, color, and signal icons, especially in combination with each other, improve warning noticeability (Laughery *et al.*, 1993) and behavioral compliance (Godfrey, Rothstein, and Laughery, 1985). However, consumers must understand clearly the meaning of a symbol in order for it to be effective. Symbols that do not have well understood meanings may communicate the opposite of what is intended (Lerner and Collins, 1980).

Unfortunately, some of the research findings on message design factors are contradictory. One study reported that pictorials had no impact on behavioral compliance (Wogalter *et al.*, 1993). Also, the use of borders to set off a warning appears to have little impact

(Laughery *et al.*, 1993). Moreover, one study reported that print size has an effect on noticeability (Swasy *et al.*, 1992), while another study reported that print size variations had no impact on attention levels (Magat and Viscusi, 1992).

Receiver characteristics also have an important impact on label effectiveness. For example, individuals who come in contact frequently with labels are most likely to notice warning messages. Mazis, Morris, and Swasy (1991) and Graves (1993) reported that the alcohol warning label was especially recognized by men, younger adults, and heavier drinkers, and Scammon, Mayer, and Smith (1991) found that recall was higher in Utah among non-Mormons than among Mormons. Consumers also tend to notice and to accept warnings more readily when such messages are consistent with their beliefs (Andrews, Netemeyer, and Durvasula, 1990).

In addition, attention to warning labels is affected by the personal relevance of the risk information. For example, Mormons are not likely to attend to alcohol warnings, in part, because such information is irrelevant to them. If a product is perceived to be relatively safe, consumers are likely to believe that information about the product's uses is more relevant than information about its dangers (Strawbridge, 1985). Thus, willingness to read a product warning is related to the perceived hazardousness of the product (Wogalter, Brelsford, Desaulniers, and Laughery, 1991). Perceived hazardousness is in turn negatively associated with product familiarity (Godfrey, Allender, Laughery, and Smith, 1983). Thus, greater familiarity may result in less attention to a warning (Stewart and Martin, 1994). When individuals have used a product many times without an accident, they may believe that the product is safe and may fail to pay attention to risk information on the product label.

New information is also more likely to be effective in gaining the consumer's attention than familiar information. Studies on the effect of hazard warning labels for chemical products indicate that information that simply serves as a reminder and does not convey new knowledge about the risk will not alter prior beliefs. Similarly, information that the recipients do not regard as convincing will not alter prior beliefs about the product (Viscusi and O'Connor, 1984).

6.4.3 Signs

Signs and posters disclosing risk information are used for a variety of purposes. For example, posters are used at the pre-purchase stage to educate or to remind people about risky behavior. Posters educating drug users about needle exchange programs and about other AIDS-preventive behavior and reminding bar patrons about designated-driver programs have been used. Signs are common also at the point-of-sale. A 1988 consent agreement mandated that ATV dealers post signs about driving hazards, precautions, and age restrictions. In addition, saccharin warnings are posted in all supermarkets, and pregnancy warnings must be disclosed in some jurisdictions in establishments serving alcoholic beverages. Finally, signs are most common at the product use stage. Examples included traffic signs, swimming pool signs, x–ray equipment signs, and signs at mines and at other industrial sites.

Several general principles have been established to maximize the effectiveness of safety signs. First, getting signs noticed is crucial; they may not be seen or read since potential exposure often occurs when people are shopping for food, driving an automobile, or operating equipment. Even traffic signs frequently are ignored (Johansson and Backlund, 1970; Shinar and Drory, 1983). In addition, signs are most likely to attract attention if they are seen at critical times during purchase or product use and if they are directly integrated into the task being performed at the time the hazard is present (Lehto and Miller, 1988).

Second, the signs must be conspicuous. As with labels and advertising, the larger signs tend to be noticed more readily than smaller signs (Godfrey, Rothstein and Laughery, 1985; Wogalter et al., 1987). Also, the shape of signs affects conspicuity. Pointed shapes, such as diamonds or triangles pointed downward, tend to have a higher hazard association than rectangles or circles (Cochran, Riley, and Douglass, 1981; Collins, 1983). In addition, there appears to be an association between color and the perceived amount of danger. Industrial workers associated the colors red and yellow with a greater degree of danger than the colors green and blue (Bresnahan and Bryk, 1975).

Third, since viewers often have only a short time to read a sign containing a hazard warning, the amount of visual clutter surrounding the sign affects its noticeability. For example, the presence of commercial signs has been found to reduce attention to traffic signs (Boersema and Zwaga, 1985), and compliance with a warning sign telling participants in a chemistry experiment to wear gloves and a mask was lower in a cluttered environment than in a non-cluttered environment (Wogalter et al., 1993).

Fourth, signal wards, symbols, and pictographs may be useful in calling attention to the hazard and in communicating key concepts to the viewer. Signal words 'danger' and 'caution' are particularly useful in communicating differential levels of hazardousness. However, it is unclear whether the signal word 'warning' connotes an intermediate level of danger as intended by the FMC labeling system (Bresnahan and Bryk, 1975).

Safety symbols, particularly in combination with a short verbal message, can be helpful in communicating a warning clearly to the target audience. Such signs are particularly effective in avoiding miscomprehension among non-English speakers. However, researchers have questioned the effectiveness of individual symbols standing alone or of multiple symbols to communicate the intended message (Collins, Lerner, and Pierman, 1982). Unfortunately, very few safety symbols are universally comprehended (Lehto and Miller, 1988). There are large variations in the comprehension of symbols on consumer and industrial signs (Easterby and Hakiel, 1981; Collins and Lerner, 1982; Collins, 1983).

Fifth, since typically signs are viewed for a short time, simple messages must be used. There may not be sufficient time to read and to process complex messages, and much of the information on complex signs may not be remembered.

Finally, advances in technology have the potential to improve safety sign efficacy. Wogalter, Racicot, Kalsher, and Simpson (1994) found that a personally relevant sign (displaying the participant's name) increased warning compliance compared to a more conventional impersonal sign (displaying the signal word 'caution'). Thus, new technology makes it practical to personalize signs and tailor messages to individual needs.

Collateral information

It is important to distinguish between brief warning messages and longer risk statements. Brief warnings consist typically of a few short sentences and focus on a limited number of risk factors (e.g., severity of potential risks, probability of risk occurrence, and methods of avoiding risks). Such brief warnings appear on products such as cigarettes, which rotate several warnings, alcoholic beverages, and consumer products such as lawn mowers, ladders, and vending machines.

Legally, any printed or graphic material that accompanies a product is considered labeling. Labeling may accompany the product physically, for example, the labeling located on the product's container (such as on over-the-counter drugs and pesticides) or printed on a separate piece of paper inserted into the product's container (such as the insert for tampons, birth control pills, or ibuprofen pain relievers). Labeling also may be distributed separately from the product (such as table-top promotional displays placed on cosmetic counters in department stores).

This information may cover many topics and often is more complex than the warnings on product labels. In crafting such longer forms of labeling, motivations and capabilities of the target audience must be considered.

While brief warnings must overcome attentional and comprehension barriers due to simple decoding difficulties, longer warning messages face higher-order information processing barriers. The successful processing of longer forms of information involves a series of cognitive activities that motivate or enable the consumer to: (1) read the document, (2) select out (parse) the most important information from the message, (3) encode the information in memory in a way that integrates the information with other relevant associations, (4) elaborate the information so that it is accepted, and (5) retrieve the information when relevant decisions and actions are necessary. Message developers must keep each of these processes in mind when considering the optimal design for longer warning messages. Design elements that may influence one element of the processing chain positively may be ineffective for (or negatively influence) other stages of the process. For example, short sentences may have a greater likelihood of being fully parsed; however, they may not provide sufficient or compelling reasons for a reader to accept or to be persuaded by the information. In this section, we suggest message design considerations that take into account the information processing considerations associated with these five information processing stages.

6.4.4 Willingness to Read

For leaflets, exposure often is assumed because the information may physically accompany the product. Surveys indicate that many consumers state that they read product labels. For example, the Nonprescription Drug Manufacturers Association (1995) recently reported on survey research which found that over 90% of consumers report reading labels before taking over-the-counter (OTC) drugs, over half of respondents say that they always read OTC drug labels, and almost 90% indicate that they often read OTC drug labels. However, these surveys do not provide sufficient detail to understand how well the information is read. Was the information processed with a cursory glance, a full skim, read in parts, fully read, or reread? Unfortunately, these practices are difficult to measure because self-report measures may not be accurate.

Messages that provide clear signals (e.g., through graphics, headlines, or highlighting) about message importance are often read. In addition, the 'invitedness' of the document is likely to provide other important cues. Low legibility (e.g., small type size, inadequate leading, and poor contrast) and poorly structured information (e.g., without sufficient headers, white space or graphics to make the information appealing to read) are apt to be perceived as difficult to process and may be avoided to conserve cognitive resources (Felker, Pickering, Charrow, Holland, and Redish, 1981; Backinger and Kingsley, 1993).

Eye tracking research indicates that consumers often utilize an initial 'quick skim' of advertisements and then redirect their vision to novel information (Fisher *et al.*, 1989). Thus, consumers who receive collateral information also are likely to quickly glance over the presented documents.

6.4.5 Parsing

Even if consumers are willing to read collateral information, they will select only the most important portions of the material for further processing. The document can alert the reader about which information should be read in greater detail. Headlines, highlighting, or other approaches to emphasizing portions of the text provide cues to the reader about the relative importance of certain information. For example, a warning message embedded in a leaflet could be emphasized by varying the typeface, using all upper-case letters, and employing shorter words and sentences than in the rest of the document. These differences will cause the risk information to stand out and will increase the rate at which the information is learned (Felker et al., 1981). Ley (1972) also has shown that placement of information within a document can serve as a cue. He found that patients remembered the information presented first in a leaflet better than information presented at the middle or end.

6.4.6 Encoding

Encoding refers to the processes the consumer uses to decode (attempts to understand) and recode (attempts to retain) information. The reader's familiarity with the information and information processing goals (how the information will be used) influence the nature and amount of encoding.

For unfamiliar information, the consumer must create a memory representation (or schema), at least on a temporary basis, to understand the product's features, including the product's level of risk. This type of information processing is termed 'bottom-up' processing. If the unfamiliar information is highly technical or difficult to process the consumer may have difficulty decoding the message and may not process the material completely (Bartek, 1995). Labor, Schommer and Patak (1995) tested a variety of drug information leaflets that varied in message complexity (i.e., number of topics covered and the difficulty of the language used to describe the concepts). The subjects indicated that they were more 'confused, doubtful, and overwhelmed' by the longer forms of information.

To conserve cognitive resources, consumers will attempt to use existing memory schema. This type of information processing is termed 'top-down' or categorical processing. Consumers retrieve existing schema or relevant pieces of schema to 'construct' a mental model based on previous use of the product or information generalized from knowledge about similar products (Jungermann, Schutz, and Thuring, 1988; Howard, 1989). New information may be encoded in the mental model by adding new facts or beliefs about the product (accretion) or by rearranging beliefs through the addition of higher-order concepts (tuning or restructuring) (Peter and Olsen, 1996).

Bostrom and colleagues have suggested that assessing the mental model that consumers hold for a risky product (e.g., radon) can help communication designers develop collateral materials. By comparing the mental models of laypersons and experts, misconceptions and knowledge gaps can be assessed and corrected (Bostrom, Fischhoff, and Morgan, 1992; Altman, Bostrom, Fischhoff, and Morgan, 1994; Bostrom, Altman, Fischhoff, and Morgan, 1994). This procedure has the advantage of assuring that new information is informative and not simply redundant with existing knowledge. However, different leaflets may be required for consumers with varying mental models of a product.

Thus, understanding consumers' schemas that will be used as a basis for categorical processing is crucial. Such knowledge may assist those drafting leaflets to emphasize information that is consistent with the schemas of the target audiences.

6.4.7 Elaboration

Consumers often are active processors of information and may change the meaning of a risk message by adding, deleting, or reinterpreting the information. However, this elaboration is likely only when the information is relevant and the consumer is concerned about the issue (Petty, Cacioppo, and Schumann, 1983).

Selecting messages that are relevant and that motivate consumers is especially important if the purpose of the collateral information is to change beliefs or attitudes. The consumer may elaborate a message by generating supportive arguments or by generating messages that refute the presented message. For example, if a leaflet advocates that women should not drink alcoholic beverages during pregnancy because there is an increased risk of birth defects, consumers can refute the argument by envisioning 'counter-factual scenarios' in which they recall familiar instances when women drank during pregnancy but the baby was born healthy (Morris, Swasy, and Mazis, 1994). These self–generated 'counterarguments' may be encoded along with the warning information from the leaflet and they become part of the mental model. Other forms of elaboration, such as support arguments and source degradation have also important implications for what is remembered and how information influences decisions and behavior. For example, Andrews, Netemeyer and Durvasula (1993) found that cognitive responses (i.e., articulations of the thought processes consumers use when material is elaborated) could be used to explain about 75% of attitudes toward consuming alcohol after viewing a warning message. The cognitive response measure used was the number of supportive arguments minus the number of negative arguments.

Kanouse, Berry, Hayes-Roth, Rodgers, and Winkler (1981) found that longer leaflets advocating brief (as opposed to continuous) use of estrogen replacement therapy (for post-menopausal women) were less persuasive than shorter documents. The shorter documents contained only a few strong arguments advocating brief use while the longer documents contained both strong and weak arguments. The presence of weak arguments provided readers who were resistant to the persuasive message more opportunities to argue against the advocated message, thereby reducing the persuasive impact of the leaflet.

6.4.8 Retrieval

When decisions need to be made or actions must be undertaken, the consumer relies on the information retrieved from memory. Because only some message elements are selected for input, only 'bits and pieces' of the original stimulus information are retrieved. Also, the retrieved or 'reconstructed' information includes situational information available at the time of output. These situational cues should not only remind the consumer about the appropriate recommended safe behavior, but also they should include key elements of the message that may have been forgotten. For example, if the source of a message (e.g., the Surgeon General) is important to the warning's believability, this information should be included in a warning reminder system (e.g., warnings about smoking cigarettes) (Ricco, Rabinowitz, and Axelrod, 1994).

6.4.9 Face-to-Face Communication

Thus far, we have focused on the delivery of risk information primarily through impersonal sources, such as labels, leaflets, advertisements, and signs. For some products, warnings may be delivered personally, for example, when a sales clerk warns a purchaser to wear safety glasses when using a lawn mower or when a pharmacist informs a customer to be careful driving when taking medication. Face-to-face communication of warnings is particularly important when medical information is delivered as part of therapy counseling.

Unlike most other channels, face-to-face communication is two-directional. A sender of a risk message can listen to the patient's concerns and adapt the information to fit the patient's needs. In addition, interpersonal risk communication not only may serve to reduce cognitive uncertainty, but also can have a positive influence on a patient's emotional state. A physician's ability to listen and provide emotional support is an essential aspect of doctor–patient communication. There are several issues to consider in developing face-to-face communications.

First, patients may simply forget what was said during a medical encounter about a therapy's risks. Physicians structure the medical interview, which typically follows a diagnostic evaluation, first to provide diagnostic information and then to explain the therapy. Ley (1972) has shown that simply changing the order of the information provided (i.e., putting therapy information first) can significantly increase memory for risk information. Ley (1972) has shown also that using a series of techniques to simplify the medication communication (e.g., 'chunking' the information and avoiding technical words) also can increase memory for risk messages.

Second, patients and physicians vary in their interpretation of the elements in a directive that are risk messages and that supply directions for use (Morris, 1990). For example, the statement 'be careful driving until you know how the drug affects you' may be perceived as a warning by health professionals because they know the medication may cause drowsiness that can make driving hazardous. However, if patients do not infer that the reason to be careful driving is because the drug can cause drowsiness, they may not interpret the statement as a risk message. Therefore, health professionals must be sure that patients are not only able to decode risk messages, but also that they understand the information's implications. Health professionals must also provide sufficient contextual information so that their risk communications are accepted and the significance of the warnings is understood. In addition, health professionals may be willing to disclose risks when they can provide patients with directions on how to avoid or minimize the risks. As newer therapies are developed for previously untreatable conditions, health professionals may be more willing to disclose disease-related risks (Amir, 1987).

Third, health professionals may be poor judges of their 'routine' behavior. Menon, Raghubir and Schwartz (1995) have shown that people are poor judges of the frequency with which routine behavior occurs. While health professionals may believe that they disclose risks frequently, observational studies indicate that disclosure rates are far lower than reported by health professionals. For example, Svarstad (1976) performed an observational study of doctor–patient communications. She found that clear directions for use were given in about a quarter of the interactions. More recent studies, based on the self-report of patients taking prescription medications in the previous four weeks, have found somewhat higher levels of side-effect disclosure; however, the majority of patients still indicate that they do not receive sufficient side-effect information (Endlund, Vainio, Wallenius, and Poston, 1991; Morris, L., Tabak, E., and Gondek, K., unpublished).

There are several approaches to improving risk disclosure. As face-to-face communications are a two-way channel, patients' explicit or implicit demands may improve risk communications. If the percentage of patients asking for side-effect information increases, physicians might disclose risks spontaneously. However, despite reports of a greater degree of patient 'activation,' the percentage of patients asking for risk information during initial prescribing visits has not increased dramatically in the past decade (Morris, L., Tabak, E., and Gondek, K., unpublished). Another method for increasing face-to-face risk disclosure is the use of alternative risk communication channels to remind health professionals to disclose risks. Boyle (1983) found that the presence of 'sticker labels' caused pharmacists to repeat the warnings on the medication vials handed to patents.

Fourth, situational factors and patients' abilities and desires to receive risk information also affect risk disclosures. For example, Svarstad (1976) found that physicians were less likely to provide drug information to patients when there was a full waiting room and the patient previously had taken the medication than when there were either fewer patients waiting or when the patient had not taken the drug previously. Physicians' perceptions of the patient's educational level, social class, emotional state, and desire for risk information also influence the degree to which physicians disclose risks and how risk information is phrased (Morris, 1990). However, physicians' perceptions are not always accurate, and physicians may underestimate the degree to which patients want to know about risk information.

6.5 FUTURE DIRECTIONS

6.5.1 New Media

New channels for communicating risk information are under development, and channels developed in the last few years are being adapted to changing technology. The most rapidly growing channel is computer-accessed information systems (such as, CD-ROM and the Internet). In addition, systems that deliver information through cable TV, facsimile, or telephone are evolving rapidly. These new channels share elements of traditional face-to-face and mass media channels. They are individually addressable, and the information can be customized to meet individual needs; however, similar messages may be communicated simultaneously to large numbers of consumers.

The unique feature of these new channels is their large channel capacity (Gates, 1995). The Internet may be used to transfer rapidly large amounts of customized information and of 'high quality' information (e.g., color graphics and full-motion video). The potential of the information superhighway to communicate risk information will expand as the number of high capacity cables grows. Some important types of information disclosure that may be communicated effectively on the Internet are described below.

6.5.2 Counter-Advertising and Corrective Messages

If the Internet is used to convey advertising messages, it may be necessary in some cases to present 'balancing' risk information so that consumers are not misinformed or misled. For example, if a cigarette manufacturer communicates on the Internet the benefits of smoking cigarettes, an advocacy group might present a counter message at a separate web site. The counter message might, for example, discuss the dangers of smoking. The advocacy group's message might be funded, in part, by the cigarette industry (cf., FDA's 1995 proposed regulations on the marketing of cigarettes). Similarly, if a cigarette manufacturer places an advocacy message on the Internet (for example, describing the importance of citizens' rights), an advocacy group might place on the Internet an antithetical message (for example, describing the importance of preventing risks of secondary smoke inhalation).

In addition, a company that places a deceptive advertisement on the Internet could be compelled by a government agency (such as the FTC or FDA) to place a 'corrective' message on the site (or at additional sites). There are several strategies that could be used to correct for deceptive promotional messages on the Internet. A company could be required to stop running the offensive message and replace it with a 'corrected' message. This is similar to the strategy that was used by the FTC to 'correct' consumer misimpressions about Listerene mouthwash (Wilkie et al., 1984). Alternatively, the company might be required to place a 'box' within the deceptive message that would enable consumers to point and click so that the deceptive aspects of the ad are pointed out and a description is provided about why these elements are deceptive.

6.5.3 Integrating Longer Risk Information

There are several available options for companies wishing to display 'balanced' product messages on the Internet describing product benefits and risks. One approach would be to include a description of the product's risks in a separate product monograph that could be available at the end of the promotional message. This is similar to the fine print full-disclosure information that appears currently at the end of prescription drug ads. Alternatively, companies could have risk information linked to promotional messages so that when consumers click on a particular part of the ad, the risk information is displayed on the computer screen. Merck Research Laboratories currently has a direct-to-consumer product advertisement for Zocor (a cholesterol-lowering prescription drug) on its Internet site. Whenever a reader clicks on the word Zocor (which is printed in blue (denoting a linkage) as opposed to the rest of the text written in black), information appears on the screen describing product warnings and necessary precautions.

6.5.4 Integrating Shorter Risk Information

If a company wishes to display a brief risk message (analogous to shorter warnings on labels or in advertisements), there are several available options. For example, a separate warning message (similar to the cigarette warning box) could be displayed as long as the ad appeared on the screen. Alternatively, the risk message could be integrated into the product message, and the risk information could be highlighted to make it stand out from the promotional material. The risk message also could be displayed if triggered by certain search patterns. For example, if an individual screen is displayed for a certain amount of time (perhaps indicating that the viewer is reading fully the promotional material), a warning could be displayed automatically in a highlighted fashion such as in a box or in a scroll across the top or bottom of the screen. Alternatively, a 'billboard' (a separate screen) that discloses risk information might pop-up after a triggering screen (describing product benefits) was accessed. These triggered warning messages also might appear automatically based upon the consumer's search pattern. Another disclosure strategy would be to provide the reader with the option of accessing the risk information through menu-driven or hypertext linkages. Such active search risk disclosure strategies have the advantage of permitting readers to control the type of information accessed. The major disadvantage of this approach is that it does not present a fail-safe method to assure that crucial risk information is delivered to consumers who access promotional information.

6.6 CONCLUSIONS

In conclusion, risk information can be communicated through a variety of channels, and developing effective warnings in each of these channels presents a unique set of challenges. No one channel is likely to inform the target audience fully about potential product hazards. Rather, each channel has strengths that can be maximized and weaknesses that may be overcome. There are two message-delivery options for risk communicators: (1) to use a single channel in which risk information is integrated with other information presented, and (2) to develop a risk communication system (or campaign) in which warning messages are communicated through a variety of channels. The use of either of these options by the risk communicator typically depends on constraints such as time, authority, and money.

In this chapter, we have reviewed some of the message design features and receivers' personal characteristics that influence communication of risk messages. However, additional research is needed about communications effectiveness of risk information within each channel. For example, consumers' trade-offs between benefit and risk information need to be explored. Another issue concerns the 'best' way to integrate risk messages across channels. Is it best to repeat the same risk messages across several channels? Or is it best to use varied (but complementary) messages across channels? For example, brief warnings about product misuse might appear on the label, alternative types of behavior might be portrayed in a video, and an individualized warning might be delivered to consumers through face-to-face communications. As technology advances to permit easier access to risk information, traditional channels (e.g., print, audio, and video) will be merged into a single channel. These new channels will present opportunities and challenges for risk communicators. However, much remains to be learned about technology and about how consumers will use this technology.

REFERENCES

ALTMAN, C.J., BOSTROM, A., FISCHHOFF, B., and MORGAN, M.G. (1994) Designing risk communications: completing and correcting mental models of hazardous processes, part I. *Risk Analysis*, 14, 779–788.

AMIR, M. (1987) Considerations guiding physicians when informing cancer patients. *Social Science and Medicine*, 24, 741–748.

ANDREWS, J.C., NETEMEYER, R.G., and DURVASULA, S. (1990) Believability and attitudes toward alcohol warning label information: the role of persuasive communications theory. *Journal of Public Policy and Marketing*, 9, 1–15.

ANDREWS, J.C., NETEMEYER, R.G., and DURVASULA, S. (1993) The role of cognitive responses as mediatory of alcohol warning effects. *Journal of Public Policy and Marketing*, 12, 57–68.

ARMSTRONG, G.M., GUROL, M.N., and RUSS, F.A. (1979) Detecting and correcting deceptive advertising. *Journal of Consumer Research*, 6, 237–246.

BACKINGER, C.L. and KINGSLEY, P. (1993) *Write it Right: Recommendations for Developing User Instructions for Medical Devices in Home Health Care*. Rockville, MD: Department of Health and Human Services.

BARLOW, T. and WOGALTER, M.S. (1993) Alcoholic beverage warnings in magazine and television advertisements. *Journal of Consumer Research*, 20, 147–156.

BARTEK, P.A. (1995) Increasing subject comprehension of the informed consent form. *Drug Information Journal*, 29, 91–98.

BAUM, C., SCHAEFFER, M.A., WIDEMAN, M.V., REDDY, D.M., and YELLIN, A.K. (1983) *Scientific Research as Promotion: A Preliminary Investigation of Effects in Prescription Drug*

Advertisements. Report prepared for Division of Drug Advertising and Labeling. Rockville, MD: Food and Drug Administration.

BEALES, H., MAZIS, M.B., SALOP, S.C., and STAELIN, R. (1981) Consumer search and public policy. *Journal of Consumer Research*, 8, 11–22.

BEST, A. (1989) The talismanic use of incomprehensible writings: an empirical and legal study of words displayed in TV advertisements. *Saint Louis University Law Journal*, 33, 285–329.

BETTMAN, J.R. (1979) *An Information Processing Theory of Consumer Choice*. Reading, MA: Addison-Wesley.

BETTMAN, J.R. (1975) Issues in designing consumer information environments. *Journal of Consumer Research*, 2, 169–177.

BETTMAN, J.R., PAYNE, J.W., and STAELIN, R. (1986) Cognitive considerations in designing effective labels for presenting risk information. *Journal of Public Policy and Marketing*, 5, 1–28.

BHALLA, G. and LASTOVICKA, J.L. (1984) The impact of changing cigarette warning message content and format. *Advances in Consumer Research*, 11, 305–310.

BOERSEMA, T. and ZWAGA, H. (1985) The influence of advertisements on the conspicuity of routing information. *Applied Ergonomics*, 16, 267–273.

BOSTROM, A., ALTMAN, C.J., FISCHHOFF, B., and MORGAN, M.G. (1994) Designing risk communications: completing and correcting mental models of hazardous processes, part II. *Risk Analysis*, 14, 789–798.

BOSTROM, A., FISCHHOFF, B., and MORGAN, M.G. (1992) Characterizing mental models of hazardous processes: a methodology and an application to radon. *Social Issues*, 48, 85–100.

BOYLE, J. (1983) *Patient Information and Prescription Drugs: Parallel Surveys of Physicians and Pharmacists*. New York: Louis Harris & Associates.

BRESNAHAN, T.F. and BRYK, J. (1975) The hazard association values of accident-prevention signs. *Professional Safety*, 17–25.

BRUCKS, M. (1985) The effects of product class knowledge on information search behavior. *Journal of Consumer Research*, 12, 1–16.

CLARK, E.M. and BROCK, R.C. (1994) *Warning Label Location, Advertising, and Cognitive Responding, Attention, Attitude and Affect in Response to Advertising*. Hillsdale, NJ: Lawrence Erlbaum Associates.

COCHRAN, D.J., RILEY, M.W., and DOUGLASS, E.I. (1981) An investigation of shapes for warning labels. In *Proceedings of Human Factors Society, 25th Annual Meeting*. Santa Monica, CA: Human Factors Society, pp. 395–399.

COLLINS, B.L. (1983) Evaluation of mine-safety symbols. In *Proceedings of the Human Factors Society 27th Annual Meeting*. Santa Monica, CA: Human Factors Society, pp. 947–949.

COLLINS, B.L. and LERNER, N.D. (1982) Assessment of fire-safety symbols. *Human Factors*, 24, 75–84.

COLLINS, B.L., LERNER, N.D., and PIERMAN, B.C. (1982) NBSIR 82-2485, *Symbols for Industrial Safety*. Washington, DC: National Bureau of Standards.

CULLINGFORD, R., DA CRUZ, L., WEBB, S., SHEAN, R., and JAMROZIK, K. (1988) Legibility of health warnings on billboards that advertise cigarettes. *The Medical Journal of Australia*, 148, 336–338.

DAVIS, R. and KENDRICK, J.S. (1989) The surgeon general's warnings in cigarette advertising: are they readable? *Journal of the American Medical Association*, 261, 90–94.

DORRIS, A.L. and PURSWELL, J.L. (1977) Human factors in the design of effective product warnings. In *Proceedings of Human Factors Society, 22nd Annual Meeting*. Santa Monica, CA: Human Factors Society, pp. 343–346.

EASTERBY, R.S. and HAKIEL, S.R. (1981) Field testing of consumer safety signs: the comprehension of pictorially presented messages. *Applied Ergonomics*, 12, 143–152.

ENDLUND, H., VAINIO, K., WALLENIUS, S., and POSTON, J. (1991) Adverse drug effects and the need for drug information. *Medical Care*, 29, 558–564.

FELKER, D., PICKERING, F., CHARROW, V., HOLLAND, M., and REDISH, J. (1981) *Guidelines for Document Designers*. Washington, DC: American Institutes for Research.

FISCHER, P.M., KRUGMAN, D.M., FLETCHER, J.E., FOX, R.J., and ROJAS, T.H. (1993) An evaluation of health warnings in cigarette advertisements using standard market research methods: what does it mean to warn. *Tobacco Control*, 2, 279–285.

FISCHER, P.M., RICHARDS, J.W., BERMAN, E.F., and KRUGMAN, D.M. (1989) Recall and eye-tracking study of adolescents viewing tobacco ads. *Journal of the American Medical Association*, 261, 84–89.

FORD, G.T. and MAZIS, M.B. (1996) Informing buyers of risks: an analysis of the marketing and regulation of all-terrain vehicles. *Journal of Consumer Affairs*, 30, 90–123.

FOXMAN, E.R., MUEHLING, D.D., and MOORE, P.A. (1988) Disclaimer footnotes in ads: discrepancies between purpose and performance. *Journal of Public Policy and Marketing*, 7, 127–137.

FUNKHOUSER, G.R. (1984) An empirical study of consumers' sensitivity to the wording of affirmative disclosure messages. *Journal of Public Policy and Marketing*, 3, 26–37.

GATES, W. (1995) *The Road Ahead*. New York: Penguin Books.

GODFREY, S.S., ALLENDER, L., LAUGHERY, K.R., and SMITH, V.L. (1983) Warning messages: will the consumer bother to look? In *Proceedings of the Human Factors Society, 27th Annual Meeting*. Santa Monica, CA: Human Factors Society, pp. 950–954.

GODFREY, S.S., LAUGHERY, K.R., YOUNG, S.L., VAUBEL, K.P., BRELSFORD, J.W., LAUGHERY, K.A., and HORN, E. (1991) The new alcohol warning labels: how noticeable are they? In *Proceedings of Human Factors Society, 35th Annual Meeting*. Santa Monica, CA: Human Factors Society, pp. 446–450.

GODFREY, S.S., ROTHSTEIN, P.R., and LAUGHERY, K.R. (1985) Warnings: do they make a difference? *Proceedings of Human Factors Society*, 29, 669–673.

GRAVES, K.L. (1993) An evaluation of the alcohol warning label: a comparison of the United States and Ontario, Canada in 1990 and 1991. *Journal of Public Policy and Marketing*, 12, 19–29.

HOUSTON, M.J. and ROTHSCHILD, M.L. (1980) Policy-related experiments on information provision: a normative model and explication. *Journal of Marketing Research*, 17, 432–449.

HOWARD, R.A. (1989) Knowledge maps. *Management Science*, 35, 903–922.

HOY, M.G. and STANKEY, M.J. (1993) Structural characteristics of televised advertising disclosures: a comparison with the FTC clear and conspicuous standard. *Journal of Advertising*, 22, 47–58.

JAYNES, L.S. and BOLES, D.B. (1990) The effect of symbols on warning compliance. *Proceedings of Human Factors Society*, 34, 984–987.

JOHANSSON, G. and BACKLUND, F. (1970) Drivers and road signs. *Ergonomics*, 13, 749–759.

JUNGERMAN, H., SCHUTZ, H., and THURING, M. (1988) Mental models of risk assessment: informing people about drugs. *Risk Analysis*, 8, 147–155.

KANOUSE, D., BERRY, S., HAYES-ROTH, B., RODGERS, W., and WINKLER, J. (1981) R-2800-FDA, *Informing Patients about Drugs*. Santa Barbara, CA: Rand Corporation.

KINNEAR, T.C., TAYLOR, J.R., and GUR-ARIE, O. (1983) Affirmative disclosure. *Journal of Public Policy and Marketing*, 2, 38–45.

KOLBE, R.H. and MUEHLING, D.D. (1992) A content analysis of the 'fine print' in television advertising. *Journal of Current Issues and Research in Advertising*, 14, 47–62.

KRUGMAN, D.M., FOX, R.J., FLETCHER, J.E., FISCHER, P.M., and ROJAS, T.H. (1994) Do adolescents attend to warnings in cigarette advertising? An eye-tracking approach. *Journal of Advertising Research*, November/December, 39–52.

LABOR, S.L., SCHOMMER, J.C., and PATAK, D.S. (1995) Information overload with written prescription drug information. *Drug Information Journal*, 29, 1–11.

LAUGHERY, K.R., YOUNG, S.L., VAUBEL, K.P., and BRELSFORD, J.W. (1993) The noticeability of warnings on alcoholic beverage containers. *Journal of Public Policy and Marketing*, 12, 38–56.

LEHTO, M.R. and MILLER, J.M. (1988) The effectiveness of warning labels. *Journal of Products Liability*, 11, 225–270.

LERNER, N.D. and COLLINS, B.L. (1980) PB81-185647, *The Assessment of Safety Symbol Understandability by Different Testing Methods*. Washington, DC: National Bureau of Standards.

LEY, P. (1972) Primacy, rated importance and recall of medical statements. *Journal of Health and Social Behavior*, 13, 311–317.

MAGAT, W.A. and VISCUSI, W.K. (1992) *Informational Approaches to Regulation*. Cambridge, MA: The MIT Press.

MAZIS, M.B., MCNEILL, B., and BERNHARDT, K. (1983) Day-after recall of Listerine corrective commercials. *Journal of Public Policy and Marketing*, 2, 29–37.

MAZIS, M.B., MORRIS, L.A., and SWASY, J.L. (1991) An evaluation of the alcohol warning label: initial survey results. *Journal of Public Policy and Marketing*, 10, 229–241.

MENON, G., RAGHUBIR, P., and SCHWARZ, N. (1995) Behavioral frequency judgments: an accessibility–diagnosticity framework. *Journal of Consumer Research*, 22, 212–228.

MORRIS, L.A. (1990) *Communicating Therapeutic Risks*. New York: Springer-Verlag.

MORRIS, L., MAZIS, M.B., and BRINBERG, D. (1989) Risk disclosures in televised prescription drug advertising to consumers. *Journal of Public Policy and Marketing*, 8, 64–80.

MORRIS, L.A., SWASY, J., and MAZIS, M. (1994) Accepted risk and alcohol use during pregnancy. *Journal of Consumer Research*, 21, 135–144.

MURPHY, J. and RICHARDS, J. (1992) Investigation of the effects of disclosure statements in rental car advertisements. *Journal of Consumer Affairs*, 26, 351–376.

MURRAY, N.M., MANRAI, L.A., and MANRAI, A.K. (1993) Public policy relating to consumer comprehension of television commercials: a review and some empirical results. *Journal of Consumer Policy*, 16, 145–170.

NONPRESCRIPTION DRUG MANUFACTURERS ASSOCIATION (1995) *New OTC Labels: Industry's Proposal for even Easier to Use OTC Labels*. Washington, DC: author.

PETER, J.P. and OLSEN, J.C. (1996) *Consumer Behavior and Marketing Strategy*. New York: Irwin.

PETTY, R., CACIOPPO, J., and SCHUMANN, D. (1983) Central and peripheral routes to advertising effectiveness: the moderating role of involvement. *Journal of Consumer Research*, 10, 135–146.

POPPER, E.T. and MURRAY, K.B. (1989) Communication effectiveness and format effects on in– ad disclosure of health warnings. *Journal of Public Policy and Marketing*, 8, 109–123.

RICCO, D., RABINOWITZ, V., and AXELROD, S. (1994) Memory: when more is less. *American Psychologist*, 49, 917–926.

RILEY, M.W., COCHRAN, D.J., and BALLARD, J.L. (1982) An investigation of preferred shapes for warning labels. *Human Factors*, 24, 737–742.

RUSSO, J.E. and LECLERC, F. (1991) Characteristics of successful product information programs. *Journal of Social Issues*, 47, 73–92.

SCAMMON, D.L., MAYER, R.N., and SMITH, K.R. (1991) Alcohol warnings: how do you know when you have had one too many? *Journal of Public Policy and Marketing*, 10, 214–218.

SHINAR, D. and DRORY, A. (1983) Sign registration in daytime and nighttime driving. *Human Factors*, 25, 117–122.

SMITH, S.J. (1990) The impact of product usage warnings in alcoholic beverage advertising. *Journal of Public Policy and Marketing*, 9, 16–29.

STAM, A., MURRY, J.P., and LASTOVICKA, J.L. (1993) Evaluating an anti-drinking and driving advertising campaign with a sample survey and time series intervention analysis. *Journal of the American Statistical Association*, 88, 50–56.

STEWART, D.W. and MARTIN, I.M. (1994) Intended and unintended consequences of warning messages: a review and synthesis of empirical research. *Journal of Public Policy and Marketing*, 13, 1–19.

STRAWBRIDGE, J. (1985) The influence of position, highlighting and imbedding on warning effectiveness. Master's thesis, California State University, Northridge.

SUJAN, M. (1985) Consumer knowledge: effects on evaluation strategies mediating consumer judgments. *Journal of Consumer Research*, 12, 1–46.

SVARSTAD, B. (1976) Physician–patient communication and patient conformity with medical advice. In MECHANIC, D. (ed.), *The Growth of Bureaucratic Medicine*. New York: Wiley.

SWASY, J.L., MAZIS, M.B., and MORRIS, L.A. (1992) Message design characteristics affecting alcohol warning message noticeability and legibility. Paper presented at the Marketing and Public Policy Conference, May 16, Washington, DC.

VISCUSI, K. and O'CONNOR, C.J. (1984) Adaptive responses to chemical labeling: are workers Bayesian decision makers? *American Economic Review*, 74, 942–946.

WILKIE, W.L., McNEILL, D.L., and MAZIS, M.B. (1984) Marketing's 'scarlet letter': the theory and practice of corrective advertising. *Journal of Marketing*, 48, 11–31.

WOGALTER, M.S., BARLOW, T., and MURPHY, S. (1995) Compliance to owner's manual warnings: influence of familiarity and the task-relevant placement of a supplemental directive. *Ergonomics*, 38, 1081–1091.

WOGALTER, M.S., BRELSFORD, J.W., DESAULNIERS, D.R., and LAUGHERY, K.R. (1991) Consumer product warnings: the role of hazard perception. *Journal of Safety Research*, 22, 71–82.

WOGALTER, M.S., GODFREY, S.S., FONTENELLE, G.A., DESAULNIERS, D.R., ROTHSTEIN, P.R., and LAUGHERY, K.R. (1987) Effectiveness of warnings. *Human Factors*, 29, 559–622.

WOGALTER, M.S., KALSHER, M.J., and RACICOT, B.M. (1993) Behavioral compliance with warnings: effects of voice, context, and location. *Safety Science*, 16, 637–654.

WOGALTER, M.S., RACICOT, B.M., KALSHER, M.J., and SIMPSON, S.N. (1994) Personalization of warning signs: the role of perceived relevance on behavioral compliance. *International Journal of Industrial Ergonomics*, 14, 233–242.

WRIGHT, P. (1979) Concrete action plans in TV messages to increase reading of drug warnings. *Journal of Consumer Research*, 6, 256–269.

WRIGHT, P., CREIGHTON, P., and THRELSALL, S.M. (1982) Some factors determining when instructions will be read. *Ergonomics*, 25, 225–237.

CHAPTER SEVEN

Attention Capture and Maintenance

MICHAEL S. WOGALTER
North Carolina State University

S. DAVID LEONARD
University of Georgia

This chapter describes the processes involved in attention to warnings. Attention has two stages. One is the capture or switch stage in which the warning must capture attention by standing out from other stimuli in cluttered and noisy environments. Attention is more likely to be drawn to a warning if it has features that enhance its conspicuousness. The second stage, maintenance, holds attention while and until information from the warning is extracted. Features such as legibility and intelligibility are involved. Recommendations for research and application are presented.

Note: Figures that do not appear in the text of this chapter are shown in the color plate section.

7.1 INTRODUCTION

Most environments are cluttered and noisy, and frequently people's attention is divided among various stimuli. According to most modern theories of attention, people have limited pools of mental resources that are used for attending and for working (conscious) memory (e.g., Baddeley, 1986). In other words, we cannot simultaneously attend to everything around us, as it would exceed the available attention capacity. Nevertheless, we can do several tasks simultaneously if they are highly practiced, automatic procedures that consume a fraction of the available capacity. Less practiced tasks are more effortful, consume more resources, and tend to require more serial, one-at-a-time, processing that can exceed capacity and degrade performance if performed concurrently with another task.

In general, we tend to look at, listen to, or think about the most salient features of our external environment or internal thought processes. As we attend to the most salient stimuli, memories of that information are produced. As memory is formed, the stimulus becomes relatively less salient, and other stimuli or thoughts become relatively more salient. Thus, as salience diminishes for one stimulus, attention may switch to a more salient stimulus. In other words, there is an on-going, continuous process of holding and switching attention to the most salient current stimulus or thought.

As the above description suggests, there are two stages of attention. One is the capture or switch stage in which a good warning serves as an attractor that draws or captures attention away from other stimuli or thoughts. To capture attention, the warning needs to be more salient than other events in the environment or those being internally processed. The second stage of attention is maintenance. Here, attentional focus is retained on the warning message while information is extracted and memory is formed (e.g., while a person examines the stimulus material). To expedite information extraction, a visual warning needs to be easy to read and legible. Likewise, an auditory warning must be easy to listen to and intelligible. In this chapter, we will focus on factors that affect both capturing and maintaining attention to visual and auditory warnings.

7.1.1 Modalities

Most warnings are transmitted visually (e.g., signs and labels) or auditorily (e.g., tones and speech). These two sensory channels or modalities are the most frequently studied in research and used in applications and, as a consequence, they are the primary foci of this chapter. Vision and audition have somewhat different characteristics (e.g., different temporal and spatial attributes), and because of these differences, certain warning features that are effective for one sensory channel are not appropriate for the other channel and vice versa. Compared to visual warnings, relatively less research has been performed on the factors that influence attention capture and maintenance of auditory warnings. However, research in this domain is increasing rapidly (see Stanton, 1994; Edworthy, Stanton, and Hellier, 1995; Edworthy and Adams, 1996).

Hazard information can be transmitted also through other sensory modalities. Examples include the olfactory sense (e.g., the odor added to natural gas to aid detection of leaks), the gustatory sense (e.g., an extremely bitter taste added to some household cleaning products), and the kinesthetic/tactile senses (e.g., a 'stick-shaker' that vibrates aircraft control sticks to warn pilots of an impending stall). These examples show that these 'other' sensory modalities may be quite useful in communicating hazard information, and probably should be used more frequently when applicable and practical. Example situations include (a) communicating to individuals who have limited visual and auditory capabilities, and (b) providing an extra, redundant cue when other cues might be missed or not easily given. We return to the issues associated with sensory capabilities and multiple cues at a later point in this chapter.

The next section reviews factors that can influence attention capture and maintenance. An immense amount of research has been conducted on factors that influence attention. Consequently, we have had to be somewhat selective in the breadth and depth of the material covered. We refer readers to the cited references for further details.

7.2 ATTENTION CAPTURE

To attract attention while other stimuli are being processed, warnings must be adequately conspicuous relative to the particular background context in which they occur (Wogalter et al., 1987; Young and Wogalter, 1990; Sanders and McCormick, 1993). Warnings must possess characteristics that make them prominent and salient so that they stand out from background clutter and noise (Frantz and Miller, 1993; Wogalter, Kalsher, and Racicot, 1993a).

In the sections that follow we describe factors that influence attention capture to warnings. We first describe the effects relevant to visual warnings followed by those relevant to auditory warnings.

7.2.1 Vision

Visual warnings are provided in a variety of media including printed labels, posters, signs, brochures, inserts, and product manuals. Some types of visual warnings are presented electronically in the form of simple on/off lights, gauges, video displays, etc. Perceptual enhancements that increase the noticeability of warnings can facilitate attention capture, whereas deficiencies in these characteristics can cause a failure to attract attention. The following sections describe some factors that influence attention capture.

Environmental conditions

Environmental conditions can adversely impact warning detection. One common problem is low illumination. Insufficient light makes printed warnings less visible. Warning visibility can be aided by adding an artificial light source directed at the surface or by back lighting it. Another strategy is to make maximum use of the light that exists, by using, for example, a retroreflective surface coating.

Too much light also can impair visibility. Glare occurs when large amounts of light reflect off a warning surface into the eyes, overpowering the print. Glare can be caused also by intense light emanating directly from a nonwarning source, such as oncoming headlights (cf. Dahlstedt and Svenson, 1977), or certain kinds of neon sign and strobe light. Wogalter *et al.* (1993a) noted that a warning sign with an attached very bright strobe light which had been intended to capture attention caused some research participants to avoid looking in the direction of the warning sign because of the strobe's intensity. Such glare sources can cause light adaptation (or a decrement in dark adaptation) which makes it difficult to see dimmer objects. Another consideration with respect to natural lighting is that the amount and direction of light can vary with the time of day and with the seasons. Other environmental conditions that can have effects similar to low illumination include the presence of smoke, fog, rain, and humidity (see, e.g., Lerner and Collins, 1983).

Duration/flash rate

Sometimes a warning is a simple visual stimulus such as an indicator light on an automobile dashboard. Such lights usually stay 'on' until the problem is corrected or the circuit disengaged. The continued presence of an indicator light increases the likelihood that individuals will detect it, but it does not ensure detection. Better than continuous indicator lights are flashing lights. Flash rates of around 10 Hz are recommended by Sanders and McCormick (1993). The flash rate should not be greater than the critical flicker fusion frequency (i.e., 24 Hz), as this produces the appearance of continuous light. If flash rates are very slow, it is important that the 'on' time is long enough that an operator will not miss the light when glancing at the display panel during its 'off' time.

Brightness contrast

One of the factors influencing whether we can see a stimulus in a particular environment is the figure–ground relationship. In a good figure–ground relationship, the figure or object is readily discernible from the background context. Discernibility is facilitated by brightness contrast, which is a function of reflectance ratios of the figure and ground. Black print on a white background or white print on a black background provides maximum brightness contrast, while gray print on a similar shade of gray background produces little contrast. Research shows that features with greater contrast are detected and localized faster than those of lower contrast (e.g., Brown, 1991; Sanders and McCormick, 1993). Lighting conditions can also affect brightness contrast. In particular, extremely dim and extremely bright light can reduce the apparent difference in light reflectance between the figure and ground.

Color contrast

Certain color combinations produce contrast that is nearly as good as black and white (e.g., black on a saturated yellow or white on saturated red). However, certain hue combinations (e.g., dark blue on dark purple or yellow on white) do not produce distinguishable figure–ground patterns and should not be used (Sumner, 1932).

Some individuals have color-vision deficiencies. Some of these persons are unable to distinguish readily between certain colors, such as between red and green or between yellow and blue because of a genetic defect. These color combinations should be avoided as figure–ground combinations.

In recent years, fluorescent-type colors have become available. Previously, fluorescent pigments tended to fade relatively quickly from exposure to environmental elements such as sunlight. Fluorescent colors interact with ultraviolet light making them appear brighter than nonfluorescent colors. In the US, fluorescent orange is now being used in many localities in signs for road construction/work zones and strong yellow/green has been used for pedestrian-crossing signs. Recent studies show benefits of fluorescent colors in warning applications (Dutt, Hummer, and Clark, 1998; Zwahlen and Schnell, 1998). Unfortunately, not all colors are available as a fluorescent. The fluorescent red is not really red; it is pink. Additional research is needed to determine the benefit of fluorescent colors with respect to their use on product labels (Wogalter, Magurno, Dietrich, and Scott, 1999).

Concern with brightness and color contrast should not be limited to the warning itself, but consideration should be given also to the predominant colors in the environment that will surround the warning. For example, in a largely red environment or context (e.g., the walls of a building, or the main parts of a product label), a red warning will be less noticeable than other colors (Young, 1991). Fullest advantage should be taken of color contrast to distinguish the warning from other colored surrounding stimuli.

Highlighting

Research indicates that when warnings are embedded in other text some form of highlighting (usually with color) helps make them stand out. Strawbridge (1986) found that participants using a glue product were more likely to notice when the embedded warning was highlighted. Young and Wogalter (1990) found that participants who were preparing to use a gas powered electric generator or a natural gas oven were more likely to remember and understand highlighted compared to nonhighlighted warning material in product manuals.

Borders

Another way to highlight safety information is to surround the warning with a distinctive border. Some research suggests that having a border around a warning sign or label enhances figure–ground differences (e.g., Ells, Dewar, and Milloy, 1980; Rodriguez, 1991). Rashid and Wogalter (1997) found that certain border conditions (e.g., having thick, colored diagonal stripes) were rated to be more attention-capturing than other border conditions (e.g., no border or a thin black line border). Example borders are shown in the next chapter (**see Figure 8.10 in color section** from Chapter 8 by Leonard, Otani, and Wogalter). Further, Wogalter and Rashid (1998) manipulated the border of a posted warning placed at a high volume pedestrian area. Their results replicated the pattern found in the earlier rating study. However, positive results have not always been found. Laughery, Young, Vaubel, and Brelsford (1993) did not find an effect of a rectangular border around a warning in a reaction time search task. Swiernega, Boff, and Donovan (1991) observed that the presence of a border slowed performance in a rapid recognition task. The latter result may be similar to a perceptual effect called lateral masking, in which it has been found that stimulus markings presented close in time and distance to target stimuli interfere with the ability to distinguish their features (Averbach and Coriell, 1961).

Size

Large objects tend to be more salient than smaller objects, and are more likely to capture attention. Highway signs are massive to ensure that drivers will see them at distances that allow enough time to attend to them, and if necessary, react to the message. Obviously, we cannot have billboard-size warnings everywhere, but the point is that generally greater size within existing constraints is desirable.

Signal word panel and multiple feature combinations

The ANSI (1991, 1998) Standards on sign and label warnings recommend that all warnings contain a signal word panel on the uppermost portion of the display. ANSI-style warnings include a rectangular-shape signal word panel on the top section. This panel usually includes a signal word (e.g., DANGER, WARNING, CAUTION), color (e.g., red, orange, yellow), and a signal icon/alert symbol (a triangle enclosing an exclamation point) or some other shape (e.g., oval, hexagon) which together comprise a multiple-feature configuration (e.g., Westinghouse Electric Corporation, 1981; FMC Corporation, 1985). Examples are shown in **Figures 7.1 and 7.2 (see color section)**. These stimuli were tested by Wogalter, Kalsher, Frederick, Magurno, and Brewster (1998d). This research is detailed in the next chapter.

Although there has been considerable research on the panel's components, individually and in combination, most of it has concerned measurement of hazard connotation (Chapanis, 1994; Kalsher, Wogalter, Brewster, and Spunar, 1995; Wogalter *et al*., 1995b, 1998d; see also Chapter 8 by Leonard, Otani, and Wogalter). Relatively few empirical studies have investigated the attention attracting effects of the signal word and other associated components. Laughery *et al*. (1993), using reaction time measures, found that an alcohol warning printed in red with a signal icon was detected on labels significantly faster than a black warning without a signal icon. Similarly, Bzostek (1998), using pharmaceutical labels, found that warning detection was significantly faster when they contained a colored signal word (that distinguished it from other text), and/or contained one of several icons.

Generally, warnings having more prominence-type features are more salient and easier to find and more likely to be noticed than those having fewer prominence-type features. Multiple features provide several cues that individually or in combination could capture attention. Additionally, warnings with multiple salient features should benefit people with sensory or perceptual deficiencies. For example, persons who are color blind might not distinguish some of the colors but may notice the warning because of the bold printing of the signal word or shapes that are used. Additional research on the relative added value of the various prominence features, separately and together, is needed to give warnings designers a better basis upon which to make decisions.

Pictorial symbols

Another component of many multi-feature warnings is pictorial symbols (or icons). Most research on pictorial symbols concerns comprehension, a topic that will be covered in Chapter 8 by Leonard, Otani, and Wogalter. However, a frequently overlooked benefit of symbols is that they are attention getting also. Research shows that warnings with pictorial symbols are rated more noticeable (Kalsher, Wogalter, and Racicot, 1996; Sojourner and Wogalter, 1997, 1998) than warnings without them. Research also shows that a warning that includes an icon is easier to detect (Laughery *et al*., 1993). The attention-getting benefit of symbols might have little or no dependence on their understandability. Thus, even if a symbol is not highly understandable, its inclusion in a warning might still be warranted as long as the critical-confusion errors are low. According to ANSI (1991, 1998), a pictorial symbol should produce no more than 5% critical confusion in a comprehension test. See Chapter 8 by Leonard, Otani, and Wogalter for more discussion on comprehension testing and critical confusion.

Location

In general, warnings should be located so that individuals who need to see them do in fact notice them. The layout of the environment and what people do in the environment need to be considered in placing a warning properly. Determining the best location(s) may require task analyses (e.g., Lehto, 1991; Frantz and Rhoades, 1993), where the work or other tasks are broken down into cognitive and motor units and are analyzed to determine the locations where people tend to focus their visual attention as they perform the work or other activity. See Chapter 13 by Frantz, Rhoades, and Lehto for a more detailed discussion on task analysis.

In general, a warning's attention-getting power will be facilitated by placing it close to the hazard. Thus, in most cases warning noticeability will be benefited by its attachment directly to the product (or its container) as opposed to a more 'distant' placement such as in a separate instruction manual (Wogalter *et al*., 1987; Frantz and Rhoades, 1993; Racicot and Wogalter, 1995; Wogalter, Barlow, and Murphy, 1995a). Although this recommendation is reasonable in most cases, in certain circumstances a warning placed too close to the hazard can be ineffective and sometimes dangerous. An example would be a roadway work-zone sign that is first visible close to or within the work zone itself. A better placement would provide sufficient advance notice about the upcoming hazard. The warning should not be too distant, however, as it might be forgotten. Analysis of the task and foreseeable circumstances can help to reveal one or more potentially appropriate placement locations to enhance warning noticeability (see Chapter 13 by Frantz, Rhoades, and Lehto).

Most people's relaxed looking angle for straight-ahead viewing is between 15° and 35° below horizontal straight ahead of them. Warning locations considerably different from where people tend to look, such as higher (or lower) in the horizontal periphery will be less likely to be noticed (Cole and Hughes, 1984).

Sometimes warnings cannot be placed at optimal locations. For some products and environments, aesthetics need to be considered. For example, people would not like having a highly conspicuous warning displayed on the front panel of a stereo receiver in their living room entertainment system. Where else might a warning for the stereo receiver be properly placed? Some potential locations are better than others. For this example, suppose a warning is needed for hazards associated with improperly connecting peripheral components to it. Besides the front panel, other potential locations for this warning could be on the top or the rear of the receiver's case. A warning at these locations would be apparent to users connecting the cables. The rear location is better than the top because people installing the receiver probably would be looking at the back panel when performing the wiring task, whereas the top might be obscured by another stacked component (and it may be considered aesthetically displeasing). The bottom of the receiver is a poor location, because most installers would not see the warning label when doing the wiring. However, the underside could be an appropriate place to put certain other kinds of warning message (e.g., a warning intended to prevent unqualified persons from removing the cover). This would be a good location for this warning because the screws are located there. Another potential location is in the product manual. Certainly warnings belong there because people may assume that the manual contains a complete listing of all relevant hazards. However, if it is a very important warning (e.g., because of severity, frequency of occurrence, etc.), then the warning should be located also on the product itself (or container of the product), because people may not read the manual or may not have it available at all. Nevertheless, sometimes poor placement options can be compensated for when used in conjunction with a well located brief accessory warning (e.g., on a front panel of a product) that directs them to look at another specific location for more detailed information (Wogalter et al., 1995a). Because there is no guarantee that every person will look where we think they will look, placing important warnings in multiple locations (e.g., both on the product and in a product manual) will increase the chance that one of them will be seen.

7.2.2 Audition

For the purposes of this chapter, we will assume that any sound stimulus, whether simple or complex, can alert and attract attention (unless masked by other sounds). Complex sounds, like voice, also have the potential of conveying general or specific information on what the problem is. In this chapter we will not be discussing the processes involved in comprehending the intended meaning of complex sounds (see Chapter 8 by Leonard, Otani, and Wogalter).

Auditory warnings are used commonly to alert people to various problems. Even relatively simple sounds, such as sirens, tones, buzzers, bells and whistles, produce an alerting reaction and sometimes a startle response. Sounds like these are a powerful way to get people's attention. Good warning alerts will arouse people from tasks on which they are highly focused. This 'kick-in-the-head' alerting characteristic makes auditory warnings a favorable tool for attracting attention.

Another major advantage of auditory warnings is the omnidirectional nature of most sounds (Wogalter and Young, 1991; Wogalter et al., 1993a). Auditory signals spread out in all directions from the source, usually reflecting off multiple surfaces before arriving at the receiver's ears. Thus, unlike visual warnings, persons at risk need not be looking at a specific location to be alerted.

Although sounds spread out, they can give directional cues also. Generally, mid to high frequency sounds direct to the ears from the source can provide location cues based on small differences in the time of arrival and intensities of certain frequencies of the sounds between the two ears. For example, a tone coming from a speaker on a control panel can cue the operator to attend to a particular visual display on the panel so that the specific reason for the auditory signal can be determined (Eastman Kodak Company, 1983; Sorkin, 1987; Sanders and McCormick, 1993). Unfortunately, location detection is poor in some circumstances. The sirens of emergency vehicles often are hard to localize amongst walls of city buildings, and can be particularly confusing when a single window of a car is open but the sound source is actually emanating from the opposite direction.

The human auditory system is more sensitive to some sounds than others. The frequencies of the human voice are those for which the auditory system is most sensitive (1000–4000 Hz) (Coren and Ward, 1989). It might be assumed that one would want to provide the auditory warnings within this frequency range because of our increased sensitivity to them. However, warning signals within this range could interfere with the reception of relevant verbal discourse in an emergency situation (which, too, might carry warnings). Thus, an important aspect for the auditory alert signal is that it be comprised of frequencies different from the expected non-warning sounds in the environment, as well as other warning sounds, that might mask it. While the warning(s) should be different from other sounds, it should still be within the sensitive regions of the frequency spectrum.

The above discussion indicates that interference is an important consideration in the design of auditory warnings. There are three kinds of interference of concern. One is masking by noise or other signals that cover up or obscure parts or all of the sound. Background noise, such as machinery in an industrial environment and music blaring in vehicles, can vary in loudness, frequency and complexity. Where possible, the warnings designer should consider whether and how other sounds might affect the auditory warning's signaling ability.

A second type of interference is attenuation (reduction in intensity). Ear protection (e.g., plugs, muffs) is used in many industrial work environments to shield workers from loud extraneous sounds and to prevent hearing loss. Closed car windows also attenuate sounds from outside the car, including sirens from emergency vehicles. Thus, auditory warnings need to be designed to be heard distinctly above the expected background din or within sound shielded conditions. One potential solution in industrial settings is to include headphone speakers inside ear muffs to allow information to get through electronically. Similarly, in automobiles and other enclosed spaces outside signals can be transmitted to within the shielded environment.

A third kind of interference is distraction of the receiver's mental processing by the warning itself. The considerable alerting value that makes auditory warnings useful for capturing attention can be a hindrance when it gets in the way of (distracts from) a very critical task—such as making corrections to the problem the warning is signaling. A loud blaring buzzer from a cockpit warning might interfere with a pilot's ability to carry out proper emergency maneuvers. The more intrusive a sound is, the more likely it will interfere with thought processes. Further, very loud sounds can cause threshold shifts which can cause temporary or permanent reduction in the ability to detect subsequent sounds (Ward, Glorig, and Sklar, 1958; Kryter, Ward, Miller, and Eldredge, 1966).

Thus to attract attention a warning should be louder and spectrally different from the expected background noise, but also it should be given at frequencies for which we are sensitive. At the same time, it should not be so loud that it distracts the listener from performing important tasks. Therefore, numerous foreseeable conditions must be evaluated when designing an auditory warning system to attract attention.

7.3 ATTENTION MAINTENANCE

A warning does little good if it just captures attention but the person gets nothing out of it or the person immediately redirects his or her attention to something else. Once attention has been attracted to the warning, it is important that the warning retain attention so that information can be encoded (see also Rousseau, Lamson, and Rogers, 1998). During this active attention period, the message text is read and/or the pictorial is examined. The warning must hold attention for the time necessary to encode and store the message contained in the warning. The warning should prevent attention from being distracted by and to other stimuli before the message is satisfactorily encoded. These processes involved in knowledge and comprehension are covered in Chapter 8 by Leonard, Otani, and Wogalter. As we did in the section on attention capture, we discuss the visual and auditory factors involved in attention maintenance.

7.3.1 Vision

If the warning is difficult to read because individuals have difficulty making out the letters comprising the words, they are less likely to devote the time and energy necessary to decipher them. In this case, the warning fails to maintain attention. An important factor for maintaining attention to a visual warning is legibility. Legibility refers to how well the separate features or markings of letter characters and pictorials can be distinguished so that they can be identified and recognized. Some writers have mistakenly confused legibility with readability. Both are concerned with ease of reading. However, readability concerns larger groups of characters (e.g., words, sentences) in which comprehension of the material is a consideration (see Chapter 8 by Leonard, Otani, and Wogalter). Legibility concerns whether the individual characters and their features are distinguishable. It concerns the way the text looks; whereas, readability concerns its content or meaning.

Size and visual angle

Frequently legibility is tied to size or, more specifically with respect to text messages, to letter height. Underlying the visual size dimension is visual angle (Smith, 1984) which relates to the area occupied on the retina by the feature's image. With a small retinal image, fewer receptors register the individual components, resulting in poorer visual acuity. If the visual angle is very small, the viewer may see a gray blur instead of separate dark and light elements. The visual angle is a function of both the stimulus size and its distance away from the eye. At greater distances, a given stimulus produces a smaller image than if it were closer. If users are expected to hold a product while examining its label, then the size of the letter characters should be based on the expected distance from the handheld label to the eye. Letter heights for a 'Keep Out' sign at an electric utility power station should be based on the distance from the sign to the peripheral approaches to the site.

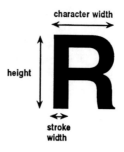

This is Horizontal Compression
This is Normal
This is Horizontal Expansion

Leading is the vertical space between lines of text.
Leading is the vertical space between lines of text.

Leading is the vertical space between lines of text.

Leading is the vertical space between lines of text.

Figure 7.3 Example typographical characteristics.

While generally large print is preferred to small print, there are limits. There cannot be monumental warnings everywhere. If people are able to read the warning under all foreseeable risk conditions, then the print does not need to be any larger. If the print is too large it will be difficult to encompass the information in a glance.

There is more to recognizing characters than simply their height. Other factors include the thickness of the character stroke, height-to-width ratio, character compression, and leading. Figure 7.3 illustrates these characteristics. See Tinker (1963) and Sanders and McCormick (1993) for more information on these and other typographical characteristics.

Sometimes warnings are printed in all upper case (capital) letters. Given the same point size, upper case letters are physically larger than lower case letters as in the following example:

Warning versus WARNING

Because of their generally smaller size, lower case letters produce smaller visual angles than larger upper case letters. By considering only character size, upper case letters might be more legible than lower case letters (Foster and Bruce, 1982). However, experts on typefaces have noted that mixed-case materials (both upper and lower case) are easier to read (Tinker, 1963; Williams, 1994). Lower case letters are more distinctive in shape, thereby making them easier to differentiate than upper case letters. Upper case letters have a block-like appearance making them highly similar and confusable with one another under low-legibility conditions (e.g., small visual angle, low illumination). Garvey,

Figure 7.4 Example of a nearly illegible pictorial symbol. It is supposed to mean 'no eating, drinking or smoking'.

Pietrucha, and Meeker (1998) compared the font Clearview to the fonts on standard highway signs. Clearview's lower case letters are 12% larger than the standard font. They found that increasing the physical size of the lowercase letters (but still using the same 'footprint' space as the standard font) produced better recognition and reaction time scores than the standard font.

Research has shown that under certain conditions reducing the space between individual letter characters enhances reading speed (Moriarity and Scheiner, 1984). When the print is above threshold legibility, closer-spacing of characters requires fewer eye movements to read. However, character spacing must be adequate for the letter components to be seen distinctly. This might account for why Anderton and Cole (1982) and Young, Laughery, and Bell (1992) found that reduced spacing between letters reduced legibility. Watanabe (1994) also found horizontally compressed characters were less legible.

Font

Font style can affect legibility particularly when highly elaborate, unusual, unfamiliar fonts are used. The ANSI (1991, 1998) Z535 Standards recommend sans serif fonts (without character embellishments) such as Helvetica over fonts with serifs (with character embellishments) such as Times Roman. Serif fonts are considered acceptable when the font size is small (as in many product labels and most manuals). Proof readers report serif fonts to be less fatiguing than sans serif fonts. Serif fonts facilitate reading under low contrast conditions because the serifs aid in letter distinguishability, and by putting more ink on the page. The presence or absence of serifs probably does not have a substantial effect as long as the font style is not extremely unusual or elaborate.

Symbols

As we have suggested earlier, the relevant features of pictorial symbols need to be legible. Too much detail can make a graphic illegible when it is reduced in size or viewed at a distance. Most design standards and guidelines recommend using large bold components in safety symbols. However, large blobs of ink can render a pictorial symbol illegible. Figure 7.4 shows a pictorial symbol with legibility problems.

A frequently used graphic shape in warnings is the prohibition or negation symbol. This symbol is a red circle with a single diagonal slash going from the top left quadrant to the bottom right quadrant. Usually the negation symbol is configured so that the slash overlays another symbol placed within the circle (but occasionally the slash is placed under the symbol or an X is used instead). The intended meaning is to prohibit whatever the internal symbol depicts. **Figure 7.5 (see color section)** shows an instance where, on the same street corner in San Francisco, both the over and under slash are used.

It is particularly important that the over slash or X does not obscure the critical details of the underlying symbol necessary for its interpretation. Dewar (1976) and Murray, Magurno, Glover, and Wogalter (1998) found that sometimes the slash can obscure critical features of symbols, decreasing their recognizability. Murray *et al.* (1998) showed that simple adjustments, such as horizontally flipping asymmetric pictorials, can aid identification performance. Examples are shown in **Figure 7.6 (see color section)**.

Figure–ground contrast

As with attention capture, figure–ground contrast is important for attention maintenance. Legibility is reduced when the contrast between the characters relative to its background is low. Ideally, the print and background should be comprised of dark print on light background (or vice versa, light print on a dark background) or of two highly distinguishable colors (e.g., red on yellow or vice versa) rather than two shades of gray or two similar shades of another single color.

Environmental conditions

The presence of smoke, fog, rain, reduced light, etc. can limit the discernibility of the individual warning features (e.g., Lerner and Collins, 1983). Another environmental-related concern is that the color red, the most important hazard color, does not maintain its hue well under dim lighting. As light is reduced, red darkens in appearance before the other hues do, thereby reducing its contrast with dark backgrounds. For expected dim lighting conditions, red printed on a light background is preferred. Another frequently used safety color, orange, can get washed out under certain kinds of artificial lighting.

Printing

Legibility can be affected adversely by poor reproduction at the printing stage where wet paint or ink may spread or 'bleed' and sometimes fill in important details that would otherwise help to distinguish the characters. A similar problem can occur with projected light displays (e.g., on computer screens). Here the stroke width of light letters on dark backgrounds generally needs to be somewhat thinner than for dark letters on a light background. Light comprising the letters spreads out making the stroke width appear wider than it is; this phenomenon is called irradiation (Sanders and McCormick, 1993).

Durability

Over time, exposure to sunlight, air pollution, dirt, grime, water, cold, and heat could cause the color and brightness contrast of the pigments and the material comprising the warning to degrade, making the warning less legible than when it was newer and in better condition. Also, colors degrade at different rates. Red and magenta pigments on outside signs fade more quickly than other colors, primarily from exposure to the sun and other environmental elements. This can create a serious problem beyond simply making the warning more difficult to detect.

Consider what can happen with negation-type symbols where the red of the circle/ slash may fade faster than the black. As a consequence, the 'inside' portion of the symbol may be seen clearly while the 'red' prohibitive portion may not. **Figure 7.7 (see color section)** shows a photograph of a 'no pedestrian crossing' symbol sign where the red circle/slash negation portion has completely faded. In this case, people might interpret the exact opposite of the intended meaning!

The conditions under which the label or sign can reasonably be expected to be used and stored must be considered when choosing materials. The warning must remain in a satisfactory condition over the expected existence of the hazard. Moisture on a paper label will cause it to disintegrate, and some glue compounds will break down with extreme temperatures. The print pigments and the materials constituting the warning should be chosen so they remain in good condition throughout the effective life of the sign or product. Therefore, one should not simply assume that a warning will hold up for the entire time that the sign or the product is in use. Hazard signs (and where applicable, product labels) should be inspected, maintained, repaired, or replaced. In commercial and industrial settings, signs and labels should be inspected periodically. The warnings should be repaired when the materials degrade, become dirty, or are vandalized. Such procedures also provide the opportunity to replace the old warning with a newer version if new materials, designs, and information have become available since its original placement. We recommend the warning designer seek professional consultation in determining the materials that will preserve it over time and in foreseeable conditions of use and abuse.

Target audience

Legibility also depends on the target audience. The persons at risk might have an assortment of vision problems, most notably uncorrected vision with acuity worse than 20/20. For example, older individuals as a group are more likely to have vision problems (Rousseau *et al.*, 1998), and are more comfortable with and prefer larger size type than younger adults (Vanderplas and Vanderplas, 1980; Zuccollo and Liddell, 1985).

Formatting

The appearance of the warning can influence whether individuals will choose to maintain attention to the material or look elsewhere. Desaulniers (1987) showed that people were more willing to read text structures arranged in an outline or list format, with spaces and bullets separating the main points, instead of continuous paragraph-type prose. We suspect that this result is due partly to people being more likely to look at and examine aesthetically pleasing material.

Location

Warnings should be placed so that people can read and examine them comfortably. A posted sign warning that is positioned at an angle, instead of straight on, can be more difficult to see and may discourage further looking. One illustration of this is the warning on one department store brand of top-load washing machines. The lid is hinged on the left side, and printed on the underside of the lid is a set of operating instructions and warnings. In order to read the horizontal print straight on while standing in front of the machine, one must cock the head sideways over the machine. Few people will make the effort to get into this awkward position to read the material.

Limited space

In many situations the types of information and feature that can be included in a warning are constrained by the space available. Limited space is a particular problem for products that have multiple hazards and are held in small containers. A complete warning of all hazards on the label would force the use of very small print, and consequently legibility would be reduced and fewer people could or would read it. Therefore, on some hazardous products one cannot print everything of relevance on labels directly attached to the product. Nevertheless, several alternative strategies can be considered in dealing with this limited space problem. One alternative is to select certain information for emphasis (Young, Wogalter, Laughery, Magurno, and Lovvoll, 1995) and exclude less important information. The abbreviated warning label could refer users to a more complete set of information in some other location (Wogalter et al., 1995a). This strategy may be acceptable if indeed complete information is actually available. Ready access to product manuals cannot be guaranteed as some are thrown away or lost after the product is first used (Wogalter, Vigilante, and Baneth, 1998c).

A second alternative is to increase the size of the label or sign to allow for more information, and/or larger print. Highway signs are sized to enable motorists to see the information legibly at a distance. Additionally, research shows consumers prefer a glue product having a container label design that increases the label's available surface area to make room for a larger warning compared to a more conventional label design with a smaller warning (Barlow and Wogalter, 1991; Wogalter, Forbes, and Barlow, 1993b; Wogalter and Young, 1994; Wogalter and Dietrich, 1995; Kalsher et al., 1996). Several alternative methods for increasing label space on small glue and pharmaceutical containers have been examined including a tag, wrap-around, and cap label designs (Wogalter and Young, 1994; Wogalter et al., 1999). Figure 7.8 has three example container label designs having additional surface area that could be used for larger print and/or additional material. Research has shown that people (particularly older adults) prefer container label designs such as those shown in Figure 7.8 and acquire more information from the label. There is also higher compliance than with conventional container label designs.

Integration or separation from instructions

Most products come with information on how to operate, maintain, and service the equipment, in addition to warning about hazards. How warnings should be presented with respect to procedural instructions and other information has been debated and frequently has been the subject of guidelines by various groups. The Environmental Protection Agency (1991) and other US agencies have suggested that precautionary statements should be in a distinct section separate from the instructions. However, research shows some conflicting results on whether warnings should be separated or integrated with the operating instructions. Friedmann (1988) noted that many individuals skipped the warning to go to the procedural/operating instructions. Venema (1989) found that twice as many individuals reported that they examined product labels for the purpose of reading the operating instructions than to read about safety instructions. Strawbridge (1986) found that more individuals read the warning on a glue label when it was placed together with the instructions. Additionally, Frantz (1992, 1994) found greater warning compliance if the warnings were included within the instructions, as compared to separate sections of warnings and instructions. Other studies have found different results. Karnes and Leonard (1986) found a positive effect of a separate warning section, but this finding is complicated by the fact

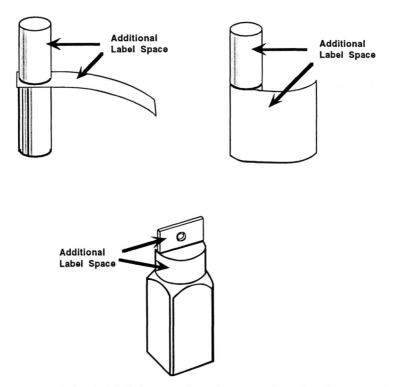

Figure 7.8 Example bottle label designs where there is additional surface space for larger print and/or important/additional material.

that the separate warning differed somewhat from the embedded version. Wogalter *et al.* (1993a) found that a warning within a set of instructions was complied with more frequently than a (larger separated) sign warning. In another study, Wogalter, Mills, and Paine (1998b) manipulated the format of risk information in the consumer portion of prescription drug advertisements. They found that a separate enhanced warning similar to the style recommended by the ANSI (1991, 1998) guidelines produced higher knowledge scores in a comprehension test than either a simple separated or integrated warning.

As the above descriptions indicate, research on integrated versus separated warnings has produced equivocal findings. Probably some of the differences are attributable to familiarity with and the complexity of the product or task and the perceived risk. Products and tasks perceived to be familiar, simple, and of low risk produce less concern than those perceived to be less familiar, complex and of high risk (Wright *et al.*, 1982). With greater familiarity, separate highly conspicuous warnings placed at strategic locations might be better than warnings integrated with the instructions. In the less familiar case, people are likely to go through the instructions step by step and, consequently, it is probably better to integrate the warnings with the operating instructions. These speculations, however, need to be verified.

7.3.2 Audition

An effective auditory warning alerts the receiver but after attention is captured, attention to the auditory stimulus may need to be maintained over time in order to process message content. This is not an issue for short duration stimuli, yet even here attention might still need to be held (e.g., to the representation in memory). With long duration auditory stimuli (e.g., a voice message), attention must be held while the message manifests itself. For the most part, attention maintenance in the auditory realm involves voice communications, as speech more frequently requires across-time processing than most nonverbal auditory stimuli.

Intelligibility

The concept of intelligibility of auditory stimuli corresponds to the concept of legibility for visual stimuli. A large body of research exists on the factors that influence intelligibility. Most of the work was done in military and aviation contexts. Some of the most important factors are described below.

A perfectly clear message at the source can be made unintelligible if played back through defective or low fidelity systems. Many people have had the experience of unsuccessfully trying to decode the speech of an order taker at a restaurant drive-through. Defective equipment (e.g., a blown loudspeaker) makes it difficult to discern the different speech components. Of course, the problem may be the order taker's enunciation, too! A person who speaks with a heavy foreign accent or with a speech impediment also decreases intelligibility.

In addition to above-noted effects attributable to the source of the message, intelligibility can be affected also by numerous other factors related to the channel, context, receiver, and the message itself. Intelligibility is reduced by (a) low level signals, (b) the presence of high levels of masking noise, (c) the receiver's low familiarity with the message, (d) a wide ranging possible vocabulary from which the message is conceived, (e) low redundancy of the sound components, (f) very fast or very slow rate of transmission, and (g) high similarity of the target voice relative to other background sounds/voices. Two excellent reviews of this literature are provided by Edworthy and Adams (1996) and Sanders and McCormick (1993). Additional information can be found in Mulligan, McBride, and Goodman (1984).

Annoyance and false alarms

As noted earlier, auditory warnings can annoy people. Highly intrusive sounds can interfere with the receiver's thought processes making some activities more effortful and error prone. Also people can become quite disturbed when too many false alarms occur. High rates of false alarms happen when the detection system's sensitivity is very high. Usually there is good reason for making a safety system highly sensitive, for example when the hazard could produce devastating results. Most people want airport bomb detectors to be highly sensitive so that no explosives make it onto passenger aircraft even though more people and baggage are searched causing delays and frustration among travellers. These efforts are worth the trouble, given the possible consequences. Besides the annoyance problem, high false alarm rates can produce the 'cry wolf' phenomenon—people ignore the warning signal because they believe that it is false. Unfortunately, the warning might actually be properly signaling a true hazard, producing tragic consequences. Frequent false alarms can increase the likelihood that people waste time searching for a way to eliminate the warning rather than trying to correct the condition which caused the warning, and purposely attempting to defeat the system. In some cases, elimination of the condition causing the signal may be simple, such as buckling lap belts or closing an unlatched car door by slamming it shut. In complex industrial environments, it may not be so simple. A means of turning off the warning might be needed. Of course the system should be designed so that if the warning is turned off, it would be automatically reset (perhaps after a short delay) so that it is available for any recurrence of the problem. Ideally, an auditory warning should always sound when it is needed and never when it is not.

Multiple voice warnings

Some systems employ multiple voice warnings. The problem is that some of these systems do not account for the possibility that they might be deployed simultaneously, a situation which could be highly confusing to the operator. How do you deal with the possibility of several simultaneous speech warnings? Some possibilities are: (a) presenting simultaneous messages in distinctly different voices that are discriminable from one another (male versus female versus synthetic voice); (b) prioritizing the order of messages so that the most important are given first; (c) having messages appear to be coming from spatially distinct locations; (d) giving the most important message(s) prominence features (e.g., loudness) based on urgency; (e) enabling playback of the message if part of it is missed the first time; and/or (e) combining a concise voice warning with a more detailed print warning (Wogalter and Young, 1991; Wogalter *et al.*, 1993a; Edworthy and Adams, 1996). In the latter case, the voice warning can serve to capture attention, concisely present the most important information, and then orient the person to a more detailed visual warning.

7.4 OTHER FACTORS AND ISSUES

7.4.1 Multi-modal warnings

As noted above, sometimes auditory and visual warnings can be combined. A benefit of having both types of warnings in a warning system is that they provide redundant cues. If one modality for the warning is blocked, information is available in the other modality. Visual and auditory cues can be combined also with cues from other sensory modalities, including smell, taste, and tactile/kinesthetic. The smell of smoke, the taste of something bitter, or the rumbling of a car over paving strips, are examples. Corrugated-pavement strips on roadways provide auditory and tactile alerting cues to reinforce the visual cues from the road and from signs indicating a reduced speed limit or imminent hazard.

Another example of multi-modal cues is interactive warnings (e.g., Hunn and Dingus, 1992; Dingus, Wreggit, and Hathaway, 1993; Frantz and Rhoades, 1993; Duffy, Kalsher, and Wogalter, 1995; Wogalter *et al.*, 1995a). Interactive warnings provide tactile/kinesthetic (touch) cues while the participant is performing a task (such as having to touch and move a warning while installing or using a product). Theoretically, interactive warnings cause a break in the performance of a familiar task by causing attention to be switched to the warning (Gill, Barbera, and Precht, 1987; Rasmussen, 1987; Lehto, 1991; Frantz and Rhoades, 1993).

7.4.2 Overloading

Overloading occurs when the amount of information is more than a person is able or willing to process. Many separate warnings or a single extensive one will be less likely to attract and maintain attention than having a few brief warnings. Prioritizing hazard communications is critical (Vigilante and Wogalter, 1997). To reduce the possibility of overloading or excessive on-product warnings, the most important information should be placed on the product and less relevant material placed in an accompanying product manual or package insert (see also Wogalter et al., 1995a).

Overloading should not be confused with overwarning. Overwarning is the notion that people encounter too many warnings in the world, and it is thought that people will be less likely to attend to warnings as a consequence of this inundation. In other words, overloading means that processing capacity is overwhelmed or exceeded by the amount of information in a given situation, whereas overwarning involves being habituated by one's overall life experience. Although overloading and overwarning are theoretically possible, research has not yet verified their occurrence clearly. Nevertheless if either occurs, it means that there should be even greater emphasis on prioritization of content, formatting, and placement.

7.4.3 Habituation

In Chapter 2 by Wogalter, DeJoy, and Laughery, the communication-human information processing (C-HIP) model was described as having a nonlinear flow of information among the processing stages. It was noted that later processing stages in the model feed back onto the attention stage (in a loop-type fashion). One example of this is habituation. Habituation is an outgrowth of the mental events described at the outset of this chapter. Initially, attention is attracted to the most salient stimulus and, while it is maintained on the stimulus, memory is formed causing the stimulus to become less salient. As a consequence of this reduction in salience, other stimuli of greater relative salience will attract attention away from the warning stimulus. Habituated warnings have inadequate salience to attract and maintain attention.

In a different, and perhaps less obvious sense habituation indicates that there is some information about the warning in memory. However, this does not mean that all of the relevant information is known. Individuals might have incomplete knowledge yet not be motivated to seek additional information.

Several design factors may help to retard or counteract habituation. The first is to incorporate the prominence features (size, color, loudness) described earlier in this chapter. Another method is stimulus variation. This can be done by modifying the warning periodically so that it looks or sounds different. Technology has now enabled control and presentation of many signs so that warnings are presented only when they are needed. One example is electronic signs on busy roadways. In the workplace and in hazardous environments, warnings could be presented at the points in time when risky behavior might be exhibited. Highly sophisticated detection and warning systems could enable personalization of the sign also (e.g., using the targeted individual's name) and varied presentation patterns (partial, irregular reinforcement) that will prevent or delay habituation (Wogalter, Racicot, Kalsher, and Simpson, 1994; Racicot and Wogalter, 1995).

Unfortunately, changing the warning is not always possible. Product manufacturers cannot visit people's homes and alter the warning label on their appliances and power tools every so often. However, some kinds of stimulus change on consumer products are possible. One is to change the styles and formats of warning labels on frequently purchased (non-durable) consumer products according to some regular schedule. For durable goods such as appliances and power tools, it may be possible to send revised warnings to consumers for previously purchased products using data bases containing purchase, rebate/coupon, warranty, and repair records.

7.4.4 Familiarity

In the last section, we described habituation as an example of a later stage of the C-HIP model that affects the 'early' stage of attention. Habituation involves memory affecting attention. Another example of feedback from a later stage of processing on attention is the effect of familiarity (see Chapter 9 by DeJoy). Numerous studies show that persons familiar with a product or task are less likely to look for or read a warning than those who are less familiar (e.g., Godfrey, Allender, Laughery, and Smith, 1983; Godfrey and Laughery, 1984; Leonard, Hill, and Karnes, 1989; Wogalter, Brelsford, Desaulniers, and Laughery, 1991; Wogalter et al., 1995a).

7.4.5 Standardization

There has been an increasing effort in recent years to produce standards that specify certain design characteristics (see Chapter 12 by Collins). An example is the ANSI (1991, 1998) Z535 format described earlier. A positive aspect of standardization is that, given its relatively constant physical characteristics, people will eventually learn what a warning looks or sounds like. In this sense, a standard warning in clutter or in noise might stand out because people will know immediately that it is a warning. A further advantage of standardization is that relatively little effort may be needed to produce a warning that conforms to the standard. However, standardization could produce problems. Unfortunately, these problems have not been thoroughly considered by advocates for standards. The purpose of standardization is to promote similarity across warnings which will exacerbate the habituation problem. If all warnings look or sound about the same, then it is quite possible that over time people will pay less attention to them, and this could have disastrous consequences. We believe that standards and guidelines are good starting points for initial warning designs. But they are minimum standards. There should be flexibility to allow the warning designer to deviate from the standards when it is useful and beneficial to do so. Testing can reveal other design variants that may be better than those specified by the standards. For example, test data might show that for a particularly important warning, the word 'DEADLY' and a diagonal stripe border capture attention better than an ANSI (1991, 1998) warning with the word 'DANGER' and a thin plain black line border. With good data to support them, modifications from the standard's specifications should not only be permitted, but encouraged.

7.4.6 Processing Mode and Relevance

A warning will more likely capture and maintain attention when individuals are in an information seeking mode than in other modes of thinking (Lehto and Miller, 1986; DeTurck and Goldhaber, 1988; Lehto, 1991). In other words, a person who is actively looking for hazard-related information, will be more likely to see, hear, and encode a warning than a person who is occupied with other tasks.

Stimuli that are personally relevant and interesting tend to elicit attentional processes. Because people's interests differ, people will look at and listen to different things. Our own name is one of the most relevant and attention-attracting stimuli. Moray (1959) found that auditory presentation of a person's name had a strong effect on attention attraction. Similarly, Wogalter et al. (1994) showed that displaying a person's first name on an electronically presented sign led to higher warning compliance with more people donning protective equipment in a chemistry laboratory situation than a generic warning signal word (CAUTION) in its place.

7.4.7 Characteristics of the Target Population

As noted earlier, an important concern in developing warnings is the intended target population. In some cases, the target population is the general population; in other cases, the population is more constrained (e.g., healthy, young military recruits). Not infrequently, broad target audiences will contain individuals having some form of limited sensory capability, such as vision or hearing impairments among older adults (Rousseau *et al.*, 1998; Wogalter and Young, 1998). The warnings designer should take care to consider the target audience's characteristics and, where applicable, specify warnings designs that compensate for potential impairments. For example, for older adults, warnings could be made larger or louder (Laughery and Brelsford, 1991; Rousseau *et al.*, 1998).

Impairments also can occur situationally. Attention to a warning can be attenuated under conditions of time stress (Wogalter, Magurno, Rashid, and Klein, 1998a), from physical or mental fatigue, alcohol or drug consumption or illness. If these conditions are likely, then consideration should be given on how they might affect attention and what might be done to compensate for the effects.

7.4.8 Testing

How can you know whether a warning attracts and maintains attention adequately? The best way to determine this capability is to test a representative sample of the target population. Other chapters in this volume provide more information about testing methods (e.g., Chapter 3 by Young and Lovvoll; Chapter 13 by Frantz, Rhoades, and Lehto). In this chapter, we mention briefly some of the most pertinent testing factors with respect to attention capture and maintenance. Some of the basic methods include: (a) having individuals rate or rank the noticeability of various prototype designs; (b) having individuals take part in legibility or intelligibility assessments that might include the warnings being presented at a distance or under degraded conditions; (c) assessing memory to determine whether participants remember seeing or hearing the warning; (d) measuring reaction time to detect and find target information in displays with and without a warning (where quicker response times indicate better noticeability); and (e) recording looking behavior to determine whether and how quickly individuals orient to the warning (e.g., eye and/or head movement), and for how long they examine it. The best evaluations are those that most closely replicate the real risk conditions and tasks. For example, measurement of looking behavior using a hidden camera is a more externally valid assessment of warning salience than subjective ratings of warnings presented in a questionnaire booklet.

7.5 IMPLICATIONS AND RECOMMENDATIONS

If people are unaware of an existing hazard, they need to be warned about it. First, attention needs to be captured and then maintained on the warning. Highly salient warnings are more likely to attract and hold attention than less salient warnings. Generally, incorporating features that add prominence to the warning is desirable. The exception to this rule is when attention to a warning adds danger to the situation. Examples include warnings that divert a pilot's attention away from highly critical displays during takeoff or a flashing dashboard light that draws a motorist's attention away from the road.

In this chapter we focused on the factors that influence the switching and holding of attention to warnings presented in the visual and auditory modalities. For visual warnings, we considered contrast, color, size and legibility, surround contours and borders, internal shapes such as pictorials and symbols, location, signal words, limited surface area, degraded environments, and durability. For auditory warnings, we considered simple and complex nonverbal signals, voice presentation, salience, and omnidirectionality, plus the problems of annoyance and false alarms. Other issues discussed included the use of multi-modal warnings (including visual and auditory presentation together, as well as other modalities), overload, habituation, interactive warnings, standardization, stimulus relevance, target population characteristics, influence by other stages of processing, and test methodology. Because attention to warnings is a function of many factors, we offer a general set of recommendations or guidelines below. The guidelines cannot be followed in every case, because in some situations they may conflict with each other and in others they may involve practical constraints.

To maximize the attention capture and maintenance, visual warnings should:

- Accentuate figure–background contrast
- Be brief
- Use large, legible print
- Include features that add prominence such as a signal word panel containing a signal word, color, and an alert symbol
- Include a pictorial symbol when possible
- Present information by way of multiple features and modalities to serve as redundant cues
- Make the formatting attractive, for example, use an outline or list format with spaces and bullets separating the main points instead of continuous, paragraph-type prose
- Be durable to endure the life of the product or hazardous condition
- Make better use of the available space on products/containers for warnings (to make the print larger or to include more information). Consider using methods that can enlarge the space for warnings. If this is not possible, refer users to another accessible source for more information
- Be located when and where the information is needed.

To maximize attention capture and maintenance, auditory warnings should:

- Be brief
- Have a high signal-to-noise ratio, but not be so loud that it badly annoys people
- Be clear and distinguishable from other sounds
- Have low false alarm rates
- Allow adjustments in detection sensitivity.

The warning development procedures should:

- Consider the sensory capabilities of the target population
- Consider the tasks and the environment in which the warning will be located
- Test a representative sample of target users.

We also recommend that, after a warning is placed into service, follow-up assessments be conducted of the warning's attentional effects. The purpose is to determine whether the warning is meeting the goals of attention capture and maintenance. If it is not working as intended, or its effectiveness has degraded over time, etc., the warning should be replaced with a better one.

By incorporating the above characteristics (and other recommendations suggested in this chapter), a warning is more likely to be successful in attracting and maintaining attention. In doing so, it paves the way for additional processing described in the next set of chapters.

We close this chapter by making a final comment. Today's increasingly sophisticated desktop publishing systems allow considerable freedom and flexibility in constructing warnings. Producers of signs and labels are free from simple typewriters that could produce only one size of type and a limited number of embellishments (i.e., all capital letters or underlining). Today's warning designers have access to word processing, graphic image processing, page layout, and document management software. Recently, some specialized sign- and label-making programs have become available. Thus, current desktop publishing capabilities make it easy to produce warnings. Similarly, computer-based sound processors can aid in the design of appropriate verbal and nonverbal auditory warnings by allowing the manipulation of loudness, frequency and complexity, rate, etc. Today's technology allows anybody with a modern computer and a color printer to construct warnings. However, it is important that warnings designers consider in their designs the factors discussed in this and other chapters.

REFERENCES

ANDERTON, P.J. and COLE, B.L. (1982) Contour separation and sign legibility. *Australian Road Research*, 1, 103–109.

ANSI (1991) Z535.1-5, *Accredited Standards Committee on Safety Signs and Colors*. Washington, DC: National Electrical Manufacturers Association.

ANSI (1998) Z535.1-5, *Accredited Standards Committee on Safety Signs and Colors*. Washington, DC: National Electrical Manufacturers Association.

AVERBACH, E. and CORIELL, A.S. (1961) Short-term memory in vision. *Bell System Technical Journal*, 40, 309–328.

BADDELEY, A.D. (1986) *Working memory*. Oxford: Oxford University Press.

BARLOW, T. and WOGALTER, M.S. (1991) Increasing the surface area on small product containers to facilitate communication of label information and warnings. In *Proceedings of Interface 91*. Santa Monica, CA: Human Factors Society, pp. 88–93.

BROWN, T.J. (1991) Visual display highlighting and information extraction. In *Proceedings of the Human Factors Society 35th Annual Meeting*. Santa Monica, CA: Human Factors Society, pp. 1427–1431.

BZOSTEK, J.S. (1998) Measuring visual search reaction time and accuracy for a product label warning as a function of icon, color, and signal word. Master's thesis. Raleigh, NC: Department of Psychology, North Carolina State University.

CHAPANIS, A. (1994) Hazards associated with three signal words and four colours on warning signs. *Ergonomics*, 37, 265–275.

COLE, B.L. and HUGHES, P.K. (1984) A field trial of attention and search conspicuity. *Human Factors*, 26, 299–314.

COREN, S. and WARD, L.M. (1989) *Sensation & perception*, 3rd Edn. San Diego: Harcourt Brace Jovanovich.

DAHLSTEDT, S. and SVENSON, O. (1977) Detection and reading distances of retroreflective road signs during night driving. *Applied Ergonomics*, 8, 7–14.

DESAULNIERS, D.R. (1987) Layout, organization, and the effectiveness of consumer product warnings. In *Proceedings of the Human Factors Society 31st Annual Meeting*. Santa Monica, CA: Human Factors Society, pp. 56–60.

deTURCK, M.A. and GOLDHABER, G.M. (1988) Consumers' information processing objectives and effects of product warnings. In *Proceedings of the Human Factors Society 32nd Annual Meeting*. Santa Monica, CA: Human Factors Society, pp. 445–449.

DEWAR, R.E. (1976) The slash obscures the symbol on prohibitive traffic signs. *Human Factors*, 18, 253–258.

DINGUS, T.A., WREGGIT, S.S., and HATHAWAY, J.A. (1993) An investigation of warning variables affecting personal protective equipment use. *Safety Science*, 16, 655–673.

DUFFY, R.R., KALSHER, M.J., and WOGALTER, M.S. (1995) Interactive warning: an experimental examination of effectiveness. *International Journal of Industrial Ergonomics*, 15, 159–166.

DUTT, N., HUMMER, J.R., and CLARK, K.L. (1998) User preference for fluorescent strong yellow-green pedestrian crossing signs. *Transportation Research Record*, 1605, 17–21.

EASTMAN KODAK COMPANY (1983) *Ergonomic Design for People at Work*, Vol. 1. Belmont, CA: Lifetime Learning.

EDWORTHY, J. and ADAMS, A. (1996) *Warning Design: A Research Prospective*. London: Taylor & Francis.

EDWORTHY, J., STANTON, N., and HELLIER, E. (1995) Warnings in research and practice. *Ergonomics*, 38, 2145–2445 (special issue).

ELLS, J.G., DEWAR, R.E., and MILLOY, D.G. (1980) An evaluation of six configurations of the railway crossbuck sign. *Ergonomics*, 23, 359–367.

ENVIRONMENTAL PROTECTION AGENCY (1991) *Pesticide Reregistration Handbook: How to Respond to the Reregistration Eligibility Document*. Washington, DC: Environmental Protection Agency, Office of Pesticide Programs.

FOSTER, J.J. and BRUCE, M. (1982) Reading upper and lower case on Viewdata. *Applied Ergonomics*, 13, 145–149.

FRANTZ, J.P. (1992) Effect of location and presentation format on user processing of and compliance with product warnings and instructions. *Journal of Safety Research*, 24, 131–154.

FRANTZ, J.P. (1994) Effect of location and procedural explicitness on user processing of and compliance with product warnings. *Human Factors*, 36, 532–546.

FRANTZ, J.P. and MILLER, J.M. (1993) Communicating a safety-critical limitation of an infant carrying product: the effect of product design and warning salience. *International Journal of Industrial Ergonomics*, 11, 1–12.

FRANTZ, J.P. and RHOADES, T.P. (1993) A task analytic approach to the temporal placement of product warnings. *Human Factors*, 35, 719–730.

FRIEDMANN, K. (1988) The effect of adding symbols to written warning labels on user behavior and recall. *Human Factors*, 30, 507–515.

FMC CORPORATION (1985) *Product Safety Sign and Label System*. Santa Clara, CA: FMC Corporation.

GARVEY, P.M., PIETRUCHA, M.T., and MEEKER, D. (1998) Effects of font and capitalization on legibility of guide signs. *Transportation Research Record*, 1605, 73–79.

GILL, R.T., BARBERA, C., and PRECHT, T. (1987) A comparative evaluation of warning label designs. In *Proceedings of the Human Factors Society 31st Annual Meeting*. Santa Monica, CA: Human Factors Society, pp. 476–478.

GODFREY, S.S., ALLENDER, L., LAUGHERY, K.R., and SMITH, V.L. (1983) Warning messages: will the consumer bother to look? In *Proceedings of the Human Factors Society 27th Annual Meeting*. Santa Monica, CA: Human Factors Society, pp. 950–954.

GODFREY, S.S. and LAUGHERY, K.R. (1984) The biasing effects of product familiarity on consumers' awareness of hazard. In *Proceedings of the Human Factors Society 28th Annual Meeting*. Santa Monica, CA: Human Factors Society, pp. 388–392.

HUNN, B.P. and DINGUS, T.A. (1992) Interactivity, information and compliance cost in a consumer product warning scenario. *Accident Analysis and Prevention*, 24, 497–505.

KALSHER, M.J., WOGALTER, M.S., BREWSTER, B., and SPUNAR, M.E. (1995) Hazard level perceptions of current and proposed warning sign and label panels. In *Proceedings of the Human Factors and Ergonomics Society 39th Annual Meeting*. Santa Monica, CA: Human Factors and Ergonomics Society, pp. 351–355.

KALSHER, M.J., WOGALTER, M.S., and RACICOT, B.M. (1996) Pharmaceutical container labels and warnings: preference and perceived readability of alternative designs and pictorials. *International Journal of Industrial Ergonomics*, 18, 83–90.

KARNES, E.W. and LEONARD, S.D. (1986) Consumer product warnings: reception and understanding of warning information by final users. In Karwowski, W. (ed.), *Trends in Ergonomics/Human Factors*, III, Part B, *Proceedings of the Annual International Industrial Ergonomics and Safety Conference*, pp. 995–1003.

KRYTER, K.D., WARD, W.D., MILLER, J.D., and ELDREDGE, D.H. (1966) Hazardous exposure to intermittent and steady-state noise. *Journal of the Acoustical Society of America*, 39, 451–464.

LAUGHERY, K.R. and BRELSFORD, J.W. (1991) Receiver characteristics in safety communications. In *Proceedings of the Human Factors Society 35th Annual Meeting*. Santa Monica, CA: Human Factors Society, pp. 1068–1072.

LAUGHERY, K.R., YOUNG, S.L., VAUBEL, K.P., and BRELSFORD, J.W. (1993) The noticeability of warnings on alcoholic beverage containers. *Journal of Public Policy & Marketing*, 12, 3856.

LEONARD, S.D., HILL IV, G.W., and KARNES, E.W. (1989) Risk perception and use of warnings. *Proceedings of the Human Factors Society 33rd Annual Meeting*. Santa Monica, CA: Human Factors Society, pp. 550–554.

LEHTO, M.R. (1991) A proposed conceptual model of human behavior and its implications for design of warnings. *Perceptual and Motor Skills*, 73, 595–611.

LERNER, N.D. and COLLINS, B.L. (1983) Symbol sign understandability when visibility is poor. In *Proceedings of the Human Factors Society 27th Annual Meeting*. Santa Monica, CA: Human Factors Society, pp. 944–946.

LEHTO, M.R. and MILLER, J.M. (1986) *Warnings*, Vol. 1, *Fundamental, Design and Evaluation Methodologies*. Ann Arbor, MI: Fuller Technical Publications.

MORAY, N. (1959) Attention in dichotic listening: affective cues and the influence of instructions. *Quarterly Journal of Experimental Psychology*, 11, 56–60.

MORIARITY, S. and SCHEINER, E. (1984) A study of close-set type. *Journal of Applied Psychology*, 69, 700–702.

MULLIGAN, B.E., McBRIDE, D.K., and GOODMAN, L.S. (1984) Special Report 84–1, *A Design Guide for Non-speech Displays*. Naval Aerospace Medical Research Laboratory, Naval Air Station, Pensacola, FL, Naval Medical Research and Development Command.

MURRAY, L.A., MAGURNO, A.B., GLOVER, B.L., and WOGALTER, M.S. (1998) Prohibitive pictorials: evaluations of different circle–slash negation symbols. *International Journal of Industrial Ergonomics*, 22, 473–482.

RACICOT, B.M. and WOGALTER, M.S. (1995) Effects of a video warning sign and social modeling on behavioral compliance. *Accident Analysis and Prevention*, 27, 57–64.

RASHID, R. and WOGALTER, M.S. (1997) Effects of warning border color, width, and design on perceived effectiveness. In DAS, B. and KARWOWSKI, W. (eds), *Advances in Occupational Ergonomics and Safety*, II. Louisville, KY: IOS Press, and Ohmsha, pp. 455–458.

RASMUSSEN, J. (1987) The definition of human error and a taxonomy for technical system design. In RASMUSSEN, J., DUNCAN, K., and LEPLAT, J. (eds), *New Technologies and Human Error*. New York: Wiley.

RODRIGUEZ, M.A. (1991) What makes a warning label salient? *Proceedings of the Human Factors Society 35th Annual Meeting*. Santa Monica, CA: Human Factors Society, pp. 1029–1033.

ROUSSEAU, G.K., LAMSON, N., and ROGERS, W.A. (1998) Designing warnings to compensate for age-related changes in perceptual and cognitive abilities. *Psychology & Marketing*, 7, 643–662.

SANDERS, M.S. and MCCORMICK, E.J. (1993) *Human factors in engineering and design*, 7th Edn. New York: McGraw-Hill.

SMITH, S.L. (1984). Letter size and legibility. In EASTERBY, R.S. and ZWAGA, H.J.G. (eds), *Information Design: The Design and Evaluation of Signs and Printed Material*. New York: Wiley, pp. 171–186.

SOJOURNER, R.J. and WOGALTER, M.S. (1997) The influence of pictorials on evaluations of prescription medication instructions. *Drug Information Journal*, 31, 963–972.

SOJOURNER, R.J. and WOGALTER, M.S. (1998) The influence of pictorials on the comprehension and recall of pharmaceutical safety and warning information. *International Journal of Cognitive Ergonomics*, 2, 93–106.

SORKIN, R.D. (1987) Design of auditory and tactile displays. In SALVENDY, G. (ed.), *Handbook of Human Factors*. New York: Wiley-Interscience.

STANTON, N. (1994) *Human Factors in Alarm Design*. London: Taylor & Francis.

STRAWBRIDGE, J.A. (1986) The influence of position, highlighting, and imbedding on warning effectiveness. In *Proceedings of the Human Factors Society 30th Annual Meeting*. Santa Monica, CA: Human Factors Society, pp. 716–720.

SUMNER, F.C. (1932) Influence of color on legibility of copy. *Journal of Applied Psychology*, 16, 201–204.

SWIERNEGA, S.J., BOFF, K.R., and DONOVAN, R.S. (1991) Effectiveness of coding schemes in rapid communication displays. In *Proceedings of the Human Factors Society 35th Annual Meeting*. Santa Monica, CA: Human Factors Society, pp. 1522–1526.

TINKER, M.A. (1963) *Legibility of Print*. Ames, IA: Iowa State University Press.

VANDERPLAS, J.M. and VANDERPLAS, J.H. (1980) Some factors affecting legibility of printed materials for older adults. *Perceptual and Motor Skills*, 50, 923–932.

VENEMA, A. (1989) Research Report 69, *Product Information for the Prevention of Accidents in the Home and During Leisure Activities: Hazard and Safety Information on Non-durable Products*. The Netherlands: Institute for Consumer Research, SWOKA.

VIGILANTE JR, W.J. and WOGALTER, M.S. (1997) On the prioritization of safety warnings in product manuals. *International Journal of Industrial Ergonomics*, 20, 277–285.

WARD, W.D., GLORIG, A., and SKLAR, D.L. (1958) Dependence of temporary threshold shift at 4kc on intensity and time. *Journal of the Acoustical Society of America*, 30, 944–954.

WATANABE, R.K. (1994) The ability of the geriatric population to read labels on over-the-counter medication containers. *Journal of the American Optometric Association*, 65, 32–37.

WESTINGHOUSE ELECTRIC CORPORATION (1981) *Product Safety Label Handbook*. Trafford, PA: Westinghouse Printing Division.

WILLIAMS, R. (1994) *The Non-designers Design Book*. Berkeley, CA: Peachpit Press.

WOGALTER, M.S., BARLOW, T., and MURPHY, S. (1995a) Compliance to owner's manual warnings: influence of familiarity and the task-relevant placement of a supplemental directive. *Ergonomics*, 38, 1081–1091.

WOGALTER, M.S., BRELSFORD, J.W., DESAULNIERS, D.R., and LAUGHERY, K.R. (1991) Consumer product warnings: the role of hazard perception. *Journal of Safety Research*, 22, 71–82.

WOGALTER, M.S. and DIETRICH, D.A. (1995) Enhancing label readability in over-the-counter pharmaceuticals for elderly consumers. In *Proceedings of the Human Factors and Ergonomics Society 39th Annual Meeting*. Santa Monica, CA: Human Factors and Ergonomics Society, pp. 143–147.

WOGALTER, M.S., FORBES, R.M., and BARLOW, T. (1993b) Alternative product label designs: increasing the surface area and print size. In *Proceedings of Interface 93*. Santa Monica, CA: Human Factors Society, pp. 181–186.

WOGALTER, M.S., GODFREY, S.S., FONTENELLE, G.A., DESAULNIERS, D.R., ROTHSTEIN, P.R., and LAUGHERY, K.R. (1987) Effectiveness of warnings. *Human Factors*, 29, 599–612.

WOGALTER, M.S., KALSHER, M.J., FREDERICK, L.J., MAGURNO, A.B., and BREWSTER, B.M. (1998d) Hazard level perceptions of warning components and configurations. *International Journal of Cognitive Ergonomics*, 2, 123–143.

WOGALTER, M.S., KALSHER, M.J., and RACICOT, B.M. (1993a) Behavioral compliance with warnings: effects of voice, context, and location. *Safety Science*, 16, 637–654.

WOGALTER, M.S., MAGURNO, A.B., CARTER, A.W., SWINDELL, J.A., VIGILANTE, W.J., and DAURITY, J.G. (1995b) Hazard association values of warning sign header components. In *Proceedings of the Human Factors and Ergonomics Society 39th Annual Meeting*. Santa Monica, CA: Human Factors and Ergonomics Society, pp. 979–983.

WOGALTER, M.S., MAGURNO, A.B., DIETRICH, D., and SCOTT, K. (1999) Enhancing information acquisition for over-the-counter medications by making better use of container surface space. *Experimental Aging Research*, 25, 27–48.

WOGALTER, M.S., MAGURNO, A.B., RASHID, R., and KLEIN, K.W. (1998a) The influence of time stress and location on behavioral compliance. *Safety Science*, 29, 143–158.

WOGALTER, M.S., MILLS, B., and PAINE, C. (1998b) Direct-to-consumer advertising of prescription medications in the print media: assessing the communication of benefits and risks. Unpublished manuscript. Raleigh, NC: Department of Psychology, North Carolina State University.

WOGALTER, M.S., RACICOT, B.M., KALSHER, M.J., and SIMPSON, S.N. (1994) The role of perceived relevance in behavioral compliance in personalized warning signs. *International Journal of Industrial Ergonomics*, 14, 233–242.

WOGALTER, M.S. and RASHID, R. (1998) A border surrounding a warning sign text affects looking behavior: a field observational study. Poster presented at the Human Factors and Ergonomics Society Annual Meeting, Chicago, IL.

WOGALTER, M.S., VIGILANTE, W.J., and BANETH, R.C. (1998c) Availability of operator manuals for used consumer products. *Applied Ergonomics*, 29, 193–200.

WOGALTER, M.S. and YOUNG, S.L. (1991) Behavioural compliance to voice and print warnings. *Ergonomics*, 34, 79–89.

WOGALTER, M.S. and YOUNG, S.L. (1994) Enhancing warning compliance through alternative product label designs. *Applied Ergonomics*, 24, 53–57.

WOGALTER, M.S. and YOUNG, S.L. (1998) Using a hybrid communication/human information processing model to evaluate beverage alcohol warning effectiveness. *Applied Behavioral Sciences Review*, 6, 17–37.

WRIGHT, P., CREIGHTON, P., and TRELFALL, F.M. (1982) Some factors determining when instructions will be read. *Ergonomics*, 25, 225–237.

YOUNG, S.L. (1991) Increasing the noticeability of warnings: effects of pictorial, color, signal icon and border. In *Proceedings of the Human Factors Society 35th Annual Meeting*. Santa Monica, CA: Human Factors Society, pp. 580–584.

YOUNG, S.L., LAUGHERY, K.R., and BELL, M. (1992) Effects of two type density characteristics on the legibility of print. In *Proceedings of the Human Factors Society 36th Annual Meeting*. Santa Monica, CA: Human Factors Society, pp. 504–508.

YOUNG, S.L. and WOGALTER, M.S. (1990) Effects of conspicuous print and pictorial icons on comprehension and memory of instruction manual warnings. *Human Factors*, 32, 637–649.

YOUNG, S.L., WOGALTER, M.S., LAUGHERY, K.R., MAGURNO, A., and LOVVOLL, D. (1995) Relative order and space allocation of message components in hazard warning signs. In *Proceedings of the Human Factors Society 39th Annual Meeting*. Santa Monica, CA: Human Factors Society, pp. 969–973.

ZUCCOLLO, G. and LIDDELL, H. (1985) The elderly and the medication label: doing it better. *Age and Ageing*, 14, 371–376.

ZWAHLEN, H.T. and SCHNELL, T. (1998) Visual detection and recognition of fluorescent color targets versus nonfluorescent color targets as a function of peripheral viewing angle and target size. *Transportation Research Record*, 1605, 28–40.

CHAPTER EIGHT

Comprehension and Memory

S. DAVID LEONARD
University of Georgia

HAJIME OTANI
Central Michigan University

MICHAEL S. WOGALTER
North Carolina State University

The factors related to the comprehension and memory of warnings are described. One of the main purposes of warnings is to inform. Warnings can help to fill in gaps of knowledge about potential 'hidden' hazards. Warnings should be understandable and sufficiently explicit so that persons at risk will be informed about the hazards and potential consequences if they fail to comply with its directives. The design of prototype warnings based on research and guidelines is described. Prototypes need to be tested to assure that the intended target population attains adequate understanding. Features comprising the warning's physical characteristics can provide an overall hazard impression which is important if the warning's specific message is not completely understood. Additionally, factors that affect comprehension and memory of warning information (e.g., familiarity, habituation, and training) are described.
Note: Figures that do not appear in the text of this chapter are shown in the color plate section.

8.1 INTRODUCTION

Chapter 7 by Wogalter and Leonard concerned the factors that affect the capture and maintenance of attention. The present chapter concerns the next stage of processing, involving the factors that produce comprehension (understanding) and memory.

This chapter covers many topics, and we have organized its presentation in the following way. It begins with a description of how information processing theory can be used to describe the processes involved in acquiring hazard/warning knowledge, and how memory is activated and reactivated based on cues provided by warnings and their context. The next three sections describe the factors affecting the comprehension of printed textual messages, symbols, and auditory warnings. In these sections, the literature is reviewed from which general principles are drawn. These general principles may be applied in the initial design stages to produce a set of preliminary prototype warnings. We then describe why it is necessary to test the warning prototypes for comprehension. We then discuss the factors that influence acquisition of information and skills (learning and training). We then turn to a discussion on how the warning's physical characteristics can provide a general hazard impression. Finally, the topics of familiarity, habituation, and prospective memory are discussed. The chapter closes with a brief summary and recommendations.

8.2 KNOWLEDGE AND AWARENESS

One principal purpose of warnings is to inform people about hazards. If the hazards are already known and people are aware of them, a warning containing the same information probably is not necessary. For example, it is not necessary to label a kitchen knife that it could cut skin or a pencil that it could puncture the eye. These are common, simple items for which the hazards are obvious to nearly everyone. The exception, of course, is young children's lack of understanding of these facts, but we assume that care givers will protect them until they are old enough to learn this information and handle these tools properly.

Nevertheless, people do injure themselves with knives and pencils—even though the injured person knew an event of this type could occur. In the USA, industrial maintenance personnel are trained to disengage the power when repairing equipment. Yet many workers are injured every year when they fail to lock out the power properly to the machinery that they are repairing. Most of the injured persons had been trained on lock out procedures and 'knew' what to do (i.e., the information was in memory), but they 'forgot.' These instances of 'forgetting' are due to people not being conscious or aware of the relevant hazards at the proper moment. Therefore, warnings not only serve to inform (to get the information into memory), but they can also serve as reminders, or cues, that activate existing knowledge in memory so that people are more likely to be aware of the hazard at the time they are at risk. We will have more to say about this reminder or cuing role at various points in the chapter.

8.2.1 Knowledge Gap and Acquisition Theory

Before the technological revolution, most hazards associated with the tools that people used were obvious and apparent. Technology has brought to the market new products that are not obvious to set up/install, use, maintain, or repair. They have 'hidden' (nonapparent or latent) hazards. For example, one can not tell the effects of a white pill just by its appearance. Also, it is not obvious that automobile air bags could cause serious injury or death if a person sits too close to them during their deployment. In both examples, warnings are needed to inform people about latent hazards, allowing them the opportunity to exercise appropriate precautions. **Figure 8.1 (see color section)** shows a warning sticker sent to registered owners of vehicles with air bags.

For a person to be adequately informed, he or she must comprehend the hazards, know how to avoid them, and know the potential consequences of unsafe behavior. Frequently the person's knowledge is incomplete, particularly for nonapparent hazards.

If there is a gap in knowledge, the warning should be designed to induce formation of new memory structure so that the person's knowledge becomes consistent with the hazard-related knowledge needed. Generally the assimilation of new information will be easier if extensive relevant memory already exists (i.e., a smaller knowledge gap) than if very little related memory already exists (i.e., a larger knowledge gap). Formation of new memory generally requires effort, and such effort uses mental resources. Because people may be occupied with other tasks that are absorbing some of their mental resources at any given time, the warning should be constructed so that the information being transmitted is easy to grasp and does not overload the system.

When only a relatively small portion of the warning information does not match, the gap in knowledge is relatively small. Here there is less information to assimilate than if a large knowledge gap existed. Producing new memory when there is a small gap requires less mental resources. The processing to reduce or close a small gap can be accomplished more readily than when a large gap exists; in the latter case the formation (accommodation) of considerable quantities of new information into memory is needed.

Generally, a warning that is easier to process is more effective than a warning that is more difficult to process. Thus, warnings that have as their basis information that the person already knows are almost always better than warnings that contain large amounts of incongruous (nonmatched) information with respect to the person's existing knowledge. Warnings that contain easy-to-grasp information are particularly beneficial when warning exposure time is short, when quick reactions are necessary, or when the individual is fatigued or stressed.

Another reason why warnings should be well matched with what people already know is that people are 'cognitive misers.' They do not want to expend considerable amounts of time and energy to fill large knowledge gaps.

Some of the above description is related to the concept of schema (or script). Schema theory suggests that individuals can develop complex mental structures regarding various topics (e.g., Rumelhart, 1980). A schema includes information from previous experiences that guides people's expectations about various situations. A schema about cooking dinner using a charcoal grill might include such things as cutting up the food, starting the coals, and applying sauce. This schema could also include the possibility of being burned by touching the hot grill or by flying sparks. Thus, based on a schema about the nature of the grill and how hot it gets, the cook might wear a mitt on one hand and avoid touching the grill with the other. However, there are other safety concerns that this person might need to know to accomplish the barbecuing task safely. For example, individuals might not know that burning coals produce carbon monoxide. Accumulated colorless and odorless carbon monoxide can deplete the oxygen supply to the brain. This is a lesser-known (nonobvious) hazard. The reason this information is not well known is due partly to its physical characteristics and how it affects the body and partly because most grilling occurs outdoors where there is sufficient ventilation to disperse the gas. However, to a company that manufactures grills, it should be foreseeable that some people will attempt to use their grill indoors in bad weather. Therefore, it is important to warn against using the grill in an enclosed space. A warning for the grill's heat hazard is probably less important than a good warning for the carbon monoxide hazard, because high temperatures are an obvious inherent feature of cooking on charcoal grills, but this is not true for the carbon monoxide hazard.

8.3 UNDERSTANDING TEXT

8.3.1 Language

For text-based warnings, it is obvious that in order to convey the message, the individual must have some knowledge of the language to be able to read and understand the warning.

Individuals who know only Spanish probably will not be able to understand warnings written only in English (although some 'foreign' words might be cognates of terms in a language the individual knows; Wogalter, Frederick, Magurno, and Herrera, 1997a). If the target audience in a particular geographic area is comprised of persons knowing distinctly different languages (not a common one), then the warning might need to be presented in more than one language. Obviously this would benefit these groups but there are also some negatives. One is that there may be limited space, and by giving more space to translations, the space available for the primary language is reduced, together with the print's prominence and legibility.

In Chapter 7 by Wogalter and Leonard we touched on the multi-language problem. One solution is to use label designs that increase the space available for warning materials. In a market comprised of a mixture of English-only users and Spanish-only users, one strategy is to present both languages equally sized on a label (label is split in half). Another potential method is to size the print based on the percentage of people in a population who use the language. For example, if in a particular consumer market 70% understand only English and 15% understand only Spanish (and 15% fitting other categories), then the English portion might be made somewhat larger and the Spanish portion somewhat smaller (although not necessarily proportional to exact population percentages). In many cases, it may not be practical to print other languages. At issue is whether it is acceptable to allow people to go unprotected when the safety instructions are not presented in a language that the user understands.

Another strategy is to include translations of only the most important information. Most of the label would be in the primary language, but space would be allocated for a short message in other language(s). The short message might present (a) the main dangers, (b) a statement that emphasizes that before using the product seek an accurate translation of the warnings and instructions from someone, and (c) an easy way to contact a multilingual representative of the manufacturer by phone, mail, or internet.

As we have seen, the issues involved in the multi-language problem are complex. A whole host of practical and societal issues must be considered. While there are no definitive solutions, it is clear that the warning designer must consider language-usage by persons exposed to the hazard.

Not only do people differ in terms of the specific languages they use, but also with respect to their skills and competence in using their primary language. The design should consider people with low literacy levels, with low verbal comprehension skills, or with limited education.

Warnings should be designed so that at-risk individuals will be able to acquire the necessary information to keep them safe. Unfortunately, this is not generally the case. Usually warnings are written by people who are more educated and knowledgeable about the to-be-communicated hazards than the individuals that comprise the target audience. Warnings designers can make the mistake of assuming that everybody knows what they know. This assumption may be correct some of the time, but it may be incorrect with respect to some particularly critical safety information. Incorrect assumptions can produce errors when important information is left out or terminology is not understood or misunderstood by target users. One example is material safety data sheets (MSDSs) which must be available to workers in all workplaces using hazardous chemicals as required by the US Occupational Safety and Health Administration (29 CFR 1910, Hazard Communication Act). MSDSs are intended to describe the hazardous nature of the chemicals that the workers work with. Unfortunately, often these documents are extremely technical and fail to do their intended job with this audience.

Table 8.1 Mean expert-judges' quality ratings of lay participants' explanations and recommended actions for warning terms (adapted from Leonard et al., 1991).[a]

Warning term	Explanation	Recommended actions
Flammable	1.19	1.66
Poison	1.67	1.63
Combustible	1.16	1.27
Irritant	1.20	1.26
Explosive	1.27	1.13
Corrosive	0.87	0.92
Dangerous when wet	1.00	0.80
Radioactive	0.68	0.74
Spontaneously combustible	1.05	0.58
Oxidizer	0.30	0.37

[a] Note that quality ratings are based on a scale ranging from 0 (poor) to 3 (good).

Another related problem is that some of the words in warnings can be interpreted differently by different people. This variability indicates that some people's beliefs are incorrect. The technical definition of 'flammable' is a substance that has a flash point of 100° Fahrenheit (about 38° Celsius) or lower (as defined in US Federal Regulations). Lay persons are not likely to be familiar with this technical definition and may interpret the term quite differently (cf. Leonard, Creel, and Karnes, 1991). In fact, research by Main, Frantz, and Rhoades (1993) showed that many lay persons interpret 'combustible' as being more hazardous than 'flammable' when actually the reverse is true—according to formal, technical definition (in Federal rules). Perhaps lay persons believe that 'flammable' means the substance will burn like a match, whereas 'combustible' sounds like an explosion.

Laughery (1993) describes several instances where manufacturers have made incorrect assumptions about what targets know. Vigilante and Wogalter (1996) note that domain experts and non-experts differ in their conceptions of which warning components are most important. Research by Leonard and his colleagues has shown that only a small proportion of people accurately understand some of the most commonly used terms in warnings (Leonard et al., 1991; Leonard and Digby, 1992). Table 8.1, adapted from Leonard et al. (1991), shows expert judges' ratings of a set of definitions produced by lay persons for several commonly used hazard terms. Ideally, there should be a match between the intended meaning of the term and the target population's understanding of it. As can be seen in the table, the match is not very good for many of these terms. For example, the term 'oxidizer' is poorly understood. However, this term may be rich in information for technically trained persons such as chemists or firefighters, activating a considerable body of existing knowledge about the nature of the hazard, and what to do to avoid negative consequences. To trained individuals, the term 'oxidizer' is probably an adequate warning. The problem is that most lay persons do not know what this term means. It does not cue knowledge on the kinds of precaution needed to prevent harm. Thus, a warning intended for the lay public needs to include terms that cue knowledge that ultimately leads to proper hazard avoidance. In the case of 'oxidizer' hazards, such information might include the potential for fire, explosion, and extreme injury. As noted earlier, technically trained persons can mistakenly believe that they know what the target audience knows. It is therefore critical to determine whether the target audience interprets the intended message properly. Also attention needs to be focused on the possibility that some people may misinterpret the message, and if so, to determine how the misinterpretation can be reduced.

8.3.2 Vagueness of Terms

Another terminology-related problem is vagueness (cf. Kreifeldt and Rao, 1986). The commonly used warning phrase 'Use in a well ventilated area' can be interpreted in many ways. The problem is that it does not tell what specific conditions are safe and unsafe. If the product is used inside, how big should the room be? Is a room with one open window adequate? Should you use a fan? A respirator? Clearly more explanation is needed than this statement provides.

Consider another phrase found on many consumer-product labels: 'Do not use near an open flame.' By itself, this phrase is inadequate for three reasons. First, it does not tell what 'near' means. Second, people may not realize that pilot lights in gas stoves, furnaces, and water heaters are 'open flames' in the technical sense. Third, people may not realize that the pilot light can be 'on' even if the appliance is 'off.' Extremely serious injuries have occurred because people did not think about pilot lights. Some people may not even realize they exist. Others may know of their existence but do not think of them as 'open flames' as usually they are located inside an enclosure and not readily visible. Other potential spark sources include common electric devices such as on–off switches, telephones, and electric motors. We suspect that people often do not think of them as sources of ignition.

Both example statements mentioned above need improvement. In particular, they need to be more explicit (i.e., to give more specific information) so that people have the opportunity to be made fully aware of the hazards, consequences, and what they need to do. Explicit warnings are better able to fill the knowledge gaps that we discussed earlier. A warning for a flammable product used by the general public needs to provide information about (a) the potential of fire and explosion hazards when used in the vicinity of ignition sources, (b) where potential ignition sources might be located, and the possibility that there might be more spark-producing sources in their environment, (c) the distance from these ignition sources that is safe, (d) how vapors may accumulate and travel, and (e) what kind of ventilation conditions are appropriate and inappropriate. Explicitness can apply to describing the nature of the hazard, instructing what to do and not do to avoid the hazard, and telling about the consequences if the instructions are not followed. Obviously this is a lot of information, but the amount could be reduced by determining what people already know. Which parts of the warning need to be explicit depends on the hazard and foreseeable use situation. While the consequences associated with oxidizers and flammables need to be explicit, the consequences associated with a wet floor hazard do not. In the latter case, an abbreviated sign is sufficient to cue most people's existing memory that slippery floors can cause falls. Unfortunately, many potentially hazardous products have warnings containing vague, non-explicit information that fail to reduce knowledge gaps adequately.

One reason that has been given for the lack of explicitness in most consumer product warnings is that manufacturers avoid them because of the belief that explicit warnings will deter people from purchasing their product, compared to a competitors' product with a less explicit warning. However, published evidence supportive of this assumption is not strong, and indeed, equivocal at best. Some studies have found that explicit warnings produced negative attitudes (Morris and Kanouse, 1981; Vaubel, 1990; Vaubel and

Brelsford, 1991), whereas other studies have found no effect (Laughery and Stanush, 1989; Silver, Leonard, Ponsi, and Wogalter, 1991) or even a positive effect (e.g., Ursic, 1984) on product preferences and purchase intentions. In finding a positive effect of explicitness on perceptions, Ursic (1984) suggests that explicit warnings reduce people's uncertainty, making people feel safer. The effects of explicitness probably depend on the type of product, the consumer, and the specific warning. Explicit warnings for dangerous tools like gas powered lawn mowers or wood shredder-mulchers are less likely to reduce sales than explicit warnings for a product that is expected to be safe, like hair spray. Also, it is not unreasonable to expect that certain large segments of the population might prefer products with more explicit warnings so that others in their household (e.g., their spouse, older children, caretakers) would be informed.

8.3.3 Inferences

As indicated earlier, development of understandable text should consider the kinds of inference that must be made by readers (Kintsch and van Dijk, 1978). Incorrect inferences could cause comprehension and recall to suffer (Britton, Van Dusen, Glynn, and Hemphill, 1990). Therefore, the warnings designer should avoid text that requires extensive inferential processing because the inferences might be wrong.

Nevertheless, some inferences will need to be made based on the need to keep warnings brief. Activation of existing information in memory through good warning cues will increase the likelihood that correct inferences will be made. The problem is that vague, highly technical, or incomplete warnings may not activate the appropriate kinds of knowledge, leaving the individual to make use of more limited information, leading to the possible production of incorrect inferences.

8.3.4 Underlying Concept

People may also have difficulty with the underlying concept to be communicated. For example, to understand the concepts of radiation and biohazard fully, extensive education and training are required. Therefore, it is unlikely that a brief textual message will be able to convey all of the hazard-related ramifications. Obviously trying to convey concepts about which people have little existing knowledge (i.e., large gaps of knowledge) is difficult. A related issue here concerns how much information should be communicated. Despite knowing that there is a large knowledge gap, and knowing that one would like to communicate to users all relevant information about the hazard, consequences, and instructions, there is, however, another important issue to consider: people might not be willing to read an extremely long message, but would be willing to read a shorter message. Therefore some tradeoff decisions may need to be made which could involve the use of a shorter on-product label, and a longer, more complete set of warnings and instructions elsewhere. Later in this chapter, we discuss alternative ways that a warning by its physical characteristics can relay some hazard information just by the way it looks (or sounds).

8.3.5 Readability

Another factor associated with warning text comprehension is readability. Readability refers to the ease with which one can extract information from a text message. This term, however, is sometimes used inappropriately to mean legibility. As described in Chapter 7 by Wogalter and Leonard, legibility refers to the ability to discriminate the component elements of the printed alphanumeric characters or parts of pictorial symbols. The goal of making a warning 'readable' is to make the text message simple, direct, and easy to understand.

Readability can be determined in several ways. One is to use a readability index, such as the one developed by Flesch (1948). Klare (1976) and Duffy (1985) review other similar indices. Most give a numerical estimate of the approximate grade-level reading skill required to understand the material (or the percentage of the native English-speaking audience that would be expected to understand it) based on such factors as the number of letters in the words and the number of words in sentences. If the textual material is intended to reach a large percentage of the general public, some authors have suggested that it should be written somewhere between 4th and 6th grade level. How these particular grade-level guidelines were determined is not known, but probably they are fairly good guideposts when producing text warnings for the general public.

The readability indices were not developed for evaluating warning text. There are three problems associated with using the currently available readability indices in evaluating warnings. One is that most readability indices require text samples of at least 100 words, whereas most posted warning signs and product labels are usually fewer than 100 words. However, there are many kinds of warning material with more than 100 words that would not have this problem, e.g., employee-safety training manuals, product manuals, and package inserts. A second problem is that warning text frequently lacks the punctuation necessary for the readability indices to parse clauses and sentences. Silver *et al.* (1991), however, demonstrated a method for compensating for these problems by duplicating the shorter warning text until it exceeded 100 words and then adding punctuation. The third, and perhaps the most serious problem with readability indices, is that they do not fulfill their main intended purpose of measuring how well people understand the material at the grade-levels or percentages they supposedly predict. For example, most of the readability formulae assume that all shorter words and sentences are more understandable than longer ones. This is true in a general sense, but using shorter words and sentences will not automatically enhance understanding. It is possible for a readability index to indicate that some sample of text is understandable by fourth graders, yet the specific words or syntax used might render the message quite difficult to understand (cf. Chapanis, 1965). Even if the individual words of a warning message are understandable, the words as a group might not be. Note that scrambling the words within phrases and sentences would give the same readability score as real sentences. Until readability indices become more sophisticated, in particular by being capable of evaluating the semantic content and context, they should be used only as a rough guide in evaluating textual warnings, and should not be treated as a reliable or valid tool for assessing warning comprehension. As we will discuss later, the only good way to know whether the material is understandable is to test it on the target population.

8.3.6 Organization

Another approach to facilitating the understanding of warning text is through the material's organization. Information structured coherently is better than a random organization. Organization can be produced in several ways, such as a hierarchical structure, in a network, and/or based on mental models. Kozminsky (1977) found that providing titles (or headings) consistent with the to-be-learned information improved memory. Desaulniers

(1987) has shown that the use of outlines, lists, and hierarchical arrangement of concepts improves its perceived organization. However, research on which kinds of organization maximally benefit knowledge acquisition has been to date quite limited and is a research area that could provide useful design guidance.

8.3.7 Guidelines

Some attempts have been made to provide rules and guidelines for producing clear writing. Hartley (1994) describes a set of guidelines for designing instructional text that appear to have applicability to warnings. Also, general rules for constructing warning text can be found in the ANSI (1991, 1998) Z535 warning standards, and in the Westinghouse Electric Corporation (1981) and FMC Corporation (1985) guidelines. The guidelines usually specify, for example, that the text should be brief and written in common, nontechnical terms when the target audience is the general public. These guidelines are probably good starting points when designing warnings. However, they are not always applicable to the specific situation or message to be communicated. Moreover, there is such an abundance of rules and guidelines, that applying them may be difficult (Wright, 1985). Compromises frequently have to be made between different design rules. For example, consider the guideline that we gave earlier regarding explicit wording: research has shown that explicit warnings can benefit comprehension. However, following this rule produces longer warnings and conflicts with another guideline that warnings should be brief. Currently, there is very little research that delineates which guidelines are relatively more important and how conflicts should be resolved.

8.3.8 Limitations of Guideline Factors

Thus far we have discussed several factors that can influence the understanding of warning text. These factors included: explicitness, readability indices, inferential analysis, organizational approaches, and use of guidelines. They can be used to develop initial prototype warnings designs. However, their use as the sole method of establishing and constructing a warning has an important limitation, which is that they can allow the production of warnings that are not as understandable as they could be. The only way to know the extent of understanding is to actually test a representative sample of the target audience. We will have more to say about this topic later in this chapter. Further information on comprehension testing is also given in Chapter 3 by Young and Lovvoll. In the next section, we consider another way of conveying hazard information using nonverbal (non-textual) symbolic representations.

8.4 UNDERSTANDING PICTORIAL SYMBOLS

An increasingly common approach in warnings design is to use pictorial symbols as an addition or substitute for words (Dewar, 1999). Symbols might depict the hazard, the consequences, actions to take or not take, or some combination of these. Well designed symbols can be useful for illiterates as well as literates who are not skilled in the particular language of the printed text message, for example travelers in foreign countries. Symbols can serve not only to enhance comprehension but also, as described in Chapter 7 by Wogalter and Leonard, symbols can also help to capture people's attention.

8.4.1 Symbol Processing

Generally, pictures are easier to remember than words; sometimes this is called the picture superiority effect (Nelson, 1979). One explanation of this effect comes from dual-code theory (Paivio, 1971, 1986). Two types of coding system in memory are hypothesized: verbal and imaginal (visuo-spatial). Words are assumed to be coded verbally, and pictures are assumed to be imaged. Each code can evoke the other code, but the translation from code to code varies in difficulty. Some words can easily evoke specific concrete images (e.g., 'gloves,' 'goggles') while other words representing abstract concepts (e.g., 'protective equipment,' 'security') are not as readily translated into mental images. High imagery words may activate both codes, which makes encoding into memory more effective and subsequent retrieval easier. This information is easier to retrieve because theoretically more 'paths' are created in memory, making the information more accessible (more likely to be cued) at later times. Thus, according to dual-code theory and the picture superiority effect, warnings with symbols should be more effective in terms of encoding and retrieval.

8.4.2 Comprehension Testing

Therefore, symbols appear to have considerable potential in communicating hazard-related information. The best symbols can convey concepts quickly and readily activate considerable pre-existing hazard knowledge. Ideally, symbols should be understood by everyone, but in most cases they are not. The research literature shows symbol comprehension rates can vary from very high to very low, depending on the symbols, the test methods, and the population tested (e.g., Laux, Mayer, and Thompson, 1989; Calitz, 1994; Leonard, 1994). Since we know that some pictorials will not be understood by some percentage of the at-risk population, a question that could be asked is: what level of comprehension is acceptable? Published standards have attempted to address this issue by quantitatively defining what constitutes an acceptable symbol. The American National Standards Institute (ANSI Z535.3, 1991, 1998) requires that 85% or more of the answers from at least 50 people should identify correctly the referent concept with no more than 5% critical confusion. Critical confusion comprises answers opposite to the intended concept or wrong answers that could lead to behavior that could result in injury or property damage. Obviously, critical confusion errors are to be avoided. As an example consider the symbol in Figure 8.2. It shows a side-view, outline shape of a pregnant woman with a circle-slash prohibition. The intended meaning is that women should not take the drug while pregnant or, if they are not pregnant, to take precautions to avoid getting pregnant while taking it. However, this symbol by itself can also be interpreted to mean birth control protection, a potentially disastrous error. Obviously, it is very important to limit the number of cases of critical confusion. Avoiding them is even more important than high comprehension scores.

According to ANSI Z535.3 (1991, 1998), when a safety symbol can not be developed that reaches the 85% criterion, then text must accompany it. In the unification of countries comprising the European Union (EU), people will freely traverse borders where they do not know the primary language. The EU's international symbols are intended to be displayed without any accompanying text. The Organization of International Standards' 3461-1 standard (ISO, 1988) requires 67% correct identification in a comprehension test.

Attainment of the benchmark comprehension score does not mean that the symbol in question is adequate, nor does a lower score indicate that it should not be used. The

Figure 8.2 Warning symbol meant to indicate that a drug for severe acne should not be taken by pregnant women. Some women have apparently interpreted this symbol to mean that the drug acts as a contraceptive, illustrating a critical confusion. This symbol appeared in the *FDA Consumer* (**22**(8), p. 26), US Food and Drug Administration, Rockville, MD.

ultimate criterion is: does the symbol/pictorial improve safety? If no better pictorial can be made and the critical confusion rate is low, it is better to use the symbol than not to use it.

The validity of the above-mentioned criterion notwithstanding, relatively few symbols in use today have been tested. In the tests that have been conducted, the results often show that many pictorial symbols in current use have low comprehension rates. Even symbols that we would expect to produce high identification rates are not understandable by substantial numbers of people. Leonard (1994) found that both college and English as a second language (ESL) students failed to recognize many symbols used commonly in transportation. Collins and Lerner (1982) found that many individuals had difficulties understanding some of the symbols used for fighting fires. Calitz (1994) found that symbols used in South Africa by legal mandate were poorly recognized. In addition, less than half of a set of automotive-related symbols were named correctly by 60% or more of the respondents in a study by Jack (1972). MacBeth and Moroney (1994) found that college students had difficulty in correctly interpreting several ISO symbols. Laux *et al.* (1989) and Ringseis and Caird (1995) both found that although some industrial safety and pharmaceutical symbols are comprehended well, others are not. In the next several sections, we discuss some of the reasons why symbol comprehension can be poor.

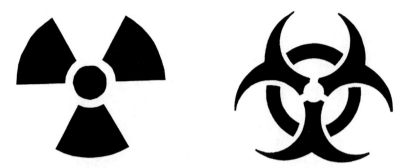

Figure 8.3 Symbols for (a) radiation and (b) biohazard are abstract representations; they do not directly represent the specific physical nature of the hazards.

8.4.3 Underlying Concept

As with text messages, people can have difficulty understanding the underlying concept that a pictorial symbol is supposed to represent. The symbols in Figure 8.3 for radiation and biohazard are two examples. Symbols for these concepts are abstract and are depictions that do not closely represent the actual physical entities. Even if the people are trained (or learn) the short referent definition associated with a symbol, the individual might not really know the nature of the intended concept, its implications for health and safety, and what he or she needs to do to avoid being hurt. Thus, as was true for textual messages, it cannot be expected that a symbol by itself will be able to convey all of the safety ramifications for complex, abstract concepts.

8.4.4 Visualizability

When the depiction bears resemblance to the actual visual objects or procedures, then in general understandability will be greater. However, some phenomena are not visible to the eye and are difficult to depict as visual images. Examples include two hazards that we have already mentioned, radiation and biohazard (Figure 8.3). Other difficult-to-depict examples include carbon monoxide, time, and beach undertow. Dewar and Arthur (1999) describe some of the difficulties of creating a pictorial display that illustrates the concept of undertow. The pictorial symbol was intended to convey that there may be strong water currents that could pull people under water. Another (usually) invisible concept, electricity, does have a reasonably successful symbol (the jagged arrow shock symbol shown in Figure 8.4), but undoubtedly it is based on (a) the shape being similar to one visible form of electricity, lightning, (b) that most people have received some home and school education that electricity can be dangerous, and (c) the symbol's frequent use.

Lack of visibility is important also with respect to the symbol itself. As mentioned in Chapter 7 by Wogalter and Leonard, sometimes the slash in the circle/slash prohibition symbol can reduce identification rates by obscuring critical components (Dewar, 1976; Murray, Magurno, Glover, and Wogalter, 1998). Figure 8.5 shows a prohibition symbol that hides a critical component necessary to understand the symbol. Does it mean 'Do not walk' or 'Do not stand'? This prohibition symbol was taken from an actual warning intended to caution people about an 'automatic' door that can close unexpectedly when the system does not detect a person in the doorway threshold. Therefore it was intended

Figure 8.4 Symbol for electricity.

Figure 8.5 Ambiguous prohibition symbol. Does it mean 'Do Not Walk' or 'Do Not Stand'? The slash obscures a critical portion of the symbol, producing the possibility of a critical confusion. (It was used for a sign on a set of doors that open automatically when a person approaches them. Unfortunately the sensor system did not work well when a person was standing in the threshold and the door could close unexpectedly. In this case, the symbol was intended to mean 'do not stand').

to mean 'Do not stand' (to keep moving) when it can also be interpreted to have the opposite meaning, 'Do not walk'—a clear cut critical confusion. Bruyas (1997) has explored several methods of measuring the relative importance of components comprising graphic symbols. Such procedures could be useful in determining which types of component can be deleted without loss of comprehension performance, and at the same time, reducing clutter and increasing legibility. Also, critical components of complex pictorials could be highlighted to distinguish them from less important details (Brantley and Wogalter, 1998).

8.4.5 Quality and Form of the Depictions

The quality of the artwork can affect comprehension levels. Numerous design guidelines exist on how to produce professional quality symbols (FMC Corporation, 1981; Westinghouse Electric Corporation, 1985; ISO, 1988; ANSI, 1991, 1998; Sanders and

McCormick, 1993). Many of the basic design guidelines derive from the perceptual principles of Gestalt psychology (see, e.g., Coren and Ward, 1989; Sanders and McCormick, 1993). These characteristics include figure–ground, simplicity, contiguity, boldness, similarity, among others. Generally, pictorial symbols with good contrast and comprised of simple forms are preferred. See Dewar (1999) and Sanders and McCormick (1993) for more information on this topic.

For any concept, many drawings are possible. The depiction of a warning concept might focus on the hazard, the consequence, the compliance instructions, or some combination of these. In addition, the objects in the symbols can be variously depicted in different perspectives, by different amounts of detail and emphasis, etc. Sometimes a minor change to a single component of a symbol can change its meaning dramatically. Several authors describe some of the issues involved in creating and refining symbols (e.g., Wolff and Wogalter, 1993, 1998; Dewar, 1999; Magurno, Wogalter, Kohake, and Wolff, 1994). Zwaga (1989) and Brugger (1999) discuss methods for reducing a large set of depictions to a few good potential symbols that can be verified by comprehension testing. We will have more to say about symbol comprehension testing later.

8.4.6 Literal Interpretation

A common pictorial symbol for 'No open flames' is a circle/slash overlaying a lit match. The literal meaning of the symbol is that no matches should be lit in the area. However, this same symbol has also been used as a signal that *all* ignition sources (including spark generating devices) must be extinguished because flammable substances are present. It is apparently being assumed that people will extract from this and other flame symbols the broader concept of ignition source. The problem is that although some people may make this extended interpretation, many people will make only the literal interpretation. It really cannot be expected that everyone will generalize to the broader concept without additional accompanying information or specific training.

8.4.7 Complexity

As noted previously, simple symbols are preferred. However, one cannot always follow this guideline when trying to produce an understandable symbol. Consider again the concept of 'no ignition source.' Possible depictions of this concept include a lit match (as described before) or just a simple flame overlaid by a red circle/slash prohibition symbol. However, neither of these symbols conveys the full concept. Therefore, it might be necessary to include other forms in the symbol, such as perhaps, electrical switches and telephones—two common devices that produce sparks. However, including these or other objects increases the detail and complexity of the symbol which could have negative effects (e.g., decreased legibility). Thus, while we would like to make simple symbols, we can not always do so and still be confident that all persons will understand the full intent of the message.

8.4.8 Single versus Multiple Panels

For some concepts, a single symbol panel may not be enough. Some symbol designers (e.g., Dewar and Arthur, 1999; USPC, 1997) have used sequences of symbols or multiple

Figure 8.6 Example symbols requiring multiple panels to convey a concept ('Take morning, noon and night' and 'Undertow').

panels to convey certain concepts. Figure 8.6 shows symbols meant to represent the concept of 'Take morning, noon, and night.' The other symbol is for 'Undertow.' One can see that adequate depiction requires more than one panel.

8.4.9 Language Accompaniment

Symbols and language can complement each other (e.g., Cairney and Sless, 1980; Morrell, Park, and Poon, 1990; Young and Wogalter, 1990; Sojourner and Wogalter, 1997, 1998). People who do not understand a pictorial might be able to read accompanying text. Then, after reading the descriptive text, they might in all subsequent exposures understand and remember the symbol's meaning. Moreover, the accompanying words provide additional, more specific information that otherwise would be very difficult to convey by pictorial symbols alone. Of course, adding text will not directly help those lacking the skill to read it. Research also shows that relative to younger adults, older adults seem to prefer and are possibly more adept with textual instructions than symbolic instructions, possibly because of age-related differences in familiarity with the two kinds of media (Morrell *et al.*, 1990; Morrow, Leirer, and Andrassy, 1996; Sojourner and Wogalter, 1997, 1998).

8.4.10 Review of Pictorial Symbol Comprehension

In the preceding sections, general symbol-design principles based on research and guidelines were presented. While symbols can be helpful in communicating hazard-related information, symbols for many concepts may fail to produce high levels of comprehension, particularly for concepts that are complex or not easily visualized. Even without high levels of comprehension, pictorials may be useful components of warnings because of their ability to attract attention, to reinforce an accompanying verbal message, and to cue knowledge in memory. One must, however, purposely look for and avoid critical confusions.

8.5 UNDERSTANDING AUDITORY WARNINGS

In this section, we discuss factors that relate to the understanding and memory of auditory warnings. Chapter 7 by Wogalter and Leonard described various characteristics of sound-based warnings that affect attention capture and maintenance. Like visual warnings, auditory warnings can carry information and affect people's hazard comprehension.

8.5.1 Nonverbal Sounds

Simple auditory warnings (e.g., common tones and beeps) usually carry less information than complex sounds. Simple auditory warnings announce the fact that there is a problem but provide little additional information. Other more complex nonverbal auditory warnings can signal (or code) specific hazards by using different frequencies or temporal arrangements of sounds. However, for different sounds to be effective in signaling specific hazards, the receiver must be able to associate specific sounds with their meanings. Research indicates that only a limited number of nonverbal auditory signals should be used in any one system; having too many will make them difficult to discriminate, learn, and remember (Cooper, 1977; Banks and Boone, 1981). Even after the set is learned, retraining and practice may be necessary to ensure the signals' meanings are not forgotten (Patterson and Milroy, 1980).

Research indicates that the design and selection of nonverbal auditory sounds be based on existing stereotypical knowledge in the target population or, in other words, having high association value with the referent (Edworthy, Loxley, and Dennis, 1991; Hellier, Edworthy, and Dennis, 1993; Edworthy, Hellier, and Hards, 1995; Haas and Casali, 1995). For example, if one wanted a sound that would give information to an operator about the slowing down or the speeding up of some industrial process, one might use an auditory signal that modulates in accord with the speed of the machine. In this case, the sound might be of a higher pitch or with a quicker beat rate when the system is operating on 'high' and a lower pitch or a slower beat rate when the system is operating on 'low.' When the system is not working perfectly the sounds could be distorted to reflect the degree of machine malfunction. In other words, nonverbal sounds can be designed to reflect the state of the system, making them useful information displays. To avoid the need for extensive training to learn associations between an arbitrary set of auditory codes and their referents, it is wise to use what people already know as a partial basis for selection and design of sound cues.

8.5.2 Voice

Complex auditory messages also can be transmitted via voice (speech). Unlike complex nonverbal messages that require specific training to know the various meanings of the coded sounds, the use of speech makes available an extensive repertoire of pre-existing language skills that can be used to decode the meaning of the sound-based information. Thus, the number of different voice messages that can be conveyed without extensive training is considerably larger than that for nonverbal auditory signals.

Some messages are not conveyed as readily auditorily as they are visually because of the nature of the two senses. With voice presentation, the message is presented across time and one's ability to review earlier-presented material is limited without a mechanism for playback. Generally with visually presented material, the section can be reread if it is not understood the first time. Difficult material requiring complex surface-to-deep structure transformations tends to tie up large amounts of working memory capacity, and tends to be less well understood when presented by voice than by print (Penney, 1989). At the same time, if the warning message is short and relatively simple, presentation by voice can be very effective at capturing attention, making it more likely that the information will be conveyed and processed further. Indeed, research shows that short voice warnings can be more effective in producing behavioral compliance than the same message in print (Wogalter and Young, 1991; Wogalter, Kalsher, and Racicot, 1993).

Voice warnings are now commonly conveyed in mass-media broadcasts and in automated cockpit systems. Technological advances have made available inexpensive miniaturized digital recording and playback devices (as in phone answering machines and in some greeting cards). These inexpensive systems combined with detection devices (e.g., photoelectric and motion sensors) may be used in a variety of applications not previously considered practical. We expect that auditory-warning systems will become more sophisticated in the future, particularly in the area of selective presentation to avoid annoyance and habituation (see also Chapter 7 by Wogalter and Leonard).

8.6 COMPREHENSION TESTING

In the previous sections, we described factors that can improve comprehension. However, none of the preceding methods is definitive because they will not necessarily produce the best warnings. The process of developing warnings should not stop after applying a set of guidelines. Guidelines are useful in helping to form an initial set of designs or prototypes that then should be put through formal comprehension testing to assure that the warnings are understood as intended (i.e., whether they activate or produce the necessary information in memory).

Direct measurement of comprehension is the ultimate determinant of whether persons are properly informed about the hazard. As we noted earlier, compromises have to be made when following warnings guidelines (e.g., between brevity and explicitness). However, one does not know before testing is carried out whether the right tradeoffs have been made. The only way to know is to test it. Testing can determine whether comprehension is adequate and whether there is a need for more design improvements. Testing can also be used to gather feedback on potential design improvements.

Another reason for comprehension testing is that sometimes warnings designers and domain experts can misrecognize that they know what the target population knows, and that their particular experiences and beliefs may not be adequate to produce understandable warnings for the general public. Frantz, Miller, and Main (1993) investigated the ability of two different groups, engineering students and law students (potential designers of warnings) to estimate the effectiveness of a set of warnings that had already been measured in previous research. Less than half of them selected the most effective warnings.

In this chapter we do not describe the specific procedures involved in testing warning comprehension. Chapter 3 by Young and Lovvoll describes some of these methods in more detail. Also, the appendix of the most recent version of the ANSI (1998) Z535.3 safety symbol standard includes some suggested methods for testing symbol understandability. Although the Z535 standard does not describe explicitly how to test text messages, the test procedures can be adapted readily from the symbol testing methods (Wolff and Wogalter, 1998).

8.6.1 Participants

Obtaining an appropriate set of participants for comprehension testing can be difficult and time consuming. Frequently, college students are used because they are convenient to university-based researchers. A possible problem with student participants is that their knowledge and education may differ from the warning's target audience. Note though, that college students are not always inappropriate research participants for warning research; sometimes they are the target audience of concern, e.g., for warnings associated with alcoholic beverage consumption. In other situations, students can give a general indication how a different group of persons might perform (e.g., Leonard and Cummings, 1994; Magurno et al., 1994). That is, if college students have difficulty understanding a warning, then surely less educated people will do no better. However, if college students perform very well, we will not know how the general public will respond. Fortunately, comparisons between the data of college students and other groups (e.g., participants solicited at flea markets) frequently show the same basic pattern of results, although college students generally produce more consistent (less variable) and sometimes higher scores than ordinary citizens (e.g., LaRue and Cohen, 1987; Silver et al., 1991; Wogalter, Kalsher, Frederick, Magurno, and Brewster, 1998). Nevertheless, despite the concordance, a warning that will be used in actual applications should be tested using an appropriate sample of the target audience when possible (Laughery and Brelsford, 1991; Young, Laughery, Wogalter, and Lovvoll, 1998).

When a representative sample of the target audience cannot be obtained because of economic, technical, or logistical reasons, testing should focus on using a sample of persons at the lower end of the distribution of language skills, education, and socio-economic status of the target population. Note that the data gathering process should not be directed at 'average' users because this would omit the most critical persons in terms of those who might not understand the material. Thus, if the warning is to appear on a consumer product (making the public at large the target audience), and if a fully representative sample of target users can not be secured, one should at least employ a sample including persons who have limited reading and language skills. Adult literacy programs and English as a second language programs (ESL) are good sources of such persons. Flea markets and community centers may be good sources of participants possessing a range of language skills and demographic characteristics.

8.6.2 Focus Groups

Another method of getting feedback on warning designs is focus groups. Groups of participants fitting some specific (usually demographic) criteria are brought together to discuss, with the aid of a facilitator, a set of issues that are considered relevant to the content and format of the warning being evaluated. The focus group participants must be given adequate background information about the hazard as well as potential injury scenarios. They are then encouraged to express their ideas and opinions and pros and cons about potential warning designs. However, the basic focus group method alone is inadequate for assessing comprehension. Some of its deficiencies include: (a) the use of small sample sizes; (b) the fact that one or two individuals may drive the entire group's ideas—thereby making the sample smaller than its nominal count; (c) these one or two individuals informing others in the group about hazards that they would not have recognized; and (d) only opinions are collected not actual knowledge or behavior. The primary problem stemming from these issues is that the best warning, one that informs and produces superior levels of safe behavior, may not be derived from the focus group method. Like the other methods discussed earlier, focus groups can be beneficial in the process of developing initial prototype warnings that later could be included in a formal comprehension test.

8.6.3 Open ended versus Multiple Choice

Although it is quite common for multiple-choice tests to be used to assess knowledge, often they are inappropriate for testing warnings-related comprehension. The main reason is that it is very difficult to develop plausible sets of alternative answers for a multiple-choice test that assess comprehension fairly. Wolff and Wogalter (1998) have shown that multiple-choice tests can produce erroneous results. The best comprehension tests involve open-ended questions, in which people are simply presented with a prototype warning (or a component of warning) and are asked what they understand about it. Open-ended tests are more difficult to score than multiple-choice tests. Judges (graders/scorers) must subjectively assess whether or not the responses should be counted as correct, which can be difficult when participants' answers are ambiguous/unclear. A standard procedure is to have more than one judge score the comprehension responses to obtain a measure of inter-judge agreement (reliability). See Collins (Chapter 12; this volume). Wolff and Wogalter (1998) and ANSI (1998) for more discussion on these points.

8.6.4 Context

Unfortunately, many studies evaluate warnings comprehension in contexts that are different from the eventual real-world situations in which a warning (or warning component) will appear. Providing the appropriate contextual information during the comprehension test not only makes the test more realistic, but it can also enhance understanding by cuing related knowledge which, in turn, could yield higher test scores than without context. Without contextual cues, the test may yield low comprehension scores which, in turn, would indicate the need for additional, and frequently costly, redesign and testing procedures. The warning might have performed much better had participants known where it would be located. Also, without an explicit context, participants may supply their own implicit context which may or may not reflect the actual context in which the warning will appear. For example, in a test where no specific context is provided, a pictorial symbol depicting a boot may produce two or more interpretations depending on the context inferred, e.g., that safety shoes must be worn or that a shoe store or repair shop is present. However, had a context been provided (e.g., showing it with a photograph of a construction site or a marketplace), the number of incorrect responses would be reduced.

Therefore, supplying context during comprehension testing should facilitate the finding (and reduce the cost of finding) an adequately understandable warning (Wolff and Wogalter, 1998; Leonard and Karnes, 1998).

8.6.5 Potential Shortcuts

Sometimes the development of adequate warnings can require considerable work. First, a set of prototype warnings based on research results and guidelines is produced, then the most likely candidates are put through a comprehension test. If the testing shows that many persons in the target audience do not understand the hazard and its important ramifications, then the prototype(s) should be modified based on feedback from the earlier test participants, followed by another comprehension test with another sample of people. The process continues iteratively (design, test, redesign, test, redesign, test) until a satisfactory level of comprehension is reached (Wolff and Wogalter, 1993; Dewar and Arthur, 1999; Magurno *et al.*, 1994). Several shortcut methods for testing pictorial symbols are described by Zwaga (1989) and Brugger (1999). This work shows that subjective ratings of understandability correlate with comprehension scores. Further research on the factors that predict comprehension should reduce some of the work involved in testing. For further information on comprehension testing, see Chapter 3 by Young and Lovvoll.

8.7 TRAINING

It is excessively optimistic to assume that people will encode and integrate large numbers of warnings simply by seeing them when they appear on equipment, on a product label, or in a manual. People engage in different degrees of 'reading'. If everyone grasped all of the information that they 'read,' all students would get near-perfect scores on tests at school and people, in general, would be better informed. As we know, people vary in how much they read and how much they comprehend. To ensure that people learn safety-related information, training may be necessary. Many large companies use training to ensure that employees know specific safety skills and procedures. Sometimes training can be quite brief, and other times, months or years of training (apprenticeship or schooling) are required. Because critical, potentially hazardous events tend to occur infrequently, periodic retraining may be necessary. To determine whether the training is producing knowledge and skills, some sort of follow-up test is necessary.

The basic premise of training is that it will promote the use of appropriate knowledge and skills when they are needed or, in other words, the effects of training transfers to actual real-world tasks. Transfer from training to actual use conditions can be positive (facilitating subsequent performance) or negative (retarding subsequent performance). One example of positive transfer is represented in Figure 8.7. Most people have learned the general concept of the circle/slash prohibition symbol, perhaps from seeing it in various other symbols such as those for 'No dogs' or 'No bicycles,' etc. From this prior experience, people are likely to transfer the knowledge that the circle/slash means 'No' to other prohibitions such as 'No bobsled' or 'Do not touch' symbols—even though they might not have seen the exact symbols before.

A situation involving negative transfer occurs when information learned earlier makes it more difficult to learn new material later. Consider the symbols shown in Figure 8.8. Suppose that a person initially learned that a flame symbol means fires are permitted (as in the symbol on the left indicating 'campfires allowed in area'), and then later tries to

Figure 8.7 Example of positive transfer. Knowing the meaning of the prohibitive symbol (e.g., as in 'No Dogs' or 'No Bicycles') can transfer so that one understands the meaning of the prohibitive symbol in newly seen symbols (e.g., 'No Bobsled' or 'Do Not Touch').

Figure 8.8 Example of negative transfer. Symbols for (a) campfires allowed in area, and (b) flammability—where fire is not permitted. Learning one symbol first can have negative transfer effects on learning a second similar symbol with a different meaning. In this particular example, there is the possibility of critical confusion.

learn that another flame symbol (on the right) indicates flammability ('fire is not permitted'). This second association to the flame symbol (flammability) might be more difficult to learn compared to one in which no earlier learned association had been formed to a symbol with a flame. Thus in negative transfer, prior learning interferes with learning a different association for a similar picture or concept.

The processes of encoding information into memory and its subsequent retrieval from memory are intimately linked. The particular encoding operations or study strategies used at the time of initial stimulus exposure determine whether and how well the information will be retrieved at a later time (Craik and Lockhart, 1972; Tulving and Thompson, 1973). In general, the best cues for retrieval are those that were present when the stimuli were encoded earlier. Thus to increase the likelihood of correct remembering, one should maximize the similarity between the conditions in which retrieval of the information is desirable and those that are employed during the study/training session.

Training can involve many methods. In the following sections, we describe three: (a) modeling, (b) simulation, and (c) paired-associate learning.

8.7.1 Modeling

Modeling involves exposing individuals to another person who demonstrates how to perform the pertinent tasks correctly and safely. The desired outcome is that the persons exposed to the model will reproduce the model's behavior. Research shows that modeling increases warning compliance after participants see a videotape presentation or a live model carrying out the proper safety procedures (Wogalter, Allison, and McKenna, 1989; Racicot and Wogalter, 1995). There are a wide variety of safety training video tapes currently available from assorted vendors. Frequently these videos employ modeling, but their effectiveness is largely unknown.

8.7.2 Simulation

A second training method, simulation, provides the opportunity to practice critical procedures under (safe) conditions that mimic actual conditions. During practice sessions, feedback is given for improving performance. For example, pilots practice in realistic cockpit simulators similar to the aircraft they will fly. A focus of this training is to put the pilots through a series of potential emergency scenarios under controlled conditions. Because emergency events occur infrequently, the proper skills might not otherwise be learned. Simulation provides the opportunity to learn and practice emergency procedures and responses. In addition, airline pilots undergo periodic refresher courses to ensure that they will not forget what to do when certain incidents and warnings are presented. Simulation is used also for training various other kinds of safety critical work including nuclear power plant operators responding to a potential accident, lifeguards practicing rescues, and nurses administering medications, among others.

8.7.3 Paired-associate Learning

Paired-associate learning has a long and extensive research history in the psychology literature. Numerous studies (see e.g., Deese and Hulse, 1967; Ashcraft, 1989) have documented the parameters of such training. Typically, pairs of stimulus items are

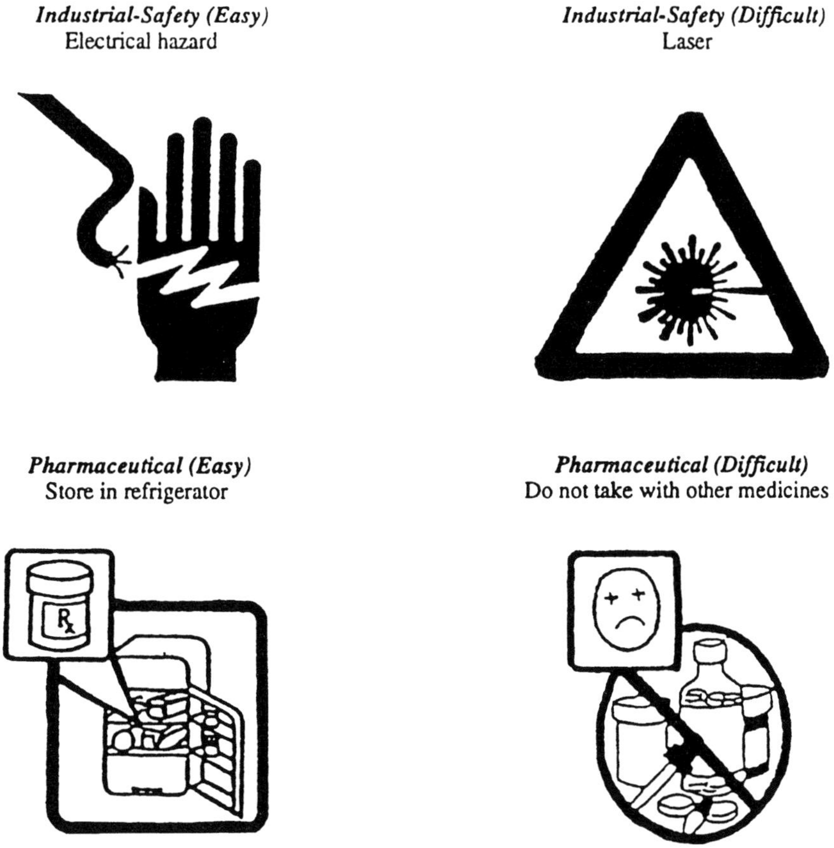

Industrial-Safety (Easy)
Electrical hazard

Industrial-Safety (Difficult)
Laser

Pharmaceutical (Easy)
Store in refrigerator

Pharmaceutical (Difficult)
Do not take with other medicines

Figure 8.9 Easy and difficult pictorial symbols used in Wogalter *et al.* (1997b).

studied together, and later when one of the two is shown, memory is activated allowing information about the other stimulus to be retrieved. One application of this process is in learning the meaning of abstract symbols such as for biohazard and radiation. Wogalter, Sojourner, and Brelsford (1997b) have demonstrated the utility of paired-associate learning for symbol comprehension. They examined the effects of training on the comprehension of 'easy' and 'difficult' industrial-safety and pharmaceutical symbols in a pre- and post-test paradigm. The easy/difficult distinction was based on earlier comprehension test performance. As Table 8.2 shows, the easy items were comprehended well in the pretest, although not all of them reached the ANSI (1991, 1998) 85% correct comprehension criterion. The difficult items produced much poorer comprehension scores on the pre-test. Following a single trial of paired-associate presentation, both types of pictorial were understood at higher levels. The increase was particularly dramatic for the more difficult symbols, about 40% better after a single training trial. These gains were maintained over time, as shown by a delayed post-test administered to another group of participants one week after training. A smaller-scale follow-up study showed only a small drop in performance six months later.

Table 8.2 Proportion correct as a function of easy versus difficult pharmaceutical and industrial safety symbols before training, immediately following training, or 7–10 days following training (adapted from Wogalter et al., 1997b).

	Pharmaceutical		Industrial	
	Easy	Difficult	Easy	Difficult
Initial test (before training)	.89	.47	.68	.33
Immediate post-training test	.96	.85	.89	.78
Delayed test (7–10 days later)	.97	.82	.90	.72

8.8 HAZARD IMPRESSION

Most of this chapter has dealt with attaining or activating knowledge for specific hazard-related concepts, including the meaning of text and pictorial symbols. In this section, we describe a different kind of information processing. This processing is *less* content-specific and more general, and it concerns the overall impression formed from a warning. The impression produced is a general feeling of danger (dangerousness)—that something bad is possible. In the warnings research literature, this dimension has been given various labels such as perceived hazardousness, arousal strength, and urgency. This impression can be formed regardless of whether the individual understands the specific content of the warning message.

The production of an appropriate hazard impression can reduce some of the problems cited earlier regarding comprehension difficulties experienced by persons who, for example, have lower-level language skills, but will not eliminate the problem entirely. Hazard impression can be helpful to individuals who fail to understand parts of the warning, either because of their personal limitations or when suboptimal conditions exist. If one or more cues of a multi-feature warning cannot be seen or interpreted accurately, then the remaining cues might compensate by providing an overall hazard impression. Also, the formation of an overall impression can serve as a redundant cue along with the specific message content of the text or symbols. We discuss other cues in the sections that follow.

8.8.1 Color

Certain colors such as red, orange, and yellow are used commonly to indicate different levels of hazard (from greater to lesser, respectively) (Bresnahan and Bryk, 1975; Westinghouse Electric Corporation, 1981; Collins, 1983; FMC Corporation, 1985; ANSI, 1991, 1998; Chapanis, 1994). Research has consistently shown that people in western cultures understand that red connotes hazard (Braun and Silver, 1995; Griffith, 1995; Wogalter et al., 1997a, 1998). Two other colors, orange and yellow, connote lower hazard than red, but people do not readily differentiate between the two on the perceived hazard dimension (Chapanis, 1994; Griffith, 1995; Wogalter et al., 1998). Besides the above-mentioned three colors and black, most of the other common colors connote little or no hazard.

8.8.2 Surrounding Shape

Sometimes warnings are enclosed in differently-shaped surround borders. The conventional 'STOP' sign is recognizable by its octagon shape. Through past experience we have learned an association among the features that comprise its octagon shape, its color red, and the word STOP. The triangular yield sign is perhaps almost as well recognized. Riley, Cochran, and Ballard (1982) examined 19 different symbol shapes with regard to hazard association. The shape most associated with hazard was the triangle (particularly with one point aimed downward). Also highly rated were a diamond, an octagon, a hexagon, and a pentagon; rounded shapes received lower ratings.

While people may understand the above-mentioned shapes, surround shape probably serves a minor role in hazard impressions relative to other potential cues. Because surround shapes do not carry much meaning in and of themselves, some sort of training or experience is required for people to recognize the intended meanings. Another problem is that surround shapes are used inconsistently across warning systems (Dewar, 1999).

On a related matter, research (Wogalter, Laughery, and Barfield, 1997c) suggests that some container shapes (e.g., the outline shape of a paint can or of an industrial-type barrel) connote greater hazard than other container shapes (e.g., the outline shape of a soda bottle or of a milk carton). This result suggests that in addition to what the label looks like and says, the container shape can provide a cue about how hazardous the substance is inside.

Also, research shows that physical characteristics of different designs of rectangular borders around a warning can influence hazard perceptions. Participants rated 51 borders that differed in color, width, and design on the dimensions of attention-capture, willingness to read the warning, and perceived hazard on 9 point scales (0 = 'not at all' to 8 = 'extremely' on the dimension). **Figure 8.10 (see color section)** shows some of the border stimuli examined by Rashid and Wogalter (1997). Tables 8.3 and 8.4 show summary statistics of the resulting hazard ratings. The tables show that the thicker borders with a red solid border or with black and red or yellow diagonal stripes produced the highest perceived hazard ratings. In a follow-up study, Wogalter and Rashid (1998) showed that the borders that received high ratings in the earlier study also were more likely to be looked at when on a sign posted in a public area.

8.8.3 Internal Shapes

Sometimes certain kinds of geometrical/configural information are included within the warning. The ANSI (1991) Z535.2 standard for environmental warning signs includes shape configurations as part of the topmost header panel containing the signal word. For example, the signal word DANGER is enclosed in an oval shape and WARNING is enclosed within an elongated hexagon shape. Some of these shape components do not carry much hazard association value by themselves (Wogalter et al., 1998). Jaynes and Boles (1990) used some of Riley et al.'s (1982) shapes and showed no effect on behavioral compliance rates. Other research has shown that diagonal stripes, the signal icon (the alert symbol with a triangle enclosing an exclamation point), and a simple skull symbol are perceived to indicate moderate to high levels of hazardousness (Wogalter et al., 1998).

Table 8.3 Mean ratings of attention capture, willingness to read warning, and perceived hazard for 51 borders comprised of combinations of color, width, and design configuration (from Rashid and Wogalter, 1997).[a]

Configuration	Width	Attention	Read	Hazard	Configuration	Width	Attention	Read	Hazard
No border	NA	0.50	1.33	NA	Black line	I	1.38	2.21	2.83
Yellow line	I	1.54	2.33	2.21	Black parallel lines	III	1.71	2.92	2.63
Yellow parallel lines	III	1.92	2.71	2.54	Green line	I	2.08	2.38	2.71
Blue line	I	2.29	2.54	2.13	Green parallel lines	III	2.42	2.83	2.54
Black line	II	2.58	3.33	3.33	Red line	I	2.58	3.13	3.50
Blue parallel lines	III	2.63	2.79	2.46	Red parallel lines	III	2.96	3.88	4.38
Black line	III	3.04	3.83	4.04	Blue line	II	3.08	3.42	3.08
Black jagged line	III	3.08	3.75	4.13	Yellow jagged line	III	3.08	3.13	3.67
Black/white stripes	II	3.08	3.83	4.00	Green line	II	3.17	3.50	3.25
Black 7 lines	III	3.25	3.33	3.54	Yellow line	II	3.33	3.79	3.63
Black/white stripes	III	3.58	4.25	5.04	Blue line	III	3.58	4.42	2.92
Blue 7 lines	III	3.58	3.88	3.21	Green jagged line	III	3.71	4.13	4.21
Yellow 7 lines	III	3.75	3.75	3.25	Black inward arrows	III	3.83	4.75	3.96
Blue jagged line	III	4.00	4.33	3.79	Green line	III	4.08	4.38	4.13
Red line	II	4.13	4.88	5.42	Yellow saw-tooth	III	4.17	4.08	4.83
Yellow line	III	4.20	4.46	4.13	Black saw-tooth	III	4.21	4.58	4.58
Green 7 lines	III	4.21	4.42	3.25	Black/green stripes	II	4.38	4.46	4.88
Black and blue stripes	II	4.46	4.46	4.38	Blue saw-tooth	III	4.46	5.17	4.67
Red 7 lines	III	4.58	5.13	5.54	Red jagged line	III	4.75	4.83	5.79
Black/red stripes	II	4.75	5.42	6.50	Black/blue stripes	III	4.92	5.29	4.71
Black/green stripes	III	5.04	5.50	5.17	Green inward arrows	III	5.08	5.13	4.54
Red line	III	5.13	6.04	6.13	Green saw-tooth	III	5.50	5.21	5.38
Yellow inward arrows	III	5.58	5.86	5.04	Blue inward arrows	III	5.58	5.13	4.25
Black/yellow stripes	II	5.63	5.63	5.88	Red inward arrows	III	5.83	5.83	6.00
Red saw-tooth	III	6.04	6.33	6.63	Black/red stripes	III	6.08	6.17	6.58
Black/yellow stripes	III	6.25	6.71	6.71					

Note: border widths: I = 0.07 cm; II = 0.35 cm; and III = 0.71 cm; NA = not applicable.

[a] Ratings were based on Likert-type scales anchored at end points with (0) 'not at all' and (8) 'extremely.'

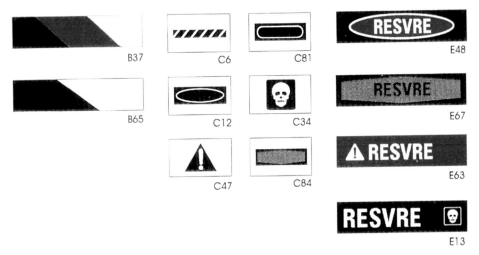

Figure 7.1 Colors and shape elements in ANSI Z535 warning standard used in a study by Wogalter et al. (1998). Used with permission from John Wiley & Sons.

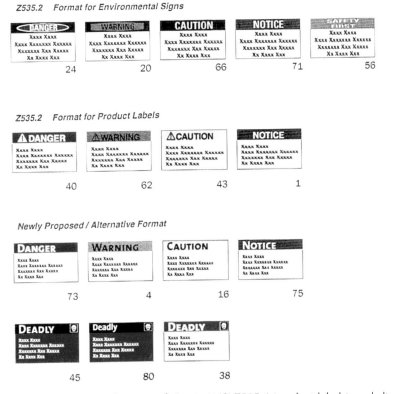

Figure 7.2 ANSI Z535.2 (environmental signs), ANSI Z535.4 (product labels), and alternative set of warnings tested by Wogalter et al. (1998). Used with permission from John Wiley & Sons.

Figure 7.5 On the same street corner in San Francisco both the over and under slash are used.

Figure 7.6 The over slash can obscure the critical parts of the internal portion of the symbol. Sometimes flipping a nonsymmetric symbol horizontally may help.

Figure 7.7 The red of the circle/slash negation symbol has faded in this sign. The two adjacent signs help to avoid a potentially dangerous critical confusion.

Figure 8.1 Air bag warning label.

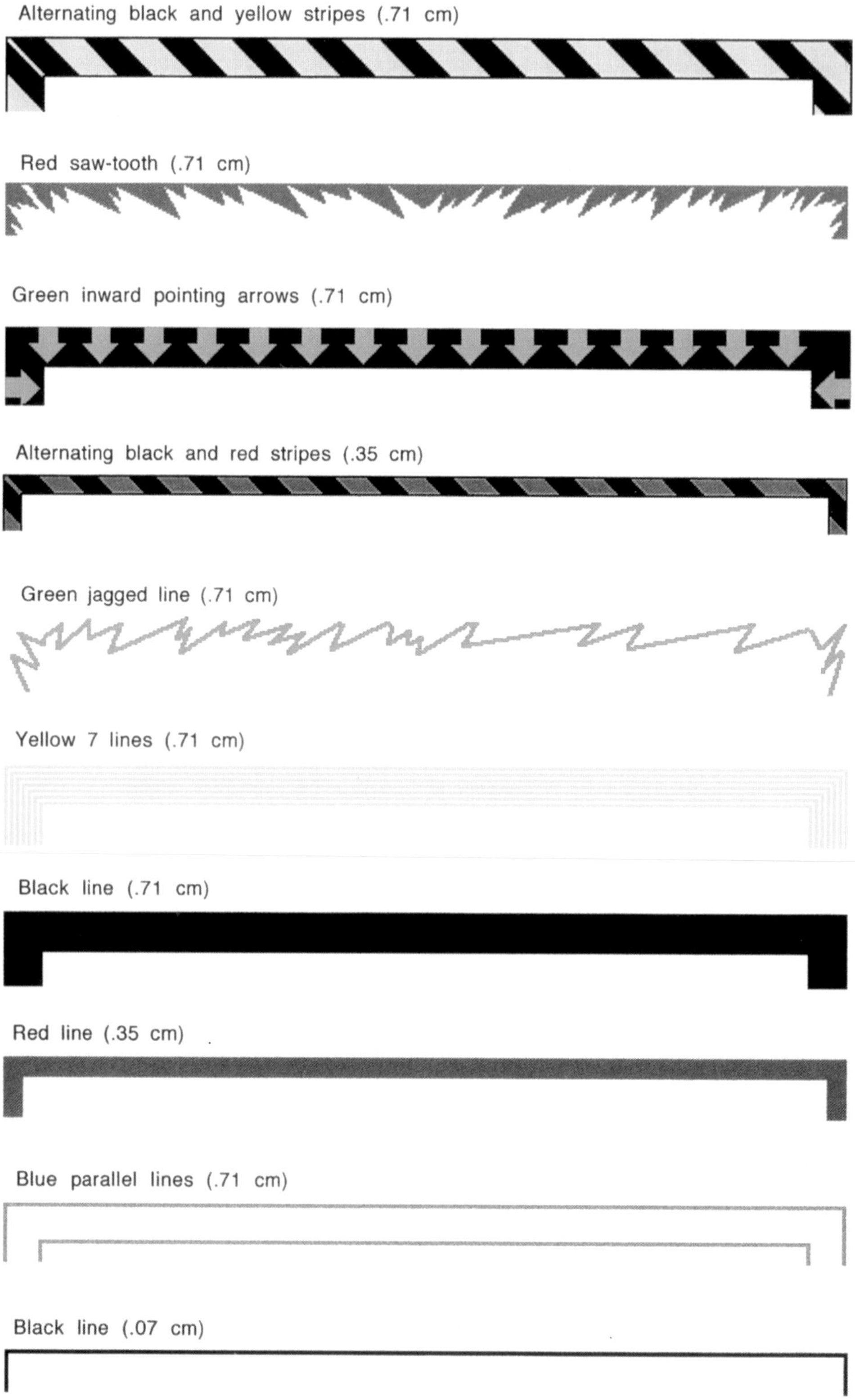

Alternating black and yellow stripes (.71 cm)

Red saw-tooth (.71 cm)

Green inward pointing arrows (.71 cm)

Alternating black and red stripes (.35 cm)

Green jagged line (.71 cm)

Yellow 7 lines (.71 cm)

Black line (.71 cm)

Red line (.35 cm)

Blue parallel lines (.71 cm)

Black line (.07 cm)

Figure 8.10 Example warning borders rated by participants in Rashid and Wogalter (1997).

Figure 12.5 Example of biohazard symbol as prescribed by OSHA. Note the distinctive design combined with the use of bright orange color.

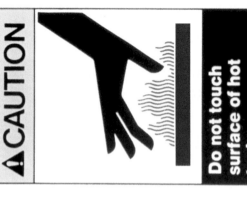

Figure 12.7 US approach for hazard warning signs, showing the use of the signal word (Danger, Warning or Caution), symbol, color (red, orange, or yellow) linked to the signal word, and short word message, using general format prescribed by the Z535 series of standards.

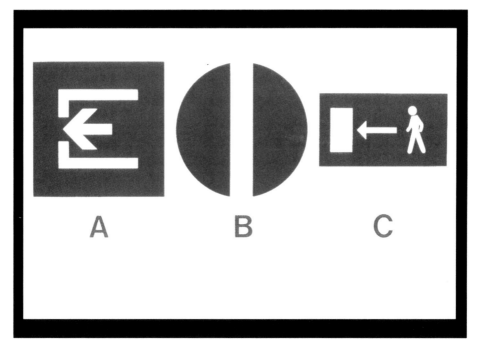

Figure 12.9 Symbols tested by NBS/NIST researchers that were not effective in communicating the message 'exit' or visible in simulated smoke conditions.

Figure 12.10 Symbol tested by NBS/NIST researchers for 'exit' that was found to communicate the message effectively and be visible in simulated smoke conditions.

Figure 12.11 Sign designed according to the Z535 criteria but using bilingual word message with one symbol. It can be imagined how complicated a sign using more than two languages would be.

Table 8.4 Mean ratings of attention capture, willingness to read warning, and perceived hazard for borders differing in color, width, and design configuration in which data are collapsed across conditions (from Rashid and Wogalter, 1997).[a]

Configuration	Attention	Read	Hazard	Configuration	Width	Attention	Read	Hazard
COLOR				WIDTH and DESIGN				
Red	4.68	5.16	5.64	No border	NA	0.50	1.33	NA
Yellow	3.95	4.25	4.19	single line	I	2.52	1.98	2.67
Green	3.97	4.19	4.00	single line	II	3.78	3.26	3.74
Blue	3.86	4.14	3.56	single line	III	4.62	4.01	4.27
Black	2.97	3.68	3.81	Parallel lines	III	3.03	2.32	2.90
				Seven lines	III	4.10	3.88	3.78
				Jagged lines	III	4.03	3.72	4.32
				Saw-tooth	III	5.01	4.88	5.22
				Inward arrows	III	5.34	5.18	4.76
				Colored stripes	II	4.46	4.76	5.13
				Colored stripes	III	5.58	5.12	5.64

[a] Note: I = 0.07 cm; II = 0.35 cm; and III = 0.71 cm widths; NA = not applicable.
Ratings were based on Likert-type scales anchored at end points with (0) 'not at all' and (8) 'extremely'.

8.8.4 Pictorial Symbols

Pictorial symbols differ from the other shapes that we have discussed in that they tend to be more detailed. We have already discussed the contribution of pictorial symbols to warning noticeability (Chapter 7 by Wogalter and Leonard) and comprehension (earlier in this chapter). In addition, symbols may convey or produce a hazard impression. This function would be important for difficult-to-depict concepts (e.g., those that are less visible, abstract, and highly technical) that people might not understand without accompanying verbal material or training. Consider the two cancer symbols in Figure 8.11. The one on the right seems to give a greater sense of hazard. A person who has never learned an association between the referent and its symbol might grasp the general gist of danger (hazard impression) just by looking at this form (though the person may not know that 'C' indicates cancer).

Figure 8.11 Two cancer symbols. Both are abstract but the one on the right gives a greater impression of hazard than the one on the left.

Figure 8.12 The skull and crossbones symbol and the Mr. Yuk symbol. Permission to reprint Mr. Yuk symbol granted by Children's Hospital of Pittsburgh.

Casey (1993) presents an interesting story about a grain shipment sent to Kurd villages in northern Iraq that was specifically meant for planting as a crop (not for direct consumption). The seeds were dyed red to indicate that they were not safe for eating. That year there was a major drought and the land remained parched and could not be seeded. This brought famine and starvation. Shortly afterwards, people began entering hospitals with severe neurological symptoms, such as inability to control their limbs. After considerable painstaking investigation, it was discovered that the neurological symptoms were due to mercury poisoning and this was subsequently tied to the grain which had been sprayed with a preservative that contained a form of mercury. The grain had been dyed to indicate that it was unfit for consumption. The on-site investigators discovered that all of the grain cases and bags prominently displayed the skull and crossbones symbol for poison (see left side of Figure 8.12). When the Kurd villagers were asked what this symbol meant, almost no one knew. They thought it was just another American design (perhaps a company logo) with no particular significance. They did wonder why the grain was red but it did not stop them from scrubbing it off with water (but unfortunately still leaving some mercury in the grain which later was made into foodstuffs and eaten). The point is that due to cultural differences not everyone understands what we might think would be one of our better danger-connoting symbols.

There's another interesting story related to the skull and crossbones symbol. The Mr. Yuk symbol shown on the right side of Figure 8.12 was developed as a substitute for the skull and crossbones symbol, because young children did not understand that the original skull and crossbones symbol indicated poison.

When designing a symbol, it may be possible to communicate the presence and level of a hazard by its inherent shape, even though individuals may not know the identity of the specific referent concept. It might be possible to redesign the biohazard and radiation symbols to enhance their perceived danger even to persons who do not know their specific meaning. It may be possible to do this by re-forming 'soft' curves within the existing symbols into 'sharp,' 'hard,' 'cutting' angles and making the appendages bolder/ fatter (as opposed to thinner).

8.8.5 Signal Words

Warnings often contain specific words intended to alert people to the presence of a hazard and the level of danger involved (severity and probability). In the USA the ANSI (1991, 1998) Z535 standards recommend the specific terms DANGER, WARNING, and

Table 8.5 Mean carefulness ratings of signal words known by 95% or more of the fourth and fifth graders and by 80% or more of the non-native English speakers. Also shown are college student and elderly participant ratings (Wogalter and Silver, 1995).[a]

	Study 1		Study 2	
	4th and 5th graders	ASU college students	Elderly	Non-native English speakers
NOTICE	5.39	4.01	5.00	3.64
CAREFUL	5.86	4.76	5.23	5.88
ALARM	6.16	5.01	6.09	4.87
IMPORTANT	5.95	5.06	5.59	5.64
CAUTION	6.64	5.22	5.91	4.75
DON'T	6.12	5.24	5.93	4.54
NO	5.63	5.60	5.81	4.68
SERIOUS	6.90	5.73	6.43	5.52
NEVER	6.09	5.93	6.27	5.34
WARNING	6.52	6.13	6.49	5.58
HOT	6.00	6.21	6.61	4.40
STOP	6.11	6.43	6.95	6.55
DANGER	7.12	6.49	7.00	7.63
DANGEROUS	7.18	6.64	7.04	7.66
POISON	7.49	7.00	7.57	7.93

[a] Note: selection of terms based on missing-value indicators of understandability.

CAUTION to be used as signal words to connote high to low levels of hazard, respectively (see also FMC Corporation, 1985). DANGER is intended for immediate hazards that will result in severe personal injury or death; WARNING is intended for hazards that could result in severe personal injury or death; and CAUTION is intended for hazards which could result in minor personal injury or damage to apparatus (FMC Corporation, 1985). Because most people do not know the formally assigned definitions and cannot accurately assign the definitions to the words when they are provided (Drake, Conzola, and Wogalter, 1998), their effect is mainly to alert people to the presence of a hazard and to produce an overall impression of the level of hazard. While some studies have shown little or no difference between DANGER, WARNING or CAUTION (Ursic, 1984; Leonard, Mathews and Karnes, 1986; Wogalter et al., 1987, 1998; Griffith, 1995), others have shown a fairly strong difference between DANGER compared to WARNING or CAUTION on perceived hazard (Bresnahan and Bryk, 1975; Dunlap, Granda, and Kustas, 1986; Leonard, Hill, and Karnes, 1989; Wogalter and Silver, 1990, 1995; Chapanis, 1994). Although sometimes statistically significant differences between WARNING and CAUTION are found, the mean differences usually are practically insignificant. Research has also investigated other potential signal words that may cover the range of the hazard dimension more effectively. One term, DEADLY, consistently produces greater levels of perceived hazard than the term DANGER (Leonard, Karnes, and Schneider, 1988; Wogalter and Silver, 1990, 1995). DEADLY could be used only for the most extreme hazards, and in this way avoid people discounting the seriousness associated with the ubiquitous DANGER signal word (Wogalter and Silver, 1990, 1995).

Table 8.5 shows a set of signal words that are understandable by 95% or more of young grade-school children (fourth and fifth graders) and by 80% or more of non-native

English readers. Also shown in this table are the ratings from college student and older adult participants (Wogalter and Silver, 1995). The words cover a much larger range of hazard than the three conventional ANSI terms, and probably they could be used as alternative terms to reduce habituation.

8.8.6 Multiple Features

The various features described can be used in combination to help cue hazards. The header panels recommended in ANSI Z535 combine multiple features, including a signal word, a colored surround, and a graphic (either the signal icon or a geometric shape). Wogalter et al. (1998) examined various individual components and combinations of components in header panels. Examples were shown in Chapter 7 by Wogalter and Leonard in **Figures 7.1 and 7.2 (see color section)**. Tables 8.6 and 8.7 show the mean hazard ratings. Probably the combined presence of multiple redundant features is most useful when seen under suboptimal conditions in which portions of the warning are not visible or not understood. The inclusion of multiple features provides alternative/redundant cues that one can hope will be adequate to provide the hazard information under suboptimal conditions.

8.8.7 Auditory Urgency

The idea of systematically matching (mapping) warning stimuli to actual hazards has been a major topic in the auditory warning literature. Edworthy and her colleagues (Edworthy and Adams, 1996; Edworthy et al., 1995b) have described factors that influence the perceived urgency of nonverbal auditory warnings. Because hazards vary in degree, it makes sense that the sound itself (ignoring the content of the word) provides a sense of urgency consonant with actual hazard level. Research has shown that sounds having

Table 8.6 Mean hazard perception ratings (overall and by participant group) and standard deviations for overall ratings.

Set-#	Stimulus	Ratings			
		Overall		Undergrad.	Comm. Vol.
		Mean	SD	Mean	Mean
Set A	Solid colors				
A-31	Red	3.2	1.1	3.2	3.1
A-58	Yellow	2.2	1.1	2.2	2.2
A-25	Orange	2.0	1.1	2.0	1.9
A-68	Black	1.7	1.5	2.0	1.5
A-60	Purple	0.8	1.0	1.0	0.6
A-64	Green	0.8	1.0	1.1	0.4
A-70	Blue	0.7	0.9	0.8	0.6
A-46	White	0.6	0.9	1.0	0.2
Set B	Multi colors				
B-69	Black/yellow	2.3	1.4	2.3	2.2
B-37	Black/red/white	2.1	1.2	2.0	2.2

Table 8.6 (cont'd)

Set-#	Stimulus	Ratings			
		Overall		Undergrad.	Comm. Vol.
		Mean	SD	Mean	Mean
B-50	Red/white	2.1	1.2	1.9	2.2
B-49	Black/orange	2.0	1.1	2.0	1.9
B-79	Black/white/red	1.9	1.2	1.7	2.1
B-65	Black/white	1.4	1.2	1.2	1.6
Set C	Shape and color configurations				
C-34	White skull in black square	3.8	0.6	3.9	3.7
C-83	Red oval in black rectangle	2.6	1.1	2.4	2.8
C-80	Black/yellow diagonal stripes	2.6	1.0	2.6	2.6
C-47	White ! in black triangle	2.3	1.1	2.4	2.1
C-84	Orange elongated hexagon in Black rectangle	2.0	1.2	1.8	2.2
C-6	Black/white diagonal stripes	1.7	1.0	1.5	1.9
C-21	Black triangle	1.4	1.0	1.3	1.5
C-51	Black elongated hexagon in black rectangle	1.1	1.1	0.9	1.4
C-12	Black oval in black rectangle	1.1	1.1	0.8	1.4
C-81	Black capsule (lozenge shape) in black rectangle	1.0	1.1	0.7	1.4
C-23	Black square	1.0	1.0	0.8	1.1
C-55	Black circle	0.9	1.0	0.8	1.1
Set D	Signal words				
D-32	DEADLY	3.8	0.6	4.0	3.6
D-53	DANGER	3.4	0.6	3.4	3.5
D-76	WARNING	2.6	0.9	2.6	2.6
D-35	CAUTION	2.3	0.8	2.5	2.0
D-11	SAFETY FIRST	1.4	1.1	1.1	1.6
D-39	NOTICE	1.2	0.8	1.2	1.2
Set E	Nonsense word headers				
E-52	White print and skull on red background	3.7	0.6	3.9	3.6
E-13	White print and skull on black background	3.6	0.8	3.7	3.5
E-63	White print and triangle / ! on red background	2.7	1.1	3.0	2.5
E-48	White print in red oval on black background	2.5	0.9	2.3	2.6
E-22	White print and triangle / ! on yellow background	2.4	0.9	2.6	2.2
E-67	Black print in orange elongated hexagon on black background	2.1	1.0	2.0	2.2
E-44	Yellow print on black background	2.0	0.1	1.9	2.1

Table 8.7 Mean hazard ratings, within-set rankings, and noticeability ratings for ANSI Z535.2, ANSI Z535.4, and alternative formats.

Rating		Hazard rating			Hazard ranking	
#	Signal word	Overall	Undergrad.	Comm. Vol.	Overall	Noticeability
ANSI Z535.2						
sign format						
24	DANGER	3.2	2.9	3.3	1.4	3.1
20	WARNING	2.7	2.5	2.7	2.4	2.7
66	CAUTION	2.4	2.5	2.4	2.4	2.8
71	NOTICE	1.2	1.2	1.2	4.2	1.4
56	SAFETY FIRST	1.1	1.4	1.0	4.6	1.4
ANSI Z535.4						
product label format						
40	DANGER	3.4	3.1	3.5	1.1	3.4
62	WARNING	2.5	2.2	2.6	2.5	2.6
43	CAUTION	2.3	2.2	2.3	2.5	2.6
01	NOTICE	1.1	1.1	1.1	3.9	1.3
Proposed formats						
73	DANGER	3.1	2.8	3.2	4.0	3.1
04	WARNING	2.4	2.2	2.5	5.3	2.4
16	CAUTION	2.1	2.2	2.1	5.3	2.2
75	NOTICE	1.4	1.3	1.4	6.9	1.7
45	DEADLY	3.8	3.8	3.9	1.4	3.8
80	Deadly	3.7	3.7	3.7	2.0	3.6
38	DEADLY (reversed color)	3.6	3.3	3.6	3.1	3.3

certain characteristics (e.g., higher frequency/pitch, faster beat rate) connote greater urgency levels (Edworthy et al., 1991, 1995a; Hellier et al., 1993; Haas and Casali, 1995).

More recently, research has begun to investigate the effects of voicing style on signaling urgency/hazard. Signal words presented in an emotionally charged female voice connote greater hazardousness judgments than the same words presented in a monotone male voice (Barzegar and Wogalter, 1998a,b; Edworthy, Clift-Matthews, and Crowther, 1998).

8.9 FAMILIARITY AND HABITUATION

The old adage, 'familiarity breeds contempt,' has some truth. A substantial body of research shows that familiarity with a product is associated strongly with lower hazard perceptions and a reduced tendency to look for warnings (e.g., Wright, Creighton, and Threlfall, 1982; Godfrey, Allender, Laughery, and Smith, 1983; Godfrey and Laughery, 1984; Leonard and Hill, 1989; Wogalter, Brelsford, Desaulniers, and Laughery, 1991). A problem related to familiarity is habituation. Habituation refers to the tendency for individuals to ignore stimuli after repeated exposure to the same stimulus (see also Chapter 2 by Wogalter, DeJoy, and Laughery, and Chapter 7 by Wogalter and Leonard). The occurrence of habituation indicates that at least some of the stimulus information is in memory.

Unfortunately, this memory may be a fraction of the total content of a warning. In other words, people might stop noticing and looking at a warning before they know all of its content. Ideally, one would like to present reliably a warning only at the times necessary to prevent unsafe behavior that would otherwise occur. However, in practice this is not possible and, consequently, warnings will be seen and heard when no unsafe behavior would potentially occur. Nevertheless, to decrease habituation one might want to alter or change warning stimuli to capture attention, like variable-information signs currently found on some major urban highways. To be on the safe side one would still probably want to present warnings more often than not, even if there is some possibility of habituation. On some non-durable consumer products purchased on a fairly consistent basis (e.g., cigarettes, beverage alcohol), a rotating-type presentation method could be used (Wogalter and Brelsford, 1994). Varying the look and the content of the warning will help to counteract habituation as well as increase knowledge (e.g., von Restorff, 1933; Wogalter and Brelsford, 1994).

8.10 PROSPECTIVE MEMORY

Most of the cognitive processes discussed thus far have dealt mainly with retrieval of items or events from the past. This is called retrospective memory, and involves recall of events that have already occurred. Prospective memory refers to remembering in advance of performing some task or, in other words, remembering to do an activity at some appropriate time in the future (Einstein and McDaniel, 1990; Einstein, McDaniel, Richardson, Guynn, and Cunfer, 1995). One plans at time A to carry out a task at time B and, if successful, one actually does remember to do the task at time B. One example is a worker remembering at the necessary time to lock-out or tag-out industrial equipment before commencing maintenance or repairs (so that the machine is not accidentally started). In using medicines one needs to remember when to take the medication and/or the specific conditions for its consumption (e.g., instructions to take twice a day an hour after eating dairy and calcium-containing products, or two hours before drinking alcohol). Automatic timers with auditory signals are available to aid prospective memory. Prospective memory can be aided also by content of the warning material. Suppose an individual wished to spray a flammable pesticide in a living-room area. The printed label instructions might state to cover furniture and other objects, followed by a directive to extinguish any pilot lights. Can the individual remember to turn off the pilot light after covering all of the furniture? Clearly, any damage to the furniture is less important than an explosion. Because prospective memory, like other types of memory, can fail, the warning instructions should direct users to turn off the pilot lights first. Prospective memory is particularly important when the compliance behavior is to be performed some time after warning exposure. Like other types of memory, it helps to have a cue (a reminder) at the time the compliance behavior needs to be performed. This relatively new area of research is likely to provide more knowledge on how to facilitate retrieval at the appropriate time in the future.

8.11 SUMMARY AND RECOMMENDATIONS

A number of warning guidelines can be put forward from this review of comprehension and memory factors. To facilitate warning comprehension, the designer should:

- Use simple language
- Verify that the text and symbols convey the intended meaning to the target population at risk
- Describe carefully and explicitly the nature of the hazard, the instructions on how to avoid the hazard, and the consequences of failing to avoid the hazard
- Design prototypes based on existing research and guidelines
- Test the best prototypes with at-risk individuals who may be least knowledgeable about the hazard
- Redesign a warning when testing reveals the target audience does not acquire the message intended.

To make warnings more memorable, one should:

- Use textual and pictorial materials that are meaningful and organized
- Provide cues to assist retrieval
- Provide training when considerable amounts of hazard-related information need to be learned
- Change the warnings occasionally so that the effects of habituation are reduced.

People should not be expected to expend substantial amounts of effort to understand warning messages. If the process is effortful, people are less likely to encode the material in the first place, but even if they do they may stop encoding the information before processing all of it. In short, warnings should be designed to convey safety messages quickly and adequately.

REFERENCES

ANSI (1991) Z535.1–5, *Accredited Standard on Safety Colors, Signs, Symbols, Labels, and Tags*. Washington, DC: National Electrical Manufacturers Association.

ANSI (1998) Z535.1–5 (revised), *Accredited Standard on Safety Colors, Signs, Symbols, Labels, and Tags*. Washington, DC: National Electrical Manufacturers Association.

ASHCRAFT, M.H. (1989) *Human Memory and Cognition*. Glenview, IL: Scott, Foresman.

BANKS, W.W. and BOONE, M.P. (1981) NUREG/CR-2147, *Nuclear Control Room Enunciators: Problems and Recommendations*. Springfield, VA: National Technical Information Service.

BARZEGAR, R.S. and WOGALTER, M.S. (1998a) Effects of auditorily-presented warning signal words on intended carefulness. In HANSON, M.A. (ed.), *Contemporary Ergonomics*. London: Taylor & Francis, pp. 311–315.

BARZEGAR, R.S. and WOGALTER, M.S. (1998b) Intended carefulness for voiced warning signal words. In *Proceedings of the Human Factors and Ergonomics 42nd Annual Meeting*. Santa Monica: Human Factors and Ergonomics Society, pp. 1068–1072.

BRANTLEY, K. and WOGALTER, M.S. (1998) Highlighting important elements in multi-panel pictorial symbols. Unpublished manuscript. Raleigh, NC: North Carolina State University.

BRAUN, C.C. and SILVER, N.C. (1995) Interaction of warning label features: determining the contributions of three warning characteristics. In *Proceedings of the Human Factors Society 39th Annual Meeting*. Santa Monica, CA: Human Factors Society, pp. 984–988.

BRESNAHAN, T.F. and BRYK, J. (1975) The hazard association values of accident-prevention signs. *Professional Safety*, January, 17–25.

BRITTON, B.K., VAN DUSEN, L., GLYNN, S.M., and HEMPHILL, D. (1990) The impact of inferences on instructional text. In GRAESSER, A.C. and BOWER, G.H. (eds), *The Psychology of Learning and Motivation*, Vol. 25, *Advances in Research and Theory*. San Diego: Academic Press, pp. 53–70.

BRUGGER, C. (1999) Public information symbols: a comparison of ISO testing procedures. In ZWAGA, H.J.G., BOERSEMA, T., and HOONHOUT, H.C.M. *Visual Information for Everyday Use: Design and Research Perspectives*. London: Taylor & Francis, pp. 305–313.

BRUYAS, M.P. (1997) Recognition and understanding of pictograms and road signs. Unpublished doctoral dissertation. University of Lumiere Lyon 2, Institut de Psychologie, Laboratoire d'Analyse de la Cognition et des Modeles, Lyon, France.

CAIRNEY, P. and SLESS, D. (1980) Understanding symbolic signs: design guidelines based on user responses. In *Proceedings of the 17th Conference of the Ergonomics Society of Australia and New Zealand*, pp. 51–58.

CALITZ, C.J. (1994) Recognition of symbolic safety signs. In *Proceedings of the 12th Triennial Congress of the International Ergonomics Association*, Vol. 4. Mississauga, Ontario, Canada: Human Factors Association of Canada, pp. 354–356.

CASEY, S. (1993) *Set Phasers on Stun: And other True Tales of Design, Technology and Human Error*. Santa Barbara, CA: Aegean.

CHAPANIS, A. (1965) Words, words, words. *Human Factors*, 7, 1–17.

CHAPANIS, A. (1994) Hazards associated with three signal words and four colours on warning signs. *Ergonomics*, 37, 265–275.

COLLINS, B.L. (1983) Evaluation of mine-safety symbols. In *Proceedings of the Human Factors Society 27th Annual Meeting*. Santa Monica CA: Human Factors Society, pp. 947–949.

COLLINS, B.L. and LERNER, N.D. (1982) Assessment of fire-safety symbols. *Human Factors*, 24, 75–84.

COOPER, G.E. (1977) NASA-CR-152071, *A Survey of the Status and Philosophies Relating to Cockpit Warning Systems*. NASA Ames Research Center, CA.

COREN, S. and WARD, L.M. (1989) *Sensation and Perception*, 3rd Edn. San Diego: Harcourt Brace Jovanovich.

CRAIK, F.I.M. and LOCKHART, R.S. (1972) Levels of processing: a framework for memory research. *Journal of Verbal Learning and Verbal Behavior*, 11, 671–684.

DEESE, J. and HULSE, S.H. (1967) *The Psychology of Learning*, 3rd Edn. New York: McGraw-Hill.

DESAULNIERS, D.R. (1987) Layout, organization, and the effectiveness of consumer product warnings. In *Proceedings of the Human Factors Society 31st Annual Meeting*. Santa Monica, CA: Human Factors Society, pp. 56–60.

DEWAR, R.E. (1976) The slash obscures the symbol on prohibitive traffic signs. *Human Factors*, 18, 253–258.

DEWAR, R.E. (1999) Design and evaluation of public information symbols. In ZWAGA, H.J.G., BOERSMA, T., and HOONHOUT, H.C.M. *Visual Information for Everyday Use: Design and Research Perspectives*. London: Taylor & Francis, pp. 285–303.

DEWAR, R. and ARTHUR, P. (1999) Warning of water safety hazards with sequential pictographs. In ZWAGA, H.J.G., BOERSEMA, T., and HOONHOUT, H.C.M. *Visual Information for Everyday Use: Design and Research Perspectives*. London: Taylor & Francis, pp. 111–117.

DRAKE, K.L., CONZOLA, V.C., and WOGALTER, M.S. (1998) Discrimination among sign and label warning signal words. *Human Factors/Ergonomics in Manufacturing*, 8, 289–301.

DUNLAP, G.L., GRANDA, R.E., and KUSTAS, M.S. (1986) Research Report No. TR 00.3428, *Observer Perceptions of Implied Hazard: Safety Signal Words and Color Words*. Poughkeepsie, NY: IBM.

DUFFY, T.M. (1985) Readability formulas: what's the use? In DUFFY, T.M. and WALLER, R. (eds), *Designing Usable Texts*. Orlando, FL: Academic Press, chapter 6.

EDWORTHY, J. and ADAMS, A. (1996) *Warning Design: A Research Prospective*. London: Taylor & Francis.

EDWORTHY, J., CLIFT-MATTHEWS, W., and CROWTHER, M. (1998) Listener's understanding of warning signal words. *Contemporary Ergonomics 1998*. London: Ergonomics Society, pp. 316–320.

EDWORTHY, J., HELLIER, E., and HARDS, R. (1995a) The semantic associations of acoustic parameters commonly used in the design of auditory information and warning signals. *Ergonomics*, 38, 2341–2361.

EDWORTHY, J., HELLIER, E., and STANTON, N. (1995b) Warnings in research and practice. *Ergonomics*, 38, 2145–2445 (special issue).

EDWORTHY, J., LOXLEY, S., and DENNIS, I. (1991) Improving auditory warning design: relationships between warning sound parameters and perceived urgency. *Human Factors*, 33, 205–231.

EINSTEIN, G.O. and MCDANIEL, M.A. (1990) Normal aging and prospective memory. *Journal of Experimental Psychology: Learning, Memory, and Cognition*, 16, 717–726.

EINSTEIN, G.O., MCDANIEL, M.A., RICHARDSON, S.L., GUYNN, M.J., and CUNFER, A.R. (1995) Aging and prospective memory: examining the influence of self-initiated retrieval processes. *Journal of Experimental Psychology: Learning, Memory, and Cognition*, 21, 966–1007.

FLESCH, R.F. (1948) A new readability yardstick. *Journal of Applied Psychology*, 32, 221–233.

FMC CORPORATION (1985) *Product Safety Sign and Label System*. Santa Clara, CA: FMC Corporation.

FRANTZ, J.P., MILLER, J.M., and MAIN, B.W. (1993) The ability of two lay groups to judge product warning effectiveness. In *Proceedings of the Human Factors Society 37th Annual Meeting*. Santa Monica, CA: Human Factors Society, pp. 989–993.

GODFREY, S.S., ALLENDER, L., LAUGHERY, K.R., and SMITH, V.L. (1983) Warning messages: will the consumer bother to look. In *Proceedings of the Human Factors Society 27th Annual Meeting*. Santa Monica, CA: Human Factors Society, pp. 950–954.

GODFREY, S.S. and LAUGHERY, K.R. (1984) The biasing effects of product familiarity on consumers' awareness of hazard. In *Proceedings of the Human Factors Society 28th Annual Meeting*. Santa Monica, CA: Human Factors Society, pp. 483–486.

GRIFFITH, L.J. (1995) The role of color in risk perception. Unpublished masters thesis. Athens, GA: University of Georgia.

HAAS, E.C. and CASALI, J.G. (1995) Perceived urgency of and response to multi-tone and frequency-modulated warning signals in broadband noise. *Ergonomics*, 38, 2313–2326.

HARTLEY, J. (1994) *Designing Instructional Text*, 3rd Edn. London: Kogan Page/East Brunswick, NJ: Nichols.

HELLIER, E.J., EDWORTHY, J., and DENNIS, I.D. (1993) Improving auditory warning design: quantifying and predicting the effects of different warning parameters on perceived urgency. *Human Factors*, 35, 693–706.

ISO (1988) ISO 3461-1, *General Principles for the Creation of Graphical Symbols*. Geneva: International Organization for Standardization.

JACK, D.D. (1972) SAE paper 720203, *Identification of Controls, a Study of Symbols*. Warrendale, PA: Society of Automotive Engineers.

JAYNES, L.S. and BOLES, D.B. (1990) The effect of symbols on warning compliance. In *Proceedings of the Human Factors Society 34th Annual Meeting*. Santa Monica, CA: Human Factors Society, pp. 984–987.

KINTSCH, W. and VAN DIJK, T.A. (1978) Toward a model of text comprehension and production. *Psychological Review*, 85, 363–394.

KLARE, G.R. (1976) A second look at the validity of readability formulae. *Journal of Reading Behavior*, 8, 129–152.

KOZMINSKY, E. (1977) Altering comprehension: the effect of biasing titles on text comprehension. *Memory and Cognition*, 5, 482–490.

KREIFELDT, J.G. and RAO, K.V.N. (1986) Fuzzy sets: an application to warnings and instructions. In *Proceedings of the Human Factors Society 30th Annual Meeting*. Santa Monica, CA: Human Factors Society, pp. 1192–1196.

LARUE, C. and COHEN, H.H. (1987) Factors affecting consumers' perceptions of product warnings: an examination of the differences between male and female consumers. In *Proceedings*

of the *Human Factors Society 31st Annual Meeting*. Santa Monica, CA: Human Factors Society, pp. 1068–1072.
LAUGHERY, K.R. (1993) Everybody knows: or do they? *Ergonomics in Design*, July, 8–13.
LAUGHERY, K.R. and BRELSFORD, J.W. (1991) Receiver characteristics in safety communications. In *Proceedings of the Human Factors Society 35th Annual Meeting*. Santa Monica, CA: Human Factors Society, pp. 1068–1072.
LAUGHERY, K.R. and STANUSH, J.A. (1989) Effects of warning explicitness on product perceptions. In *Proceedings of the Human Factors Society 33rd Annual Meeting*. Santa Monica, CA: Human Factors Society, pp. 431–435.
LAUX, L., MAYER, D.L., and THOMPSON, N.B. (1989) Usefulness of symbols and pictorials to communicate hazard information. In *Proceedings of Interface '89*. Santa Monica, CA: Human Factors Society.
LEONARD, S.D. (1994) How well are warning symbols recognized? In *Proceedings of the 12th Triennial Congress of the International Ergonomics Association*, Vol. 4. Mississauga, Ontario, Canada: Human Factors Association of Canada, pp. 349–350.
LEONARD, S.D., CREEL, E., and KARNES, E.W. (1991) Adequacy of responses to warning terms. In *Proceedings of the Human Factors Society 35th Annual Meeting*. Santa Monica, CA: Human Factors Society, pp. 1024–1028.
LEONARD, S.D. and CUMMINGS, J.B. (1994) Influences on ratings of risk for consumer products. In *Proceedings of the Human Factors Society 38th Annual Meeting*. Santa Monica, CA: Human Factors Society, pp. 451–455.
LEONARD, S.D. and DIGBY, S.E. (1992) Consumer perceptions of safety of consumer products. *Advances in Industrial Ergonomics & Safety*, Vol. IV. London: Taylor & Francis.
LEONARD, S.D. and HILL IV, G.W. (1989) Risk perception is affected by experience. In *Proceedings of the Human Factors Society 33rd Annual Meeting*. Santa Monica, CA: Human Factors Society, pp. 1029–1033.
LEONARD, S.D., HILL IV, G.W., and KARNES, E.W. (1989) Risk perception and use of warnings. In *Proceedings of the Human Factors Society 33rd Annual Meeting*. Santa Monica, CA: Human Factors Society, pp. 550–554.
LEONARD, S.D. and KARNES, E.W. (1998) Influence of context on warnings. In KUMAR, S. (ed.), *Advances in Occupational Ergonomics and Safety*. Amsterdam: IOS Press, pp. 104–107.
LEONARD, S.D., KARNES, E.W., and SCHNEIDER (1988) Scale values for warning symbols and words. In *Trends in Ergonomics/Human Factors*, Vol. V. New York: North-Holland, pp. 669–674.
LEONARD, S.D., MATTHEWS, D., and KARNES, E.W. (1986) How does the population interpret warnings signals? *Proceedings of the Human Factors Society 30th Annual Meeting*. Santa Monica, CA: Human Factors Society, pp. 116–120.
MACBETH, S.A. and MORONEY, W.F. (1994) Development and evaluation of automobile symbols: the focus group approach. In *Proceedings of the Human Factors Society 38th Annual Meeting*. Santa Monica, CA: Human Factors Society, p. 978.
MAGURNO, A., WOGALTER, M.S., KOHAKE, J., and WOLFF, J.S. (1994) Iterative test and development of pharmaceutical pictorials. In *Proceedings of the 12th Triennial Congress of the International Ergonomics Association*, Vol. 4. Toronto: Human Factors Association of Canada, pp. 360–362.
MAIN, B.W., FRANTZ, J.P., and RHOADES, T.P. (1993) Do consumers understand the difference between 'flammable' and 'combustible'? *Ergonomics in Design*, July,14–17, 32.
MORRELL, R.W., PARK, D.C., and POON, L.W. (1990) Effects of labeling techniques on memory and comprehension of prescription information in young and old adults. *Journal of Gerontology*, 45, 166–172.
MORRIS, L.A. and KANOUSE, D.E. (1981) Consumer reactions to the tone of written drug information. *American Journal of Hospital Pharmacy*, 38, 667–671.
MORROW, D.G., LEIRER, V.O., and ANDRASSY, J.M. (1996) Using icons to convey medication schedule information. *Applied Ergonomics*, 27, 267–275.

MURRAY, L.A., MAGURNO, A.B., GLOVER, B.L., and WOGALTER, M.S. (1998) Prohibitive pictorials: evaluations of different circle–slash negation symbols. *International Journal of Industrial Ergonomics*, 22, 473–482.

NELSON, D.L. (1979) Remembering pictures and words: appearance, significance, and name. In CERMAK, L.S. and CRAIK, F.I.M. (eds), *Levels of Processing in Human Memory*. Hillsdale, NJ: Lawrence Erlbaum Associates.

PAIVIO, A. (1971) *Imagery and Verbal Processes*. New York: Holt, Rinehart and Winston.

PAIVIO, A. (1986) *Mental Representations: A Dual Coding Approach*. New York: Oxford University Press.

PATTERSON, R.D. and MILROY, R. (1980) *Auditory Warnings on Civil Aircraft: The Learning and Retention of Warnings*. Civil Aviation Authority Contract 7D/S/0142. Cambridge: MRC Applied Psychology Unit.

PENNEY, C.G. (1989) Modality effects and the structure of short-term verbal memory. *Memory and Cognition*, 17, 398–422.

RACICOT, B.M. and WOGALTER, M.S. (1995) Effects of a video warning sign and social modeling on behavioral compliance. *Accident Analysis and Prevention*, 27, 57–64.

RASHID, R. and WOGALTER, M.S. (1997) Effects of warning border color, width, and design on perceived effectiveness. In DAS, B. and KARWOWSKI, W. (eds), *Advances in Occupational Ergonomics and Safety*, Vol. II. Louisville, KY: IOS Press, and Ohmsha, pp. 455–458.

RILEY, M.W., COCHRAN, D.J., and BALLARD, J.L. (1982) An investigation of preferred shapes for warning labels. *Human Factors*, 24, 737–742.

RINGSEIS, E.L. and CAIRD, J.K. (1995) The comprehensibility and legibility of twenty pharmaceutical warning pictograms. In *Proceedings of the Human Factors Society 39th Annual Meeting*. Santa Monica, CA: Human Factors Society, pp. 974–978.

RUMELHART, D.E. (1980) Schemata: the building blocks of cognition. In SPIRO, R., BRUCE, B.C., and BREWER, W.F. (eds), *Theoretical Issues in Reading Comprehension*. Hillsdale, NJ: Lawrence Erlbaum Associates.

SANDERS, M.S. and MCCORMICK, E.J. (1993) *Human Factors in Engineering and Design*, 7th Edn. New York: McGraw-Hill.

SILVER, N.C., LEONARD, D.C., PONSI, K.A., and WOGALTER, M.S. (1991) Warnings and purchase intentions for pest-control products. *Forensic Reports*, 4, 17–33.

SOJOURNER, R.J. and WOGALTER, M.S. (1997) The influence of pictorials on evaluations of prescription medication instructions. *Drug Information Journal*, 31, 963–972.

SOJOURNER, R.J. and WOGALTER, M.S. (1998) The influence of pictorials on the comprehension and recall of pharmaceutical safety and warning information. *International Journal of Cognitive Ergonomics*, 2, 93–106.

TULVING, E. and THOMPSON, D.M. (1973) Encoding specificity and retrieval processes in episodic memory. *Psychological Review*, 80, 352–373.

URSIC, M. (1984) The impact of safety warnings on perception and memory. *Human Factors*, 26, 677–682.

USPC (1997) *USP Pictograms*. Rockville, MD: United States Pharmacopeial Convention, Inc.

VAUBEL, K.P. (1990) Effects of warning explicitness on consumer product purchase intentions. In *Proceedings of the Human Factors Society 34th Annual Meeting*. Santa Monica, CA: Human Factors Society, pp. 513–517.

VAUBEL, K.P. and BRELSFORD, J.W. (1991) Product evaluations and injury assessments as related to preferences for explicitness in warnings. In *Proceedings of the Human Factors Society 35th Annual Meeting*. Santa Monica, CA: Human Factors Society, pp. 1048–1052.

VIGILANTE JR., W.J. and WOGALTER, M.S. (1996) The ordering of safety warnings in product manuals. In MITAL, A., KRUEGER, H., KUMAR, S., MENOZZI, M., and FERNANDEZ, J. (eds), *Advances in Ergonomics and Safety*, Vol. 1 (2). Cincinnati, OH: International Society for Occupational Ergonomics and Safety, pp. 717–722.

VON RESTORFF, H. (1933) Uber die Wirking von Bereichsbildungen im Spurenfeld. *Psychologie Forschung*, 18, 299–342.

WESTINGHOUSE ELECTRIC CORPORATION (1981) *Product Safety Label Handbook.* Trafford, PA: Westinghouse Printing Division.

WOGALTER, M.S., ALLISON, S.T., and MCKENNA, N.A. (1989) The effects of cost and social influence on warning compliance. *Human Factors*, 31, 133–140.

WOGALTER, M.S. and BRELSFORD, J.W. (1994) Incidental exposure to rotating warnings on alcoholic beverage labels. In *Proceedings of the Human Factors and Ergonomics Society 38th Annual Meeting.* Santa Monica, CA: Human Factors and Ergonomics Society, pp. 374–378.

WOGALTER, M.S., BRELSFORD, J.W., DESAULNIERS, D.R., and LAUGHERY, K.R. (1991) Consumer product warnings: the role of hazard perception. *Journal of Safety Research*, 22, 71–82.

WOGALTER, M.S., FREDERICK, L.J., MAGURNO, A.B., and HERRERA, O.L. (1997a) Connoted hazard of Spanish and English warning signal words, colors, and symbols by native Spanish language users. In *Proceedings of the 13th Triennial Congress of the International Ergonomics Association (IEA'97)*, Vol. 3, pp. 353–355.

WOGALTER, M.S., GODFREY, S.S., FONTENELLE, G.A., DESAULNIERS, D.R., ROTHSTEIN, P., and LAUGHERY, K.R. (1987) Effectiveness of warnings. *Human Factors*, 29, 599–612.

WOGALTER, M.S., KALSHER, M.J., FREDERICK, L.J., MAGURNO, A.B., and BREWSTER, B.M. (1998) Hazard level perceptions of warning components and configurations. *International Journal of Cognitive Ergonomics*, 2, 123–143.

WOGALTER, M.S., KALSHER, M.J., and RACICOT, B.M. (1993) Behavioral compliance with warnings: effects of voice, context, and location. *Safety Science*, 16, 637–654.

WOGALTER, M.S., LAUGHERY, K.R., and BARFIELD, D.A. (1997c) Effect of container shape on hazard perceptions. In *Proceedings of the Human Factors and Ergonomics Society 41st Annual Meeting.* Santa Monica, CA: Human Factors and Ergonomics Society, pp. 390–394.

WOGALTER, M.S. and RASHID, R. (1998) A border surrounding a warning sign affects looking behavior: a field observational study. Poster presented at the Human Factors and Ergonomics 42nd Annual Meeting (October, Chicago, IL). In *Proceedings of the Human Factors and Ergonomics Society 42nd Annual Meeting.* Santa Monica, CA: Human Factors and Ergonomics Society, in press.

WOGALTER, M.S. and SILVER, N.C. (1990) Arousal strength of signal words. *Forensic Reports*, 3, 407–420.

WOGALTER, M.S. and SILVER, N.C. (1995) Warning signal words: connoted strength and understandability by children, elders, and non-native English speakers. *Ergonomics*, 38, 2188–2206.

WOGALTER, M.S., SOJOURNER, R.J., and BRELSFORD, J.W. (1997b) Comprehension and retention of safety pictorials. *Ergonomics*, 40, 531–542.

WOGALTER, M.S. and YOUNG, S.L. (1991) Behavioural compliance to voice and print warnings. *Ergonomics*, 34, 79–89.

WOLFF, J.S. and WOGALTER, M.S. (1993) Test and development of pharmaceutical pictorials. In *Proceedings of Interface 93.* Santa Monica: Human Factors Society, pp. 187–192.

WOLFF, J.S. and WOGALTER, M.S. (1998) Comprehension of pictorial symbols: effects of context and test method. *Human Factors*, 40, 173–186.

WRIGHT, P. (1985) Editing: policies and processes. In DUFFY, T.M. and WALLER, R. (eds), *Designing Usable Texts.* Orlando: Academic Press, pp. 63–96.

WRIGHT, P., CREIGHTON, P., and THRELFALL, S.M. (1982) Some factors determining when instructions will be read. *Ergonomics*, 25, 225–237.

YOUNG, S.L., LAUGHERY, K.R., WOGALTER, M.S., and LOVVOLL, D. (1998) Receiver characteristics in safety communications. In KARWOWSKI, W. and MARRAS, W.S. (ed.), *The Occupational Ergonomics Handbook*, Boca Roton, FL: CRC Press, pp. 693–706.

YOUNG, S.L. and WOGALTER, M.S. (1990) Comprehension and memory of instruction manual warnings: conspicuous print and pictorial icons. *Human Factors*, 32, 637–649.

ZWAGA, H.J.G. (1989) Comprehensibility estimates of public information symbols: their validity and use. In *Proceedings of the Human Factors Society 33rd Annual Meeting.* Santa Monica, CA: Human Factors Society, pp. 979–983.

CHAPTER NINE

Attitudes and Beliefs

DAVID M. DEJOY

University of Georgia

This chapter examines the contribution of attitudes and beliefs to warning effectiveness. Warnings attempt to influence precautionary intent, and precautionary intent is largely a function of the individual's expectations about possible consequences. Adopting a general value–expectancy framework, this chapter organizes research on attitudes and beliefs into three broad categories: threat-related expectations, outcome-related expectations, and receiver characteristics. Some of the most robust findings in the warnings literature, namely those involving perceived hazardousness, familiarity, and costs of compliance, essentially parallel predictions offered by value–expectancy theory. The results of this review suggest that attitude and belief factors (expectations) broadly influence how the individual will approach and interact with virtually any hazardous situation. Although less thoroughly researched, there is also evidence that well designed warnings can be effective in altering expectations under some circumstances. However, the task of motivating precautionary behavior is especially difficult because the receiver must first attend to and process the warning before there will be any chance that it will alter his or her expectations. For the most part, receptivity to safety-related information and guidance is a function of prior expectations. Warning features that influence perceived threat either directly or indirectly show the most promise for altering expectations.

9.1 INTRODUCTION

The basic communication-human information processing model (C-HIP) of the warnings process outlined in Chapter 2, by Wogalter, DeJoy and Laughery, suggests that the effectiveness of a warning is determined by the success at each stage of the model. If, for example, a warning is not attended to, it will not be processed any further. However, once a warning has been attended to and understood, the next major stage in the sequence concerns attitudes and beliefs (see Chapter 2, Figure 2.1). Beliefs refer to convictions about phenomena or objects that are accepted as true (regardless of actual truth), and often beliefs are viewed as the building blocks of attitudes. Rokeach (1966, p. 529) defines attitudes as: 'a relatively enduring organization of beliefs about an object or situation predisposing one to respond in some preferential manner.' Because of their similarity, attitudes and beliefs are grouped into a single stage in the current model. This stage of processing has not

Table 9.1 Percentages of subjects who noticed, read, and complied with warnings in four studies.

Study	Noticed	Read	Complied
Frantz and Rhoades (1993)	57%	42%	28%
Friedmann (1988)	88	46	27
Otsubo (1988)	64	39	26
Strawbridge (1986)	91	77	37

received as much empirical attention as the preceding stages, but it is apparent that people's attitudes and beliefs can have powerful effects on whether a warning will be effective.

Borrowing loosely from social-cognitive theory (e.g., Bandura, 1986), this chapter introduces the term 'expectation' to help organize thinking about the contribution of attitudes and beliefs to warning effectiveness. Expectations are the anticipatory outcomes of behavior, and attitudes and beliefs are important ingredients in the formation of expectations. Referring to the basic model, a warning should help to alter the individual's expectations about the hazardousness of a particular product, object, or activity. Warnings attempt to influence precautionary intent, and precautionary intent is largely a function of the person's expectations about possible outcomes or consequences.

First and foremost, warnings are devices for communicating risk, and risk communication is basically an interactive process (National Research Council, 1989). A well designed warning label or message should influence the expectations of the person who attends to it, but the interactive nature of the process suggests that considerable importance should also be attached to what the person brings to the situation and to the way in which the person interacts with warnings and related materials. Returning to the basic information processing model, it is important to recognize that expectations also can affect processing at earlier stages of the model. For example, an individual who does not believe that something is hazardous may not look for, attend to, or read a warning message. Moreover, a number of studies indicate that many people who notice and/or read a warning still fail to comply with it (see Table 9.1). It is quite likely that the flow of information through the model is not entirely linear; at a minimum, there are feedback loops at each stage of the model. Some might assign an even larger role to attitudes and beliefs, and argue that these factors broadly determine how the individual will approach and respond to warnings and related communications. This general topic will be returned to later in the chapter.

Introducing the concept of expectations in the present chapter also provides a potentially useful bridge to theory and research on health behavior and compliance with medical regimens. The literature on both of these is quite extensive. Most theoretical models of health behavior, such as the health belief model, the theory of reasoned action, and protection motivation theory, are derived from value–expectancy theory (for general reviews of these models, see Wallston and Wallston, 1984; Weinstein, 1993). Value–expectancy theory holds that people estimate the seriousness of risk, evaluate the costs and benefits of various actions, and then choose a course of action that will maximize the expected outcome (Cleary, 1987). Inherent in this perspective is the assumption that the person's subjective evaluation of the situation is more important than the objective level of risk.

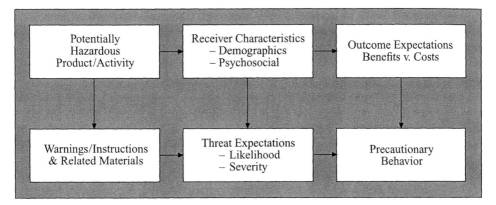

Figure 9.1 Schematic of the warnings process from a value–expectancy perspective.

Some of the most robust findings in the warnings literature, namely those related to perceived hazardousness, familiarity, and costs of compliance, essentially parallel predictions offered by value–expectancy theory (DeJoy, 1989, 1991). When a person is confronted with some type of warning message, he or she must make a decision to follow or not follow the advice provided. It seems entirely logical to assume that this decision-making involves the weighing of associated costs and benefits. This general rational-economic perspective has been applied also to the analysis of pharmaceutical and food labeling (e.g., Viscusi, Magat, and Huber, 1986; Viscusi, 1994). Figure 9.1 provides a general schematic diagram of the warnings process from a value–expectancy perspective.

This chapter organizes research on attitudes and beliefs into three broad categories: threat-related expectations, outcome-related expectations, and receiver characteristics. Threat-related expectations emphasize the individual's beliefs about the potential negative consequences associated with a particular hazard. Outcome-related expectations center around the individual's beliefs about the effectiveness of recommended precautions and the costs or barriers associated with performing these actions. For present purposes, receiver factors include any aspect or attribute of the person which facilitates or hinders self-protective behavior, including necessary skills and access to resources required to attain the behavior. Receiver characteristics contribute to the formation of expectations, either directly or indirectly. For example, users who are highly experienced in using a particular product bring a well developed set of expectations to the product use situation. This experience or familiarity can be expected to influence both threat and outcome expectations in important ways. All three categories of factors emphasize that precautionary behavior is influenced by the individual's subjective perceptions of the hazard and its associated consequences.

9.2 THREAT-RELATED EXPECTATIONS

It is generally thought that expectations about adverse consequences and the desire to avoid them is what provides the motivation for precautionary behavior. Risk perception researchers usually conceptualize perceived threat as consisting of two principal dimensions: likelihood and severity. Likelihood refers to the perceived probability of experiencing some type of adverse consequence (injury, illness, etc.); severity pertains to the perceived seriousness of the consequence in question (e.g., Lowrance, 1980;

Slovic, Fischhoff, and Lichtenstein, 1979). These two dimensions are thought to interact multiplicatively, in that a high probability of injury may not necessarily lead to precautionary behavior if the severity of anticipated injury is minimal. At the other extreme, perceived likelihood may be the primary dimension. The relative importance of the two dimensions can be expected to vary across hazards.

This section discusses research in two areas. The first area includes warning-related studies on perceived hazardousness. Perceived hazardousness is the term used most often in the warning literature to connote the threat posed by a particular product, object, or activity. Perceived hazardousness has been used as both an independent and dependent variable. In the former case, studies have examined the effects of perceived hazardousness on various measures of warning effectiveness. When treated as a dependent variable, interest has been with assessing the ability of various warning messages to influence judgements of perceived hazardousness. The second area of research deals with risk perception more generally, particularly the potential sources of bias that may distort how people evaluate and respond to various hazards.

9.2.1 Perceived Hazardousness

Perceived hazardousness and warning effectiveness

One of the strongest and most robust findings in the warnings literature is that warning effectiveness increases with the perceived hazardousness of the product. As can be seen in Table 9.2, supportive results have been obtained for a variety of outcome measures, including willingness to read warnings, warning recall, need for warnings, intent to comply, and behavioral compliance.

In terms of the likelihood × severity interaction discussed above, perceived severity appears to be the most salient dimension that people use in making judgements of perceived threat with respect to a broad array of consumer products (Wogalter, Brelsford, Desaulniers, and Laughery, 1991: Wogalter, Brems, and Martin, 1993a). Wogalter and colleagues (1991, Study 1) asked subjects to rate 72 generically named consumer products in terms of hazardousness, injury severity, familiarity, and several other dimensions. They found that perceived likelihood and perceived severity were both correlated with willingness to read warnings ($r^2 = -0.64$ and 0.89, respectively). However, regression analysis showed that severity by itself accounted for 80% of the variance in willingness to read warnings. The addition of likelihood to the model provided virtually no improvement in explanatory power. Including a severity × likelihood interaction term also failed to boost prediction. A follow-up study (1991, Study 2) found essentially similar effects using perceived hazardousness as the criterion variable. When subjects were asked to generate injury scenarios for various products (1991, Study 3), the severity of the first generated scenario was most predictive of perceived hazard ($r^2 = 0.81$).

The apparent importance of perceived severity has important implications for warnings design and practice. These findings suggest that warnings should emphasize how severely a person may be injured if recommended safeguards are not followed. Still, the inconsistency of these findings with other data in the broader risk perception literature is puzzling. Risk perception researchers have tended to emphasize the probabilistic nature of events, and perceived likelihood is often considered to be the primary determinant of perceived threat or risk (e.g., Slovic, Fischhoff, and Lichtenstein, 1980). Young, Wogalter, and Brelsford (1992) conducted an interesting study to try to resolve this inconsistency. Subjects were asked to rate either a broad set of consumer products that had been used in

Table 9.2 Summary of findings for perceived hazardousness.

Study	Product(s)	Principal effects
Beltramini (1988)	Cigarettes	Hazardousness increased warning believability
Donner and Brelsford (1988)	12 consumer products	Hazardousness increased accident scenarios, degree of precaution, and likelihood of reading warnings
Frantz (1994)	Water repellant sealer	Hazardousness increased warning reading and compliance
Friedmann (1988)	Liquid drain opener/wood cleaner	Hazardousness increased warning reading, recall, and compliance
Godfrey et al. (1983)	8 household products	Hazardousness increased likelihood to look for warnings
LaRue and Cohen (1987)	12 consumer products	Hazardousness increased willingness to read and need for warnings
Leonard et al. (1986)	12 hazards and consumer products	Hazardousness increased compliance likelihood and need for warnings
Otsubo (1988)	Circular saw/jig saw	Hazardousness increased warning reading and compliance
Silver et al. (1991)	22 pest control products	Hazardousness increased willingness to read warnings
Wogalter et al. (1987)/ Rating Experiment 2	25 hazard warning signs	Hazardousness increased perceived effectiveness of warnings
Wogalter et al. (1991)/ Study 1	72 consumer products	Hazardousness increased willingness to read warnings and need for warnings
Wogalter et al. (1991)/ Study 3	18 consumer products	Hazardousness increased precautionary intent
Wogalter et al. (1993a)/ Experiment 2	18 categories of consumer products	Hazardousness increased precautionary intent
Young, Brelsford, and Wogalter (1990)	72 consumer products	Hazardousness increased cautious intent and willingness to read warnings

previous warnings-related research (Wogalter et al., 1991) or the set of technologies and activities used by Slovic and associates (1979). The two lists are contained in Table 9.3. Subjects rated each item according to hazardousness, likelihood, severity, cautious intent, likelihood of reading warnings, familiarity, control, and catastrophe. Differences between the two lists were found on all eight of these dimensions. Most notably, Slovic's items were perceived as having more severe and catastrophic consequences, as being less controllable, and as requiring greater caution. Separate regression models for the two lists revealed that perceived likelihood accounted for 86.6% of the variance for Slovic's list, while perceived severity explained 94.7% of the variance in Wogalter's list.

Table 9.3 Listing of the 72 Wogalter et al. (1991) products and the 30 Slovic et al. (1979) items.

Wogalter et al. (1991) products

Electrical

Battery alarm clock	electric carving knife	oscillating fan	steam iron	curling iron	electric food slicer
Photoflash unit	toaster oven	desk lamp	electric hedge trimmer	pocket calculator	transistor radio
Digital watch	flashlight	quartz/space heater	trash compactor	drip coffee maker	metal detector
Sewing machine	typewriter	electric blanket	microwave oven	sun lamp	vacuum cleaner

Chemical

Antacid	cake mix	kerosene	roasted peanuts	alcoholic beverage	cough medicine
Lacquer stripper	roll-on deodorant	applesauce	drain cleaner	milk	shampoo
Artificial sweetener	nonprescription diet aid	dried cereal	skin moisturizer	aspirin	eggs
Oven cleaner	soap	baby powder	household bleach	pesticide	suntan lotion

Non-electrical tools

Bicycle	garden shears	hunting knife	rake	binoculars	garden sprinkler
Inflatable boat	screwdriver	chain saw	gas outdoor grill	ladder	scuba gear
Clothesline	golf club	lawn mower	semi-automatic rifle	dart game	hammer
Life vest	wheel barrow	football helmet	hiking boot	ping pong table	wood splitter

Slovic et al. (1979) items

Bicycle	hunting	general (private) aviation	food preservatives	H.S. & college football
Food coloring	railroads	skiing	handguns	pesticide
Commercial aviation	X-rays	motorcycles	prescription antibiotics	spray can
Motor vehicles	nuclear power	contraceptives	fire fighting	home appliances
Mountain climbing	large construction	smoking	alcoholic beverage	lawn mower
Swimming	vaccinations	surgery	electric power	police work

From Young et al. (1992)

Looking back at the items in Table 9.3, the list of consumer products contains very few items which normally would be associated with catastrophic or fatal outcomes. By contrast, Slovic's list contains a number of such items. Young and colleagues concluded that severity is likely to be the more important dimension for most consumer products. However, when severity reaches a certain level, the only remaining uncertainty is the probability of the event. For outcomes involving certain death or devastating injury or illness, perceived likelihood may indeed be the primary dimension. Thus, if the topic is warnings in the broadest sense, then both likelihood and severity have important roles to play. Still, for many (perhaps most) consumer product situations, better results in terms of influencing threat expectations might be obtained by emphasizing the severity of possible consequences. The difficulty that most people have in comprehending probabilistic data (e.g., Lichtenstein, Slovic, Fischhoff, Layman, and Combs, 1978) and in differentiating between small probabilities (e.g., Desaulniers, 1991) also favors the use of severity information.

Warnings and perceived hazardousness

A question of considerable practical significance is the extent to which warnings can influence perceived hazardousness. In an early study, Ursic (1984) found that adding warnings to products decreased rather than increased perceived hazardousness. Ursic presented subjects with display boards for several hypothetical brands of bug killers and hair dryers. Some brands of both products contained warnings, while others did not. The products with safety warnings were rated as safer and more effective than those without warnings. This unexpected finding probably occurred because product brand was confounded with the warning manipulation. Having subjects make direct comparisons of similar products with and without warnings may have inadvertently raised questions about the safety of the brands that did not carry warnings. The presence of a warning may have been interpreted as reflecting greater carefulness and responsibility on the part of the manufacturer.

Other investigators have revisited this basic question with different results. Viscusi *et al.* (1986) found that any of three warnings on two household cleaning products increased precautionary intent relative to when these same products contained no warnings. Wogalter and Barlow (1990) also found that the presence of a warning label increased the rated hazardousness of the products. In addition, products with warnings conveying higher injury severity were judged to be more hazardous than those with warnings conveying lower injury severity. No effects were found for injury likelihood. A follow-up experiment examined behavioral compliance. Compliance was higher when a warning was present (66% versus 13%); severity also impacted compliance, but only for the low likelihood condition (81% versus 44% compliance for high versus low severity). Similar findings have been obtained for alcohol warnings. Bohannon and Young (1993) reported that the presence of warning labels in alcoholic beverage advertisements reduced the tendency of young adolescents to underestimate the level of personal risk associated with alcohol consumption (relative to the absence of warnings).

Some of the same warning attributes which have been shown to attract attention may also enhance hazard perception. For example, the presence of a signal word, relative to its absence, appears to increase perceived hazard (Wogalter, Jarrard, and Simpson, 1994a). However, the ability of different signal words to influence hazard perception is less clear. Leonard, Mathews, and Karnes (1986) found no reliable differences between the terms danger, warning, and caution. Wogalter and Silver (1990) compared danger, warning,

Table 9.4 Cigarette warning content in the United States.

Year	Warning content
1965	Caution: Cigarette Smoking May Be Hazardous to Your Health.
1969	Warning: The Surgeon General Has Determined That Cigarette Smoking Is Dangerous To Your Health.
1984	1. SURGEON GENERAL'S WARNING: Smoking Causes Lung Cancer, Heart Disease, Emphysema, and May Complicate Pregnancy. 2. SURGEON GENERAL'S WARNING: Quitting Smoking Now Greatly Reduces Serious Risks to Your Health. 3. SURGEON GENERAL'S WARNING: Smoking by Pregnant Women May Result in Fetal Injury, Premature Birth, and Low Birth Weight. 4. SURGEON GENERAL'S WARNING: Cigarette Smoke Contains Carbon Monoxide.

caution, and a number of other terms. Differences were found between danger and warning, but not between warning and caution. Lirtzman (1984) also reported that workers did not consistently assign a hazard level intermediate between danger and caution. More recently, Wogalter and colleagues (1994a) found that subjects perceived differences between extreme terms, such as 'note' versus 'danger' or 'lethal', but not between the frequently used danger, warning, and caution.

Research has examined the effects of using certain colors and shapes in warnings, and with adding symbols or icons to warnings. Industrial workers associated the colors red and yellow with greater levels of hazard than the colors green or blue (Bresnahan and Bryk, 1975). Pointed shapes, such as diamonds and triangles, appear to convey greater hazard than regular rectangles or circles (Jones, 1978; Riley, Cochran, and Ballard, 1982; Collins, 1983). Results have been mixed with respect to the use of icons and symbols. Friedmann (1988) found that products containing written warnings and symbols were perceived as more hazardous than those containing only a written warning. On the other hand, Wogalter and colleagues (1994a) found that adding a signal icon (exclamation point surrounded by a triangle) to the signal word, as recommended in several warning design guidelines (e.g., Westinghouse Electric Corporation, 1981; FMC, 1985; American National Standards Institute, ANSI, 1991) did not increase the level of perceived hazard.

The specific content of the warning can influence perceived hazardousness. Leonard *et al.* (1986) found that including information about consequences that might result from failure to obey the warning increased perceived risk and intent to comply. Wogalter and Barlow (1990) reported that providing information about injury severity increased judgements of hazard. Other studies suggest that the explicitness or concreteness with which consequences are stated can increase perceived hazard (Morris and Kanouse, 1981; Loring and Wiklund, 1988; Laughery, Rowe-Hallbert, Young, Vaubel, and Laux, 1991; Laughery, Vaubel, Young, Brelsford, and Rowe, 1993). In particular, the Laughery *et al.* (1993) findings suggest that providing explicit information about consequences may elevate perceived hazard by increasing perceptions of injury severity. Explicit warnings may help to construct vivid images of severe injury and danger.

Cigarette warnings in the US have been modified several times over the past 30 years in the direction of increasing the explicitness of the consequences of smoking (see Table 9.4).

Between 1965 and 1969, the word hazardous was replaced by the less ambiguous term danger. Four rotating warnings were introduced in 1984, and two of these are very explicit in terms of consequences. The relatively new alcoholic beverage warnings in the US are similar in their explicitness to current cigarette warning labels. Trends in Gallop poll responses (Viscusi, 1992) suggest that warning messages have helped to alter lung cancer risk perceptions in the general public, but it is virtually impossible to isolate the effects of warnings *per se* from the massive public education campaign directed at cigarette smoking.

Finally, there is scattered evidence that the location and length of the warning message may influence perceived hazardousness. Frantz (1994) found that subjects exposed to a product which included precautions in the directions for use rated the product as more hazardous than those who received the product with the precautions placed in a separate section. It is possible that including warnings within instructions serves to increase the personal relevance of the warning information (Wogalter, Racicot, Kalsher, and Simpson, 1993c). Wogalter *et al.* (1991) found that subjects expected warnings to be closer to the product for products that were judged to be hazardous. Other research suggests that longer and more detailed warnings may connote greater potential hazard. In their study of pesticide labels, Silver, Leonard, Ponsi, and Wogalter (1991) found that perceived hazardousness increased with several label characteristics, including the number of pests that the product would destroy, the number of chemical ingredients it contained, and the inclusion of such statements as 'Keep out of reach of children'.

9.3 BIASES IN RISK PERCEPTION

To respond appropriately to any hazard, the individual must have a reasonably accurate appreciation of the nature and magnitude of the risk involved. Unfortunately, an impressive amount of research indicates that people have great difficulty in perceiving, structuring, and processing information in complex decision-making situations. An important development in this area has been the identification of a number of mental rules or heuristics used by people to reduce the cognitive demands of decision-making in these situations (Tversky and Kahneman, 1974). These rules are valid and useful in many situations, but they can also lead to large and persistent biases in decision-making. Most of this research has involved rather abstract laboratory tasks (see Kahneman, Slovic, and Tversky, 1982), but attempts have been made to apply this work to consumer products and warnings (e.g., Fischhoff, 1977; DeJoy, 1987).

This section discusses three sources of bias that appear to be particularly relevant to warnings: overconfidence and optimism, availability, and suppression. Overconfidence relates to the fact that people tend to have excessive and unwarranted confidence in their interpretation of events. Related to this, people are also unrealistically optimistic in judging their personal risk and vulnerability to a wide variety of life events. Availability refers to the observation that judgements of personal risk can be biased by the extent to which certain events can be imagined or recalled. Suppression involves the tendency of people to selectively discount or ignore information that conflicts with pre-existing interpretations of risk.

9.3.1 Overconfidence and Optimism

In general, people do not seem to be very sensitive to their lack of knowledge about objects and events. Research shows that people are so confident in their perceived level of knowledge about various topics that they will readily accept highly disadvantageous bets based on their confidence estimates (e.g., Fischhoff, Slovic, and Lichtenstein, 1977). People also have great difficulty in personalizing risk. In many respects, people possess fairly accurate concepts of total societal risk (Lichtenstein et al., 1978); the problem is that they do not think that these risks apply to them personally.

When asked to make comparative judgments of personal risk, people consistently overestimate the likelihood of experiencing positive life events and underestimate the likelihood of experiencing negative events (e.g., Weinstein, 1980, 1987). For example, a number of studies have asked subjects to judge their personal driving skill and safety in comparison to the average driver or other drivers in their age and gender group. Approximately 75–90% of drivers consider themselves to be safer and more skillful than other drivers (e.g., Svenson, 1981). Almost everybody appears to be optimistically biased to some extent; however, optimism tends to be more pronounced in younger than older people, and may be particularly strong in young males, especially when they are asked to make judgments about skill-based activities such as driving (DeJoy, 1992).

More specific to warnings, several studies indicate that people are able to give fairly accurate overall estimates of the risk associated with various consumer products (Brems, 1986; Martin and Wogalter, 1989; Wogalter et al., 1993a). This means that their subjective estimates of injury frequencies correlate reasonably well with objective injury data, such as the data provided by the NEISS (National Electronic Injury Surveillance System). This database provides frequency estimates of emergency room injuries associated with consumer products based on a sample of 64 statistically representative hospitals in the USA.

Comparative risk ratings have been obtained in a couple of warning studies. Rethans (1979) found that subjects were generally optimistic when judging their comparative risk of injury from a variety of consumer products. More recently, Bohannon and Young (1993), in a study of alcoholic beverage advertisements, reported that adolescents (mean age = 13.6 years) were optimistic when judging their risk compared to adolescents in general, and that the inclusion of warnings in the advertisements diminished this optimism. The older subjects (mean age = 23.3 years) in the study did not make optimistic judgments of personal risk. Other warnings-related research indicates that males tend to be more confident than females when judging their ability to avoid hazards and to use products without experiencing adverse consequences (Friedmann, 1988; Vredenburgh and Cohen, 1993; Young, Martin, and Wogalter, 1989).

9.3.2 Availability

According to the availability heuristic, people judge the likelihood of an event by considering how readily the event can be imagined or recalled. In their classic study, Lichtenstein et al. (1978) found that subjects had a highly consistent subjective scale for judging the frequencies of 41 causes of death. However, this subjective ordering differed from the objective ordering reflected in public health statistics. In general, people tended to underestimate frequent causes of death and overestimate infrequent causes of death. Lichtenstein and colleagues reasoned that these differences reflect the use of the availability heuristic. While more frequent events are often easier to recall or imagine, sometimes less frequent events can be more available because they are more dramatic, vivid, or emotionally salient. Often these same events receive a disproportionate amount of media coverage. In this study, many of the overestimated causes involved dramatic and overreported events such as homicide, accidents, fires, floods, and tornadoes. Underestimated causes tended to be more mundane killers, such as asthma, emphysema, and diabetes, all of which typically claim large numbers of people one at a time.

The tendency to overestimate infrequent causes and to underestimate frequent causes has been observed also in the warnings literature (Wogalter et al., 1993a). Moreover, attempts to 'de-bias' subjects by having them generate accident scenarios or fault trees for various products (Brems, 1986; Wogalter et al., 1993a) generally have been unsuccessful. Lichtenstein and her colleagues also found that the availability bias was quite resistant to modification. Although Wogalter and colleagues found that forcing subjects to think about or analyze the hazard did not alter their frequency estimates, it did influence their ratings of precautionary intent. Subjects who spent the least amount of time making estimates gave lower ratings of precautionary intent. The very high correlation between precautionary intent and injury severity in this study ($r = 0.97$) suggests that expectations about injury severity may be an important determinant of how much caution one is likely to demonstrate.

It seems plausible to hypothesize that availability plays some role in the injury severity–perceived hazardousness–cautious intent sequence. Providing information about injury severity may itself influence availability. A warning that states that a particular tool, if mishandled, could cause amputation or severe loss of blood is likely to evoke vivid images and emotional responses in the reader. The relative ease with which people can imagine a serious or dramatic injury (availability) should influence the level of perceived hazard which, in turn, should influence how carefully they approach a particular product or activity.

9.3.3 Suppression

Suppression involves the tendency of people to discount or ignore selectively information that conflicts with an existing interpretation of a situation (Arkes and Harkness, 1980; Freedman and Sears, 1965). Arkes and Harkness reported that once subjects had arrived at a diagnosis of a problem, they became more likely to recognize false symptoms that were consistent with their diagnosis and less likely to recognize actual symptoms that were inconsistent with the diagnosis. In terms of warnings, it is reasonable to expect that people holding preconceived notions about the risk associated with a particular product or activity will ignore or misappraise new information if it is inconsistent with their current thinking. Although this source of bias has not been explored systematically by warning researchers, suppression may be a contributing factor to the well established familiarity effect: that warning effectiveness decreases as familiarity with the product, object, or activity increases (see Table 9.5 later for a summary of relevant findings).

However, when people do not hold strong views, the opposite situation exists, and they are at the mercy of how messages are framed. Under such circumstances, presenting the same information in different ways or formats can alter people's perspectives and actions significantly (Tversky and Kahneman, 1981). The content and configuration of warnings, especially the way in which risk information is represented, can be extremely important for those who are new to the situation. Consistent with dynamic (e.g., DeJoy, 1991) and levels of performance models of the warning process (e.g., Lehto and Papastavrou, 1993), people may be at different stages in terms of how they interact with and process warning messages.

9.4 OUTCOME-RELATED EXPECTATIONS

Outcome-related expectations emphasize the individual's beliefs about the effectiveness of recommended precautions and the costs or barriers associated with performing such behavior. From a value-expectancy perspective, perceived threat establishes the motivation to engage in precautionary behavior, but generally the actual decision to do so is thought to involve the weighing of expected costs and benefits. With respect to warning messages, the benefits of complying with the indicated precautions are evaluated against the costs associated with following the recommendations. The effectiveness of a warning message should increase if either the benefits of taking the action can be heightened or if the costs can be reduced. However, if a particular hazard is perceived as representing little threat, this cost–benefit analysis becomes moot. Logically, there can be few benefits associated with following the recommended precautions under such circumstances.

This section considers three topics pertinent to the formation of outcome expectations. First, warning research, which directly examines costs of compliance, is reviewed. Next, issues related to perceived effectiveness or expected utility are discussed. Third, self-efficacy is introduced as a potentially important consideration in warnings-related behavior. The principal argument for including self-efficacy is that people must be confident that they are capable of performing the specific type or types of behavior required to produce the desired outcome (Bandura, 1986). Warnings may be more or less effective depending on the individual's self-efficacy expectations.

9.4.1 Costs of Compliance

Several studies have examined how costs of compliance influences warning effectiveness (Wogalter, Godfrey, Fontenelle, Desaulniers, Rothstein, and Laughery, 1987; Wogalter, Allison, and McKenna, 1989; Dingus, Wreggit, and Hathaway, 1993). Results are quite consistent in showing that cost is an important factor in compliance. In one field experiment (Wogalter *et al.*, 1987), warnings were placed on a broken exit door in a campus building. The warning sign directed users to an adjacent door (low cost), or another set of doors 15 meters away (moderate cost), or a third set of doors 60 meters away (high cost). The results showed that the warning was almost always obeyed in the low cost condition (94%), but totally ignored in the high cost condition (0%). Wogalter *et al.*, (1989) produced similar effects in a chemistry laboratory simulation. When subjects could access the recommended protective equipment (mask and gloves) with little effort, compliance was quite high. By contrast, very few subjects were willing to go even into the next room to retrieve the protective equipment. Dingus *et al.* (1993) examined costs of compliance in recreational and consumer products contexts. Again, relatively small increments in cost produced large decrements in compliance.

On the positive side, these same studies also suggest that quite high rates of compliance can be achieved when costs are minimal. Specific to this point Dingus *et al.* (1993, Experiment 2) manipulated cost in terms of whether protective equipment (gloves and respirator masks) were enclosed in the product packaging for an 'industrial strength tile descaler.' Providing this equipment along with the product had a dramatic effect on compliance. In fact, slightly over 80% of subjects wore the gloves and 50% used the masks in the experimental condition in which there was no specially crafted warning message urging them to do so. Compliance rates were even higher when the safety equipment was combined with various warning message configurations. Racicot and Wogalter

(1995), in a recent study of video warnings, found that 85% of subjects who reported seeing the safety equipment (gloves and masks) complied with the warning to use them.

Compliance costs certainly are not limited to the amount of effort that must be expended to obtain needed safety equipment or to behave in a safe manner. Studies of helmet usage among ATV operators (e.g., Lehto and Foley, 1991), seat belt usage in automobiles (e.g., Fhaner and Hane, 1974), and personal protective equipment usage in industry (e.g., Acton, 1977; Cleveland, 1984), all suggest that virtually any type of discomfort, restriction of movement or freedom, or other encumbrance can serve as a barrier to compliance. The simple fact that behaving unsafely is sometimes more pleasurable or rewarding than behaving safely also qualifies as a cost of compliance (McDowell, 1988). All of these costs or barriers may find their way into the personal weighing of costs and benefits.

9.4.2 Perceived Effectiveness or Expected Utility

Beliefs or expectations about the effectiveness of recommended precautions have received very little direct attention in the warnings literature. Expected benefits are often expressed as the difference in beliefs about the likelihood and severity of injury or harm (a) assuming no change in behavior, and (b) assuming the performance of some self-protective action (Weinstein, 1993). The warning literature suggests that this basic cost–benefit tradeoff can be shifted in the direction of compliance by increasing the costs of *non-compliance*. Looking at (a) and (b) above, it is quite easy to see that increasing the costs of non-compliance serves to widen the difference between (a) and (b), and thus increases the benefits associated with following the indicated precautions.

One promising way to increase the costs of non-compliance is to use very explicit wording in the consequences portion of the message. Several warning studies show that providing explicit information about the nature and severity of possible injury serves to heighten a perceived hazard (Morris and Kanouse, 1981; Loring and Wiklund, 1988; Laughery *et al.*, 1991, 1993). When the level of perceived threat (and the motivation to comply) is increased, this almost automatically increases the perceived benefits of compliance. The only situation in which this would not happen is if the indicated precautions were not perceived as being effective in reducing the threat. Somewhat surprisingly, very few warning studies have asked subjects to rate the effectiveness of the actions or measures called for in the warnings.

Presumably, any warning attribute or design characteristic that increases perceived hazard might also alter beliefs about the cost of non-compliance. In a recent study, Wogalter, Barlow, and Murphy (1995) found that placing a warning so that it actually blocked use of the equipment resulted in better compliance among experienced users. The investigators speculated that this arrangement might have helped to create the belief that this was a particularly important warning. Interactive warnings increase the noticeability of warnings (e.g., Duffy, Kalsher, and Wogalter, 1993; Frantz and Rhoades, 1993) by disrupting ongoing or 'automatic' behavior, but certain types may also alter people's expectations about the costs of non-compliance.

9.4.3 Self-efficacy

Self-efficacy has gained a prominent place in research on health behavior (Strecher, DeVellis, Becker, and Rosenstock, 1986). According to Bandura (1977), behavior change and maintenance are a function of (1) expectations about the outcomes that will result from certain behavior, and (2) expectations about one's ability to perform the behavior. It is important to remember that self-efficacy reflects beliefs about capabilities rather than actual or true capabilities. Self-efficacy is also behavior and context specific. That is, it relates to performing specific behavior in particular situations; it is not an enduring characteristic or personality trait.

Bandura argues that self-efficacy affects all aspects of behavior, including the learning of new types of behavior, the inhibition of current behavior patterns, the effort and persistence devoted to tasks, and the individual's emotional reactions to various circumstances and situations. As such, self-efficacy also should influence how people respond to warning messages. More specifically, people with low self-efficacy expectations should be less likely to follow the recommendations contained in warnings. Self-efficacy has not been examined in any systematic fashion by warning researchers, but there is at least some evidence indicating that it may be an important factor and that people differ with respect to their perceptions about the effectiveness of their own precautionary behavior (Laux and Brelsford, 1989; Celuch, Lust, and Showers, 1998). This general lack of attention to self-efficacy is not entirely surprising, in that many warnings call for very simple, discrete actions, such as donning gloves or other protective equipment. Self-efficacy is more likely to be an important factor when the behavior indicated is complex, when it involves relatively high levels of knowledge and skill, or when the behavior must be performed and maintained for long periods of time. However, it is not that difficult to imagine situations in which self-efficacy might be quite important, such as warnings in occupational settings or warnings that accompany certain drugs or medical devices.

9.5 RECEIVER CHARACTERISTICS

The term 'receiver characteristics' is borrowed from communication theory (e.g., McGuire, 1980), and is used in this chapter to represent any relevant aspect or attribute of the person to whom the warning is directed. Viewing risk communication as an interactive process means that considerable importance is assigned to what the person brings to the situation and to how the person interacts with warnings and related materials. Referring back to Figure 9.1 for a moment, receiver factors are most likely to impact precautionary behavior by influencing the formation of threat and/or outcome expectations. A diverse array of demographic, psychosocial, personality, and other variables qualify as receiver characteristics. This portion of the chapter considers four categories of receiver factor. The first three categories include: familiarity/experience, demographic factors (gender and age), and personality, respectively. The last category is labeled 'other receiver characteristics', and discusses several issues that bear on the general competence of the receiver or target audience.

9.5.1 Familiarity

A number of studies have examined the effect of product familiarity and/or experience on various measures of warning effectiveness. Familiarity is typically defined in terms of the individual's personal knowledge of and/or experience with the product, object, or activity in question. Although many investigators do not make a distinction between familiarity and experience, it is worth noting that familiarity and experience are not necessarily

Table 9.5 Summary of findings for product familiarity: 'product use simulation' studies.

Study	Product(s)	Principal effects
Andrews, Netemeyer, and Durvasula (1991)	Alcohol beverages	Familiarity decreased believability of warnings
Beltramini (1988)	Cigarettes	Familiarity did not affect believability of warnings
Gardner-Bonneau et al. (1989)	Cigarettes	Familiarity increased warning recall
Godfrey et al. (1983)	8 household products	Familiarity decreased likelihood to look for warnings for the low hazard products
Goldhaber and deTurck (1988a)	Swimming pools	Familiarity decreased warning detection and compliance likelihood
Goldhaber and deTurck (1988b)	Swimming pools	Familiarity decreased perceived risk and compliance likelihood
Johnson (1992)	Scaffolding	Familiarity decreased likelihood of reading warnings
Karnes, Leonard, and Rachwal (1986)	All terrain vehicles	Familiarity decreased perceived risk
LaRue and Cohen (1987)	12 consumer products	Familiarity decreased willingness to read and need for warnings
Otsubo (1988)	Circular saw and jig saw	Familiarity decreased warning reading and compliance
Wogalter et al. (1991, Study 1)	72 consumer products	Familiarity decreased willingness to read warnings and need for warnings
Wogalter et al. (1993a) Experiment 2	18 categories of consumer products	Familiarity decreased precautionary intent
Wogalter et al. (1995)	Computer disk drive	Familiarity decreased warning compliance
Wright (1982)	60 consumer products	Familiarity decreased likelihood of reading instructions

identical concepts. People may have some familiarity with products or classes of product that they seldom or never use themselves. In this regard, Wogalter et al. (1991) noted that ratings of product familiarity for a variety of consumer products were correlated only 0.66 with frequency of use and 0.28 with time of contact.

Studies involving product familiarity generally have taken one of two forms. First, a number of studies have employed some type of product use simulation: that is, subjects were asked (1) to interact with specific products and/or related warning messages; (2) to image themselves using or purchasing specific products, or categories of products; or (3) to indicate their general reactions and likely responses to specific products or product categories. The results from a number of these studies are summarized in Table 9.5. As

Table 9.6 Summary of findings for product familiarity: 'awareness or impact surveys.'

Study	Product(s)	Principal effects
Graves (1993)	Alcohlic beverages	Consumption/use increased awareness of warnings
Greenfield and Kaskutas (1993)	Alcoholic beverages	Consumption/use increased detection and recall of warnings
Kaskutas and Greenfield (1992)	Alcoholic beverages	Consumption/use increased detection and recall of warnings
MacKinnon, Pentz, and Stacy (1993)	Alcoholic beverages	Consumption/use increased awareness of labelling law and exposure to and memory of warnings
MacKinnon and Fenaughty (1993)	Cigarettes/smokeless tobacco/alcoholic beverages	Consumption/use increased recognition memory of warnings
Marin (1994)	Alcoholic beverages/ cigarettes	Consumption/use increased awareness of warnings
Mayer et al. (1991)	Alcoholic beverages	Consumption/use increased awareness of warnings
Mazis et al. (1991)	Alcoholic beverages	Consumption/use increased awareness and knowledge of warnings
Patterson et al. (1992)	Alcoholic beverages	Consumption/use decreased perceived risk

can be seen in this table, people who are more familiar with a product are less likely to look for, read, and comply with warnings.

The second type of study might best be described as general awareness or impact surveys. Some of these studies have involved large scale, population-based surveys of the general public (e.g., Mazis, Morris, and Swasy, 1991; Kaskutas and Greenfield, 1992), while others have focused on specific target or at-risk subgroups such as adolescents (Patterson, Hunnicutt, and Stutts, 1992), minority group members (e.g., Marin, 1994), or those affiliated with specific religious groups (e.g., Mayer, Smith, and Scammon, 1991). Primary concern in these surveys has been with assessing the impact or 'reach' of US government mandated warnings associated with alcohol and tobacco products (see Table 9.6). Typically respondents were asked about their awareness of labeling legislation, whether they have ever seen warning labels on alcohol or tobacco products or advertising, and if so whether they remember the content of the warning messages. The results of these studies also are quite consistent: people who use these products or who use them more frequently or heavily are more likely to be aware of the warnings and to be able to recall portions of the message (see Table 9.6). For the most part, these findings are probably a function of exposure. People who do not consume or serve alcoholic beverages are not likely to come into contact with the warning labels. Another possible explanation brings in the dimension of personal relevancy (Stewart and Martin, 1994). Consumers may not pay attention to certain warnings because the warnings are not of interest to them—they are not personally relevant.

The findings from these two groups of studies are quite easy to reconcile. It is hardly surprising that those who drink alcoholic beverages are more likely to have seen the warnings and are more able to recall their content. Conducting a large-scale survey on almost any consumer product (e.g., a power lawn mower) would show that product users are probably more aware of associated warnings and hazards than non-users. However, if these same people were brought into the laboratory and told to use the product in question, non-users would be more likely to read and follow warning messages. In general, familiar or experienced users can be expected to be more knowledgeable about product hazards and more confident in using the product in question. It also follows that this knowledge and confidence also means that they will probably not be motivated to seek additional information about the product. The inexperienced user, by contrast, is more likely to be in an active or information-seeking mode.

Benign experience is one obvious explanation for the familiarity effect. As people use a product without incurring any safety problems, they quite naturally become less concerned about its dangers and more confident in using the product. Findings are generally consistent with this explanation. Otsubo (1988), in her study of circular saw and jig saw warnings, found that those who read the warning were less likely to have had prior experience with the tool and were less confident in using it. She also found that those who complied with the warning had less experience with tools in general, had less experience with similar types of saws, were less confident in using the saw, and were more likely to have been injured using a similar tool. Goldhaber and deTurck (1988a,b), in two studies of swimming pool warnings, found that those who were more familiar with swimming pools were less likely to notice a warning about not diving into the pool. In addition, those with a history of performing the prohibited behavior perceived less danger in the behavior and were more likely to ignore the warning's recommendation in the future.

Script theory also has been offered as an explanation for the familiarity effect (Wogalter et al., 1995). According to script theory (e.g., Schank and Abelson, 1977), a large amount of human behavior is essentially automatic and occurs with little conscious thought. As such, it is reasonable to hypothesize that users who are quite practiced in using a particular product rely on scripts stored in memory, and devote little attention to warning labels, instructions, or other materials that accompany the product. Information seeking and more deliberate action occurs when scripts are not readily available.

Habituation may also contribute to the familiarity effect. It follows that the ability of even a well designed warning message to attract attention is likely to degrade as a function of repeated exposure. People simply become accustomed to seeing the warning and it becomes part of the background. It is also possible that people might actively ignore messages which they find to be highly intrusive and bothersome (Stewart and Martin, 1994). In an interesting qualitative study of coal mine warnings, Mallett, Vaught, and Brnich (1993) found that frequent false alarms were one possible reason why miners were slow to respond when the fire warning system was activated in the mine. Indeed, the sheer prevalence of warning messages in today's environment and the trend to standardize their appearance and content may actually contribute to the problem of habituation. A number of studies have addressed various ways to counteract habituation, including the use of technology (e.g., Wogalter, Racicot, Kalsher, and Simpson, 1994b; Racicot and Wogalter, 1995), interactive warnings (e.g., Hunn and Dingus, 1992; Frantz and Rhoades, 1993; Duffy et al., 1995), warning personalization (e.g., Wogalter et al., 1994b), and multi-modal warnings (Wogalter and Young, 1991; Wogalter, Kalsher, and Racicot, 1993b).

All of these explanations suggest that experienced users may be unlikely to attend to or comply with warnings. By extension, experienced users might also be less likely to attend to warnings when they switch products within a category of familiar products. This is a potentially serious problem since new versions of familiar products may be more dangerous than older versions, and for some products specific hazards may be detected only after a particular product has been in use for several years or even longer. Two studies bear directly on this issue. Morris, Mazis, and Gordon (1977) found that approximately 78% of women read the patient package insert when they first started using oral contraceptives, but less than 11% read the insert that accompanied subsequent prescription refills. Godfrey and Laughery (1984) surveyed women about tampons and toxic shock syndrome and found that of the women who used tampons, 73% said they noticed a warning on or in the package. By contrast, only 32% of those switching from one tampon product to another noticed a warning on the new product. Furthermore, older women (aged 27 and older) were less likely to have noticed a warning (50%) than younger women (78%).

To a considerable extent, perceived hazardousness may be an intervening mechanism for explaining the familiarity effect (Laughery and Brelsford, 1991). In other words, familiar users are less likely to comply with warnings because they perceive the product or situation as less hazardous. Several studies have examined the effects of both perceived hazardousness and familiarity (e.g., LaRue and Cohen, 1987; Otsubo, 1988; Wogalter et al., 1991, 1993a). Two of these studies, LaRue and Cohen (1987) and Wogalter and associates (1991, Study 1), provided correlation matrices. In both studies, perceived hazardousness/danger was more strongly correlated with the need for warnings than was familiarity ($r = 0.89$ and 0.95, respectively for hazardousness/danger and -0.63 and -0.62 for familiarity). There was also a fairly substantial intercorrelation between hazardousness and familiarity ($r = -0.59$ and -0.63), indicating that as people become more familiar with a product, they perceive it as being less hazardous. Using regression analysis, Wogalter and colleagues (1991) found that familiarity added very little explanatory power beyond that offered by perceived hazardousness. While the warnings literature indicates that both perceived hazardousness and familiarity influence how people approach and respond to warnings, perceived hazardousness may be the factor that motivates information seeking and precautionary action in most situations.

9.5.2 Demographic Factors

Gender

The literature on gender differences and warnings is not very extensive, but available findings suggest that warnings are likely to be more effective with females than males. For example, females appear to be more likely to look for and read warnings on products (LaRue and Cohen, 1987; Godfrey, Allender, Laughery, and Smith, 1983). In the Godfrey and colleagues study, gender interacted with product type. Male–female differences were essentially confined to the products that were judged to be hazardous by both sexes. Females also may be more likely to take appropriate actions in response to warnings (Viscusi et al., 1986; Goldhaber and deTurck, 1988b; Desaulniers, 1991; Vredenburgh and Cohen, 1993).

Gender differences also have been noted on measures related to perceived hazard and confidence. Young et al. (1989) found that for products judged to be relatively hazardous by both sexes, males assigned lower hazard ratings than females. Males were also more confident in 'knowing all the hazards' related to these products. Friedmann (1988) also found gender differences for the two products in her study: liquid drain opener and wood cleaner. Males were more confident in using both products than females. Males and females did not differ in the level of hazard assigned to the drain opener, but females considered the wood cleaner to be much more hazardous than did males. Interestingly, this last effect occurred in the absence of significant gender differences in familiarity for the two products. Females also believed that accidents were more probable if protective equipment was not worn when using the products. Vredenburgh and Cohen (1993), in a study of snow skiing and scuba diving participants, found that female participants perceived these activities as more hazardous than did males. Males considered themselves to have more ability and to be more likely to engage in high risk activities. Vredenburgh and Cohen did not find that females were more likely to read warnings, but females did indicate that they more often complied with the warnings that they read.

At least one study suggests that there may be gender differences also in the interpretation of some safety related symbols. Laux, Mayer, and Thompson (1989) gave subjects a list of 19 types of precautionary behavior and asked them to select the behavior that they would follow in response to three different pictograms from the Westinghouse Electric Corporation (1981) *Product Safety Label Handbook*: 'electric shock' (wire shocking hand); 'poison' (skull-and-crossbones), and 'flammability' (flames). Other subjects were simply asked to list four precautions they would take in response to each symbol. The results showed that the poison symbol conveyed different information concerning possible modes of injury to males and females. Although males and females may not differ in their ability to recognize certain safety symbols, mere recognition does not necessarily mean that they have equivalent knowledge or that their knowledge is sufficient to avoid the hazards addressed by the symbol.

Age

Age as a variable of direct interest has not received much systematic attention by warning researchers. Several studies have investigated how different age groups view various signal words (Leonard, Hill, and Karnes, 1989; Silver and Wogalter, 1991; Silver, Gammella, Barlow, and Wogalter, 1993). Silver and Wogalter had elementary, middle school, and college students rate 43 potential signal words on a 'carefulness' scale: 'how careful would you be after seeing each term?' Although younger students generally assigned higher carefulness ratings than did the college students, the rank order of the words was consistent across the different groups of subjects. Table 9.7 contains mean carefulness ratings for a list of words that were known by 95% or more of the fourth and fifth graders. A follow-up study with elderly subjects (Silver *et al.*, 1993) showed that the rank ordering of terms did not differ appreciably from that noted in the other age groups. Leonard *et al.* (1989) gave subjects various hazard descriptions and asked them to assign appropriate signal words. The older subjects in their sample (older than 25) tended to assign signal words that conveyed more serious consequences, for example, 'deadly' versus 'danger.'

Wogalter *et al.* (1994) had high school students, college students, and shopping mall visitors (ages 21–80) rate actual labels from various household products. Overall, high school students gave significantly higher hazard ratings than college students, who in turn gave significantly higher ratings than mall participants. However, the rank ordering of hazards across the three groups was quite consistent. Younger subjects gave higher hazard

Table 9.7 Carefulness means and standard deviations of signal words that were known by 95% or more of the fourth and fifth graders.

Word[a]	Carefulness ratings	
	Mean	SD
NOTICE	5.39	2.50
NO	5.63	2.78
CAREFUL*	5.86	2.51
IMPORTANT	5.95	2.26
HOT*	6.00	2.59
NEVER	6.09	2.43
STOP*	6.11	2.58
DON'T	6.12	2.30
ALARM	6.16	2.35
RISKY	6.25	2.14
ALERT	6.45	2.02
UNSAFE	6.47	2.18
WARNING*	6.52	2.07
HAZARD	6.57	1.99
CAUTION	6.64	2.03
BEWARE*	6.66	2.04
HARMFUL*	6.74	2.09
SERIOUS	6.90	1.74
DANGER*	7.12	7.12
DANGEROUS	7.18	1.57
POISON*	7.49	1.65
EXPLOSIVE	7.51	1.44
DEADLY*	7.57	1.58

[a] Words with asterisks were known by 99% or more of the fourth and fifth graders. From Silver and Wogalter (1991).

ratings to the various signal words in the labels (consistent with Silver and Wogalter, 1991). They also rated the labels as less likely to capture their attention.

Silver *et al.* (1991) compared college students and middle-aged adults in terms of their willingness to read warnings on household pest control products and their likelihood of purchasing these products. For the willingness to read measure, perceived hazardous accounted for over 41% of the variance for students and about 25% for adults; warning understandability and warning attractiveness also made significant contributions to this model. Familiarity was the overwhelming predictor of purchase intentions for both groups. Although different factors appear to predict precautionary versus purchasing behavior, the two age groups were quite similar in their responses. Braun, Silver, and Stock (1992) also assessed willingness to read. They compared the responses of college students and older adults to detergent labels that varied according to font type, font size, and related legibility characteristics. The results showed no main effects or interactions involving age or gender.

Goldhaber and deTurck (1989) compared middle school and high school students in terms of their responses to 'no diving' signs at swimming pools. Age interacted with gender on the compliance likelihood measure. In essence, the presence of the warning signs did not alter the intentions of the middle school students. However, among the high school students, females were much less likely to engage in the prohibited behavior (diving into shallow water) when the sign was present; high school males were actually more likely to behave unsafely when the sign was present. This type of 'boomerang effect' has been observed also with respect to young people and alcohol warnings (Snyder and Blood, 1991) and anti-drug messages (Feingold and Knapp, 1977). In one other study that investigated compliance likelihood, Desaulniers (1991) found that people aged 40 and older showed stronger intentions to comply with warnings than did younger people.

Older people may have greater difficulty in comprehending warning messages. For example, two studies (Easterby and Hakiel, 1981; Collins and Lerner, 1982) suggest that older people have greater difficulty comprehending safety signs involving pictorials. Other research indicates that the elderly also have difficulty in interpreting correctly the instructions on their own prescription medications (Zuccollo and Liddell, 1985). Zuccollo and Lidell found that 60% of elderly patients (mean age about 79) had trouble reading their medication labels and only 23% had a clear understanding of all the instructions.

In summary, there is some indication that younger people may assign higher levels of hazard to various signal words and consumer products, but only minor differences have been noted in how different age groups rank order different signal words and hazards. Compliance-related data are conspicuously absent, although a couple of studies suggest that compliance likelihood may increase with age. Also, there is scattered evidence suggesting that older adults may experience some difficulty in comprehending certain warning symbols and messages. Given the paucity of research, these conclusions should be viewed as very tentative.

9.5.3 Personality

One personality factor, risk- or sensation-seeking, has been hypothesized to affect warning-related behavior (e.g., Ayres et al., 1989). Personality research indicates that people vary in their need for and tolerance of environmental stimulation (Zuckerman, 1979). Some people actively seek out new experiences and thrills, while others become distressed when their regular activities are even marginally disrupted. It follows that people who are high in sensation seeking may be less likely to comply with warnings; in fact, they may intentionally disregard some warnings in the interest of experiencing high levels of risk. In one relevant study, Purswell, Schlegel, and Kejriwal (1986) used discriminant analysis to classify subjects into safe and unsafe groups with respect to their use of four consumer products (sabre-saw, electric knife, router, and drain cleaner). Their results indicated that sensation seeking was a useful predictor for each of the products. However, the presentation of the results did not provide information on the relative importance of sensation-seeking to classification accuracy. Still, these are provocative results that deserve follow-up.

The possible contribution of sensation-seeking to warning effectiveness raises a number of interesting questions about the general role of personality in warnings. First, does sensation-seeking make a unique contribution to warning-related behavior beyond that explained by other receiver variables such as age and gender? Second, is this general propensity to take risks related to an aspect of personality or to differences in how people judge the risk associated with various products or activities? Young males may drive in a risky fashion because they are seeking thrills, or because they underestimate the level of risk in their actions and/or because they overestimate their driving skill. Also, Wagenaar

(1992) has argued that in many situations people 'run' risks rather than consciously or intentionally take risks. Third, does sensation-seeking affect the person's response to all warnings or just those associated with certain products or activities? For example, people high in sensation-seeking may be especially likely to ignore warnings associated with controllable or skill-based activities (e.g., driving) or recreational and leisure activities. Some have argued that people are willing to accept higher levels of risk for volitional or leisure-type activities (e.g., Starr, 1969).

9.5.4 Other Receiver Characteristics

There are a number of other receiver characteristics or factors that may influence warning-related behavior but which have not been examined systematically by warning researchers. Most of these characteristics fall under the general category of receiver competence (Laughery and Brelsford, 1991). For example, a variety of sensory and behavioral limitations have the potential to affect how people detect and/or respond to warnings. Some of these are quite obvious. A visually impaired person is going to have difficulty in detecting and reading standard print warnings. In other instances, compliance may be difficult or impossible because of the particular physical skills required or because indicated equipment or supplies are unavailable or unattainable.

A second category of factors involves the level of technical information conveyed in a warning message and the ability of the intended receiver to understand and respond appropriately. Warnings that accompany pharmaceuticals and other medical products, chemicals, and complex machinery and equipment often demand specialized knowledge for full comprehension. This problem may be exacerbated by the need to provide complete factual information in a small space and the desire to reach multiple target groups using a single message. The risk communication process can be complicated further by the presence of some type of 'learned intermediary.' In the case of prescription drugs, for example, different warning messages are often directed to the physician who prescribes the drug and the patient who ultimately uses it. This can complicate matters because there may be implicit assumptions about what the physician should or will do in the way of informing or educating the patient about the drug he or she is prescribing. The pharmacist or pharmacy is also an official intermediary in this context. Some states in the US now require dispensers to provide certain types of risk and safety information to consumers at the point of purchase.

The contribution of reading ability and language skills has received some attention in the warning literature. The general recommendation is that warnings should be written at the lowest level of the intended target audience. This can be a challenging requirement to meet when the target audience includes young children, people with limited education, and/or those with little or no English language proficiency. A recent study of alcohol and tobacco warnings (Marin, 1994) found that Hispanics displayed lower rates of awareness for these warnings when compared to findings for the general population. In addition, when the study sample was divided into high and low acculturation groups (largely on the basis of English fluency), the low acculturation group had lower levels of awareness than their more acculturated peers.

Writing warnings at the lowest level may not be beneficial in all situations. Silver *et al.* (1991) examined the readability of warning labels on pest control products. Readability was assessed by the Flesch index as modified by Gray (1975) and the Coleman and Liau (1975) index. Understandability made a significant contribution to willingness to read the warnings for both the college student and adult subjects. However, contrary to expectation, subjects were more likely to read warnings that contained more sentences/statements and that were written at higher grade levels. This may be not be an isolated finding. Mazis, Morris, and Gordan (1978), in a national survey of oral contraceptive users, also found that users preferred longer and more detailed messages. This preference was most evident for younger and more educated women.

Using symbols and pictorials and explicit statements in warning messages are two promising methods for reaching people with limited reading or language skills. Pictorials can be used to illustrate hazards, consequences, and the actions necessary to avoid or reduce particular hazards. Research findings have been somewhat mixed on the benefits of using pictorials in warning messages (e.g., Friedmann, 1988; Otsubo, 1988; Jaynes and Boles, 1990; Wogalter, Kalsher, and Racicot, 1992); however, most of these studies used college students or other participants with above average language skills. Pictorials may be especially useful when the target audience has limited ability to process textual material.

The use of explicit or concrete messages has been shown in several studies to increase perceived risk and intent to comply with warnings (e.g., Morris and Kanouse, 1981; Leonard *et al*., 1986; Loring and Wiklund, 1988; Laughery *et al*., 1991, 1993; Frantz, 1994). In most instances, explicit statements are less ambiguous and provide more definitive information about appropriate and inappropriate actions and possible consequences. Pictorials can differ sometimes in terms of explicitness (see Loring and Wiklund, 1988).

9.6 SUMMARY AND CONCLUSIONS

Warnings are devices for informing people about risks and for altering their expectations about possible adverse consequences. The research reviewed in this chapter indicates quite clearly that attitudes and beliefs are important ingredients in the formation of these expectations. However, a more important question concerns how these expectations influence warning effectiveness. This issue was touched upon at the beginning of the chapter, and perhaps the best answer at this juncture is that there is a reciprocal or bi-directional relationship. On the one hand, attitude and belief factors (expectations) can be expected to influence how the individual approaches and interacts with almost any potentially hazardous situation. On the other hand, a well crafted warning message, if appropriately processed, can influence the individual's expectations regarding the product, object, or activity at issue.

The evidence for the expectations–warning linkage is quite strong. First, virtually without exception, warning effectiveness increases as a function of perceived hazardousness. The more hazardous the product or activity is thought to be, the greater the likelihood of self-protective behavior. Second, research shows that warning effectiveness decreases as familiarity or experience increases. Familiar or experienced individuals typically have well formed expectations about related hazards and do not actively seek additional information. The inverse correlation between familiarity and perceived hazardousness noted in several studies suggests that threat-related expectations also play an important role in the familiarity effect. Third, findings with respect to age, gender, risk-taking tendencies, and other receiver characteristics, although quite limited, reinforce the conclusion that individuals can be expected to differ in the ways they approach hazardous situations and related warning messages. Fourth, risk perception research indicates that people use various mental rules or heuristics to simplify complex decisions. Although quite useful in many situations, these judgmental heuristics can lead to serious and persistent biases in decision-making. For example, the well documented tendency of people to be unrealistically optimistic in judging their personal risk certainly can influence how an individual approaches a potentially dangerous product or activity.

Support also exists for the warning–expectations linkage, but here the task is more challenging because the individual must first attend to and process the warning message before there is any chance that it will alter his or her expectations. In addition, the work summarized in the previous paragraph suggests that the playing field is not always level in this respect. Some people come to the situation seeking information and guidance, while others are quite sure that they know exactly what they need to know. The warning–expectations linkage deals with the ability of a warning to persuade or motivate. Warnings are thought to motivate either by informing people about unsuspected or unknown hazards or by reminding them about known hazards. In this second instance, warnings can have impact without imparting new knowledge, to the extent that the reminder is timely and occurs at the point of unsafe behavior (Hilton, 1993).

Although behavioral compliance studies often yield fairly low overall rates of compliance, these same studies also show that the individual who reads a warning message is quite likely to comply with it. For example, Friedmann (1988) found that about 60% of subjects who read the warning also complied with it. Otsubo (1988) reported that about 70% who read the warning for the high hazard product (circular saw) complied with its recommendation. For the low hazard product (jig saw), about 50% of readers complied. Frantz (1994) reported a mean compliance rate of 73% for subjects who read the warning label. These findings provide at least some prima facie evidence that warnings can alter expectations and influence behavior. Of course, the question remains whether the warning itself actually made the difference in compliance.

In some instances, the mere presence of a warning, relative to its absence, is sufficient to increase the level of perceived hazard. Other findings suggest that various aspects of the warning message can influence perceived hazard and precautionary intent. In particular, several studies suggest that using explicit or concrete language in the consequences portion increases both perceived hazard and intent to comply. Providing explicit information about the nature and severity of possible adverse outcomes should not only elevate perceived threat, it should also increase the perceived benefits of compliance (or the costs of non-compliance). The use of certain signal words and certain colors and shapes also may heighten perceived hazard under some circumstances. Further, including precautions within the directions for use and providing longer and more detailed messages may produce this effect in some instances. Highly intrusive or interactive warnings that block use of the product may be another way to alter expectations about the importance and seriousness of the hazard.

Lehto and Miller (1988) offer a very concise statement about warning effectiveness that is apropos to summarizing this chapter. They conclude (p. 254): '... people will make decisions that are consistent with safety-related knowledge when they perceive the hazard to be high and/or when the safety-related action will require little subsequent effort or cost.' Without question, perceived threat plays a major role in warning effectiveness. This by itself is useful information, but it would even more useful to be able to identify the critical or most salient dimensions of perceived threat in the context of warning messages. In many respects, perceived injury severity may be the active ingredient. The literature is quite clear also with respect to costs. Research shows that warning effectiveness degrades rapidly as the costs of compliance increases; the imposition of even modest encumbrances produces large decreases in precautionary behavior. However, in situations where costs are minimal, rather impressive levels of compliance have been achieved.

This chapter has used a traditional value-expectancy framework to examine research on the contribution of attitudes and beliefs to warning effectiveness. To a considerable extent, this basic cost–benefit framework succeeds in organizing and explaining some of the strongest and most consistent findings in the warnings literature. It also serves to highlight the importance and interplay of threat-related expectations, outcome-related expectations, and individual or receiver characteristics. However, this perspective is not without limitations. First, it assumes that people always engage in a conscious and rational decision-making process when faced with choices about self-protective action. At a minimum, some routine behavior is automatic or script-driven; people may also simply 'run risks' without making a formal decision to behave unsafely; and in other situations, people may choose a risky course of action intentionally 'for the thrill of it' or in response to some perceived infringement of their personal freedom or control.

A second limitation is that the value-expectancy perspective is essentially a static model of self-protective behavior (Weinstein, 1993). It assumes that the relative probability of action is an algebraic function of the individual's beliefs, and that very little changes in this equation from the time the person first becomes aware of the threat to when action is taken. An alternative to this view is to conceptualize precautionary behavior as consisting of a series of qualitatively distinct stages. This type of dynamic approach has been explored with respect to warnings. Lehto and Papastavrou (1993) have proposed a levels of performance model that holds that information is processed as various levels of abstraction depending on the receiver and task being performed. This framework builds on Rasmussen's (1986) model and features four hierarchical levels of performance: skill-based, rule-based, knowledge-based, and judgment-based. DeJoy (1991) focuses on the risk appraisal process and proposes four stages with bidirectional links to warnings, instructions, and other product materials. The four stages are: awareness of the hazard, personal risk assessment, decision-making, and precautionary behavior. An important implication of these models is that the type of information needed to move people along toward self-protective action varies as a function of the stage involved. These models are quite preliminary, and they do serve to complicate the warning process. However, there is very little to indicate that precautionary behavior is simple and easy to understand.

REFERENCES

ACTON, W.I. (1977) Problems associated with the use of hearing protection. *Annals of Occupational Hygiene*, 29, 387–395.

ANSI (1991) ANSI Z535.4, *Product Safety Signs and Labels*. American National Standards Institute.

ANDREWS, J.C., NETEMEYER, R.G., and DURVASULA, S. (1991) Effects of consumption frequency on believability and attitudes toward alcohol warning labels. *Journal of Consumer Affairs*, 25, 323–338.

ARKES, H.R. and HARKNESS, A.R. (1980) Effect of making a diagnosis on subsequent recognition of symptoms. *Journal of Experimental Psychology: Human Learning and Memory*, 6, 568–575.

AYRES, T.J., GROSS, M.M., WOOD, C.T., HORST, D.P., BEYER, R.R., and ROBINSON, J.N. (1989) What is a warning and when will it work? In *Proceedings of the Human Factors Society 33rd Annual Meeting*. Santa Monica, CA: Human Factors Society, pp. 426–430.

BANDURA, A. (1977) *Social Learning Theory*. Englewood Cliffs, NJ: Prentice Hall.

BANDURA, A. (1986) *Social Foundations of Thought and Action: A Social Cognitive Theory*. Englewood Cliffs, NJ: Prentice-Hall.

BELTRAMINI, R.F. (1988) Perceived believability of warning label information presented in cigarette advertising. *Journal of Advertising*, 17, 26–32.

BOHANNON, N.K. and YOUNG, S.L. (1993) Effect of warnings in advertising on adolescents' perceptions of risk for alcohol consumption. In *Proceedings of the Human Factors Society 37th Annual Meeting*. Santa Monica, CA: Human Factors Society, pp. 974–978.

BRAUN, C.C., SILVER, N.C., and STOCK, B.R. (1992) Likelihood of reading warnings: the effect of fonts and font sizes. In *Proceedings of the Human Factors Society 36th Annual Meeting*. Santa Monica, CA: Human Factors Society, pp. 926–930.

BREMS, D.J. (1986) Risk estimation for common consumer products. In *Proceedings of the Human Factors Society 30th Annual Meeting*. Santa Monica, CA: Human Factors Society, pp. 556–560.

BRESNAHAN, T.F. and BRYK, J. (1975) The hazard association values of accident-prevention signs. *Professional Safety*, 17–25.

CELUCH, K., LUST, J., and SHOWERS, L. (1998) A test of a model of consumers' responses to product manual safety information. *Journal of Applied Social Psychology*, 28, 377–394.

CLEARY, P.D. (1987) Why people take precautions against health risks. In WEINSTEIN N.D. (ed.), *Taking Care: Understanding and Encouraging Self-protective Behavior*. Cambridge: Cambridge University Press, pp. 119–149.

CLEVELAND, R.J. (1984) Factors that influence safety shoe usage. *Professional Safety*, 29, 26–29.

COLEMAN, M. and LIAU, T.L. (1975) A computer readability formula designed for machine scoring. *Journal of Applied Psychology*, 60, 283–284.

COLLINS, B.L. (1983) Evaluation of mine-safety symbols. In *Proceedings of the Human Factors Society 27th Annual Meeting*. Santa Monica, CA: Human Factors Society, pp. 947–949.

COLLINS, B.L. and LERNER, N.D. (1982) Assessment of fire-safety symbols. *Human Factors*, 24, 75–84.

DEJOY, D.M. (1987) Judgemental heuristics in consumer products safety. In *Proceedings of Interface 87*. pp. 265–272.

DEJOY, D.M. (1989) Consumer product warnings: review and analysis of effectiveness research. In *Proceedings of the Human Factors Society 33rd Annual Meeting*. Santa Monica, CA: Human Factors Society, pp. 936–940.

DEJOY, D.M. (1991) A revised model of the warnings process derived from value-expectancy theory. In *Proceedings of the Human Factors Society 35th Annual Meeting*. Santa Monica, CA: Human Factors Society, pp. 1043–1047.

DEJOY, D.M. (1992) An examination of gender differences in traffic accident risk perception. *Accident Analysis and Prevention*, 24, 237–246.

DESAULNIERS, D.R. (1991) An examination of consequence probability as a determinant of precautionary intent. Unpublished doctoral dissertation. Rice University, Houston, TX.

DINGUS, T.A., WREGGIT, S.S., and HATHAWAY, J.A. (1993) Warning variables affecting personal protective equipment use. *Safety Science*, 16, 655–673.

DONNER, K.A. and BRELSFORD, J.W. (1988) Cuing hazard information for consumer products. In *Proceedings of the Human Factors Society 32nd Annual Meeting*. Santa Monica, CA: Human Factors Society, pp. 532–535.

DUFFY, R.R., KALSHER, M.J., and WOGALTER, M.S. (1995) Increased effectiveness of an interactive warning in a realistic incidental product-use situation. *International Journal of Industrial Ergonomics*, 15, 159–166.

EASTERBY, R.S. and HAKIEL, S.R. (1981) Field testing of consumer safety signs: the comprehension of pictorially presented messages. *Applied Ergonomics*, 12, 143–152.

FEINGOLD, P.C. and KNAPP, M.L. (1977) Anti-drug abuse commercials. *Journal of Communication*, 27, 20–28.

FHANER, G. and HANE, M. (1974) Seat belts: relations between beliefs, attitude, and use. *Journal of Applied Psychology*, 59, 472–484.

FISCHHOFF, B. (1977) Cognitive liabilities and product liability. *Journal of Products Liability*, 1, 207–220.

FISCHHOFF, B., SLOVIC, P., and LICHTENSTEIN, S. (1977) Knowing with certainty: the appropriateness of extreme confidence. *Journal of Experimental Psychology: Human Perception and Performance*, 3, 552–564.

FMC CORPORATION. (1985) *Product Safety Sign and Label System*. Santa Clara, CA: FMC Corporation.

FRANTZ, J.P. (1994) Effect of location and procedural explicitness on user processing of and compliance with product warnings. *Human Factors*, 36, 532–546.

FRANTZ, J.P. and RHOADES, T.P. (1993) A task analytic approach to the temporal placement of product warnings. *Human Factors*, 35, 719–730.

FREEDMAN, J.L. and SEARS, D.O. (1965) Selective exposure. In BERKOWITZ, L. (ed.), *Advances in Experimental Social Psychology*. New York: Academic Press, pp. 58–97.

FRIEDMANN, K. (1988) The effect of adding symbols to written warning labels on user behavior and recall. *Human Factors*, 30, 507–516.

GARDNER-BONNEAU, D.J., KABBARA, F., HWANG, M., BEAN, H., GANTT, M., HARTSHORN, K., HOWELL, J., and SPENCE, R. (1989) Cigarette warnings: recall of content as a function of gender, message context, smoking habits, and time. In *Proceedings of the Human Factors Society 33rd Annual Meeting*. Santa Monica, CA: Human Factors Society, pp. 928–930.

GODFREY, S.S., ALLENDAR, L., LAUGHERY, K.R., and SMITH, V.L. (1983) Warning messages: will the consumer bother to look? In *Proceedings of the Human Factors Society 27th Annual Meeting*. Santa Monica, CA: Human Factors Society, pp. 950–954.

GODFREY, S.S. and LAUGHERY, K.R. (1984) The biasing effects of product familiarity on consumers' awareness of hazard. In *Proceedings of the Human Factors Society 28th Annual Meeting*. Santa Monica, CA: Human Factors Society, pp. 388–392.

GOLDHABER, G.M. and DETURCK, M.A. (1988a) Effects of consumers' familiarity with a product on attention to and compliance with warnings. *Journal of Products Liability*, 11, 29–37.

GOLDHABER, G.M. and DETURCK, M.A. (1988b) Effectiveness of warning signs: gender and familiarity effects. *Journal of Products Liability*, 11, 271–284.

GOLDHABER, G.M. and DETURCK, M.A. (1989) A developmental analysis of warning signs: the case of familiarity and gender. In *Proceedings of the Human Factors Society 33rd Annual Meeting*. Santa Monica, CA: Human Factors Society, pp. 1019–1023.

GRAVES, K.L. (1993) An evaluation of the alcohol warning label: a comparison of the United States and Ontario, Canada in 1990 and 1991. *Journal of Public Policy and Marketing*, 12, 19–29.

GRAY, W.B. (1975) *How to Measure Readability*. Philadelphia: Dorrance & Co.

GREENFIELD, T.K. and KASKUTAS, L.A. (1993) Early impacts of alcoholic beverage warning labels: national study findings relevant to drinking and driving behavior. *Safety Science*, 16, 689–707.

HILTON, M.E. (1993) An overview of recent findings on alcoholic beverage warning labels. *Journal of Public Policy and Marketing*, 12, 1–9.

HUNN, B.P. and DINGUS, T.A. (1992) Interactivity, information, and compliance cost in a consumer product warning scenario. *Accident Analysis and Prevention*, 24, 497–505.

JAYNES, L.S. and BOLES, D.B. (1990) The effects of symbols on warning compliance. In *Proceedings of the Human Factors Society 34th Annual Meeting*. Santa Monica, CA: Human Factors Society. pp. 984–987.

JOHNSON, D. (1992) A warning label for scaffold users. In *Proceedings of the Human Factors Society 36th Annual Meeting*. Santa Monica, CA: Human Factors Society, pp. 611–615.

JONES, S. (1978) Symbolic representation of abstract concepts. *Ergonomics*, 21, 573–577.

KAHNEMAN, D., SLOVIC, P., and TVERSKY, A. (1982) *Judgement under Uncertainty: Heuristics and Biases*. New York: Cambridge University Press.

KARNES, E.W., LEONARD, S.D., and RACHWAL, G. (1986) Effects of benign experiences on the perception of risk. In *Proceedings of the Human Factors Society 30th Annual Meeting*. Santa Monica, CA: Human Factors Society, pp. 121–125.

KASKUTAS, L. and GREENFIELD, T.K. (1992) First effects of warning labels on alcoholic beverage containers. *Drug and Alcohol Dependence*, 31, 1–14.

LARUE, C. and COHEN, H.H. (1987) Factors affecting consumers' perceptions of product warnings: an examination of the differences between male and female consumers. In *Proceedings of the Human Factors Society 31st Annual Meeting*. Santa Monica, CA: Human Factors Society, pp. 610–614.

LAUGHERY, K.R. and BRELSFORD, J.W. (1991) Receiver characteristics in safety communications. In *Proceedings of the Human Factors Society 35th Annual Meeting*. Santa Monica, CA: Human Factors Society, pp. 1068–1072.

LAUGHERY, K.R., ROWE-HALLBERT, A.L., YOUNG, S.L., VAUBEL, K.P., and LAUX, L.F. (1991) Effects of explicitness in conveying severity information in product warnings. In *Proceedings of the Human Factors Society 35th Annual Meeting*. Santa Monica, CA: Human Factors Society, pp. 481–485.

LAUGHERY, K.R., VAUBEL, K.P., YOUNG, S.L., BRELSFORD, J.W., and ROWE, A.L. (1993) Explicitness of consequence information in warning. *Safety Science*, 16, 597–613.

LAUX, L.F. and BRELSFORD, J.W. (1989) Locus of control, risk perception, and precautionary behavior. In *Proceedings of Interface 89*. pp. 121–124.

LAUX, L.F., MAYER, D.L., and THOMPSON, N.B. (1989) Usefulness of symbols and pictorials to communicate hazard information. In *Proceedings of Interface 89*. pp. 79–83.

LEHTO, M.R. and FOLEY, J.P. (1991) Risk-taking, warning labels, training, and regulation: are they associated with the use of helmets by all-terrain vehicle riders? *Journal of Safety Research*, 22, 191–200.

LEHTO, M.R. and MILLER, J.M. (1988) The effectiveness of warning labels. *Journal of Products Liability*, 11, 225–270.

LEHTO, M.R. and PAPASTAVROU, J.D. (1993) Models of the warning process: important implications towards effectiveness. *Safety Science*, 16, 569–595.

LEONARD, S.D., HILL, G.W., and KARNES, E.W. (1989) Risk perception and use of warnings. *Proceedings of the Human Factors Society 33rd Annual Meeting*. Santa Monica, CA: Human Factors Society, pp. 550–554.

LEONARD, S.D., MATHEWS, D., and KARNES, E.W. (1986) How does the population interpret warning signals? In *Proceedings of the Human Factors Society 30th Annual Meeting*. Santa Monica, CA: Human Factors Society, pp. 116–120.

LICHTENSTEIN, S., SLOVIC, P., FISCHHOFF, B., LAYMAN, M., and COMBS, B. (1978) Judged frequency of lethal events. *Journal of Experiment Psychology: Human Learning and Memory*, 4, 551–578.

LIRTZMAN, S.I. (1984) Labels, perception, and psychometrics. In O'CONNOR, C.J. and LIRTZMAN, S.I. (eds), *Handbook of Chemical Industry Labeling*. Park Ridge, NJ: Noyes.

LORING, B.A. and WIKLUND, M.E. (1988) Improving swimming pool warning signs. In *Proceedings of the Human Factors Society 32nd Annual Meeting*. Santa Monica, CA: Human Factors Society, pp. 910–914.

LOWRANCE, W.W. (1980) *Of Acceptable Risk: Science and the Determination of Safety*. Los Altos, CA: William Kaufman.

MCDOWELL, I. (1988) Man the risk taker. *American Journal of Preventive Medicine*, 4, 172–177.

MCGUIRE, W.J. (1980) The communication-persuasion model and health-risk labeling. In MORRIS, L.A., MAZIS, M.B., and BAROFSKY, I. (eds), *Product Labeling and Health Risks* Banbury Report No 6, Coldspring Harbor Laboratory.

MACKINNON, D.P. and FENAUGHTY, A.M. (1993) Substance use and memory for health warning labels. *Health Psychology*, 12, 147–150.

MACKINNON, D.P., PENTZ, M.A., and STACY, A.W. (1993) The alcohol warning label and adolescents: the first year. *American Journal of Public Health*, 83, 585–587.

MALLETT, L., VAUGHT, C., and BRNICH, M.J. (1993) Sociotechnical communication in an underground mine fire: a study of warning messages during an emergency evacuation. *Safety Science*, 16, 709–728.

MARIN, G. (1994) Self-reported awareness of the presence of product warning messages and signs by Hispanics in San Francisco. *Public Health Reports*, 109, 275–283.

MARTIN, E.G. and WOGALTER, M.S. (1989) Risk perception and precautionary intent for common consumer products. In *Proceedings of the Human Factors Society 33rd Annual Meeting*. Santa Monica, CA: Human Factors Society, pp. 931–935.

MAYER, R.N., SMITH, K.R., and SCAMMON, D.L. (1991) Evaluating the impact of alcohol warning labels. *Advances in Consumer Research*, 18, 706–714.

MAZIS, M., MORRIS, L.A., and GORDON, E. (1978) Patient attitudes about two forms of printed oral contraceptive information. *Medical Care*, 16, 1045–1054.

MAZIS, M.B., MORRIS, L.A., and SWASY, J.L. (1991) An evaluation of the alcohol warning label: initial survey results. *Journal of Public Policy and Marketing*, 10, 229–241.

MORRIS, L.A. and KANOUSE, D.E. (1981) Consumer reaction to the tone of written drug information. *American Journal of Hospital Pharmacy*, 38, 667–671.

MORRIS, L.A., MAZIS, M., and GORDON, E. (1977) A survey of the effects of oral contraceptive patient information. *Journal of the American Medical Association*, 238, 2504–2508.

NATIONAL RESEARCH COUNCIL (1989) *Improving Risk Communication*. Washington, DC: National Academy Press.

OTSUBO, S.M. (1988) A behavioral study of warning labels for consumer products: perceived danger and use of pictographs. In *Proceedings of the Human Factors Society 32nd Annual Meeting*. Santa Monica, CA: Human Factors Society, pp. 536–540.

PATTERSON, L.T., HUNNICUTT, G.G., and STUTTS, M.A. (1992) Young adults' perceptions of warnings and risks associated with alcohol consumption. *Journal of Public Policy and Marketing*, 11, 96–103.

PURSWELL, J.L., SCHLEGEL, R.E., and KEJRIWAL, S.K. (1986) Prediction model for consumer behavior regarding product safety. In *Proceedings of the Human Factors Society 30th Annual Meeting*. Santa Monica, CA: Human Factors Society, pp. 1202–1205.

RACICOT, B.M. and WOGALTER, M.S. (1995) Effects of a video warning sign and social modeling on behavioral compliance. *Accident Analysis and Prevention*, 27, 57–64.

RASMUSSEN, J. (1986) *Information Processing and Human–Machine Interaction: An Approach to Cognitive Engineering*. New York: North Holland.

RETHANS, A. (1979) An investigation of consumer perceptions of product hazards. Doctoral dissertation, University of Oregon.

RILEY, M.W., COCHRAN, D.J., and BALLARD, J.L. (1982) An investigation of preferred shapes for warning labels. *Human Factors*, 24, 737–742.

ROKEACH, M. (1966) Attitude change and behavior change. *Public Opinion Quarterly*, 30, 529–550.

SCHANK, R.C. and ABELSON, R. (1977) *Scripts, Plans, Goals, and Understanding*. Hillsdale, NJ: Lawrence Erlbaum, Associates.

SILVER, N.C., GAMMELLA, D.S., BARLOW, A.N., and WOGALTER, M.S. (1993) Connoted strength of signal words by elderly and non-native English speakers. In *Proceedings of the Human Factors Society 37th Annual Meeting*. Santa Monica, CA: Human Factors Society, pp. 516–519.

SILVER, N.C., LEONARD, D.C., PONSI, K.A., and WOGALTER, M.S. (1991) Warnings and purchase intentions for pest-control products. *Forensic Reports*, 4, 17–33.

SILVER, N.C. and WOGALTER, M.S. (1991) Strength and understanding of signal words by elementary and middle school students. In *Proceedings of the Human Factors Society 35th Annual Meeting*. Santa Monica, CA: Human Factors Society, pp. 580–594.

SLOVIC, P., FISCHHOFF, B., and LICHTENSTEIN, S. (1979) Rating the risks. *Environment*, 21, 14–20, 36–39.

SLOVIC, P., FISCHHOFF, B., and LICHTENSTEIN, S. (1980) Facts and fears: understanding perceived risk. In SCHWING, R. and ALBERS JR, W.A. (eds), *Societal Risk Assessment: How Safe is Safe Enough*. New York: Plenum.

SNYDER, L.B. and BLOOD, D.J. (1991) Alcohol advertising and the Surgeon General's alcohol warnings may have adverse effects on young adults. Paper presented to the International Communication Association Annual Conference, Chicago, May.

STARR, C. (1969) Social benefit versus technological risk. *Science*, 165, 1232–1238.

STEWART, D.W. and MARTIN, I.M. (1994) Intended and unintended consequences of warning messages: a review and synthesis of empirical research. *Journal of Public Policy and Marketing*, 13, 1–19.

STRAWBRIDGE, J.A. (1986) The influence of position, highlighting, and imbedding on warning effectiveness. In *Proceedings of the Human Factors Society 30th Annual Meeting*. Santa Monica, CA: Human Factors Society, pp. 716–720.

STRECHER, V.J., DEVELLIS, B.McE., BECKER, M.H., and ROSENSTOCK, I.M. (1986) The role of self-efficacy in achieving health behavior change. *Health Education Quarterly*, 13, 73–92.

SVENSON, O. (1981) Are we all less risky and more skillful than our fellow drivers? *Acta Psychologica*, 47, 143–148.

TVERSKY, A. and KAHNEMAN, D. (1974) Judgement under uncertainty: heuristics and biases. *Science*, 185, 1124–1131.

TVERSKY, A. and KAHNEMAN, D. (1981) The framing of decisions and the psychology of choice. *Science*, 211, 453–458.

URSIC, M. (1984) The impact of safety warnings on perception and memory. *Human Factors*, 26, 677–682.

VISCUSI, W.K. (1992) *Smoking: Making the Risky Decision*. New York: Oxford University Press.

VISCUSI, W.K. (1994) Efficacy of labeling of foods and pharmaceuticals. *Annual Review of Public Health*, 15, 325–343.

VISCUSI, W.K., MAGAT, W.A., and HUBER, J. (1986) Informational regulation of consumer health risks: an empirical evaluation of hazard warnings. *Rand Journal of Economics*, 17, 351–365.

VREDENBURGH, A.G. and COHEN, H.H. (1993) Compliance with warnings in high risk recreational activities: skiing and scuba. In *Proceedings of the Human Factors and Ergonomics Society 37th Annual Meeting*. Santa Monica, CA: Human Factors and Ergonomics Society, pp. 945–949.

WAGENAAR, W.A. (1992) Risk-taking and accident causation. In YATES, J.F. (ed.), *Risk-taking Behavior*. New York: Wiley, pp. 257–281.

WALLSTON, B.S. and WALLSTON, K.A. (1984) Social psychological models of health behavior: an examination and integration. In BAUM, A., TAYLOR, S.E., and SINGER, J.E. (eds), *Handbook of Psychology and Health*, Vol. 4. Hillsdale, NJ: Lawrence Erlbaum Associates, pp. 23–54.

WEINSTEIN, N.D. (1980) Unrealistic optimism about future life events. *Journal of Personality and Social Psychology*, 39, 806–820.

WEINSTEIN, N.D. (1987) Unrealistic optimism about illness susceptibility: conclusions from a community-wide sample. *Journal of Behavioral Medicine*, 10, 481–500.

WEINSTEIN, N.D. (1993) Testing four competing theories of health-protective behavior. *Health Psychology*, 12, 324–333.

WESTINGHOUSE ELECTRIC CORPORATION (1981) *Product Safety Label Handbook*. Trafford, PA: Westinghouse Printing Division.

WOGALTER, M.S., ALLISON, S.T., and MCKENNA, N.A. (1989) Effects of cost and social influence on warning compliance. *Human Factors*, 31, 133–140.

WOGALTER, M.S. and BARLOW, T. (1990) Injury likelihood and severity in warnings. In *Proceedings of the Human Factors Society 34th Annual Meeting*. Santa Monica, CA: Human Factors Society, pp. 580–583.

WOGALTER, M.S., BARLOW, T., and MURPHY, S. (1995) Compliance to owner's manual warnings: influence of familiarity and the task-relevant placement of a supplemental directive. *Ergonomics*, 38, 1081–1091.

WOGALTER, M.S., BRELSFORD, J.W., DESAULNIERS, D.R., and LAUGHERY, K.R. (1991) Consumer products warnings: the role of hazard perception. *Journal of Safety Research*, 22, 71–82.

WOGALTER, M.S., BREMS, D.J., and MARTIN, E.G. (1993a) Risk perception of common consumer products: judgments of accident frequency and precautionary intent. *Journal of Safety Research*, 24, 97–106.

WOGALTER, M.S., GODFREY, S.S., FONTENELLE, G.A., DESAULNIERS, D.R., ROTHSTEIN, P.R., and LAUGHERY, K.R. (1987) Effectiveness of warnings. *Human Factors*, 29, 599–622.

WOGALTER, M.S., JARRARD, S.W., and SIMPSON, S.W. (1994a) Influence of signal words on perceived label of product hazard. *Human Factors*, 36, 547–556.

WOGALTER, M.S., KALSHER, M.S., and RACICOT, B.M. (1992) The influence of location and pictorials on behavioral compliance to warnings. In *Proceedings of the Human Factors Society 36th Annual Meeting*. Santa Monica, CA: Human Factors Society, pp. 1029–1033.

WOGALTER, M.S., KALSHER, M.J., and RACICOT, B.M. (1993b) Behavioral compliance with warnings: effects of voice, context, and location. *Safety Science*, 16, 637–654.

WOGALTER, M.S., RACICOT, B.M., KALSHER, M.J., and SIMPSON, S.N. (1993c) Behavioral compliance to personalized warning signs and the role of perceived relevance. In *Proceedings of the Human Factors Society 37th Annual Meeting*. Santa Monica, CA: Human Factors Society, pp. 950–954.

WOGALTER, M.S., RACICOT, B.M., KALSHER, M.J., and SIMPSON, S.N. (1994b) Personalization of warning signs: the role of perceived relevance on behavioral compliance. *International Journal of Industrial Ergonomics*, 14, 233–242.

WOGALTER, M.S. and SILVER, N.C. (1990) Arousal strength of signal words. *Forensic Reports*, 3, 407–420.

WOGALTER, M.S. and YOUNG, S.L. (1991) Behavioural compliance to voice and print warnings. *Ergonomics*, 34, 79–89.

WRIGHT, P. (1982) Some factors determining when instructions will be read. *Ergonomics*, 25, 225–237.

YOUNG, S.L., BRELSFORD, J.W., and WOGALTER, M.S. (1990) Judgments of hazard, risk, and danger: do they differ? In *Proceedings of the Human Factors Society 34th Annual Meeting*. Santa Monica, CA: Human Factors Society, pp. 503–507.

YOUNG, S.L., MARTIN, E.G., and WOGALTER, M.S. (1989) Gender differences in consumer product hazard perceptions. In *Proceedings of Interface 89*. pp. 73–78.

YOUNG, S.L., WOGALTER, M.S., and BRELSFORD, J.W. (1992) Relative contribution of likelihood and severity of injury to risk perceptions. In *Proceedings of the Human Factors Society 36th Annual Meeting*. Santa Monica, CA: Human Factors Society, pp. 1014–1018.

ZUCCOLLO, G. and LIDDELL, H. (1985) The elderly and the medication label: doing it better. *Age and Ageing*, 14, 371–376.

ZUCKERMAN, M. (1979) *Sensation Seeking: Beyond the Optimal Level of Arousal*. Hillsdale, NJ: Lawrence Erlbaum Associates.

CHAPTER TEN

Motivation

DAVID M. DEJOY

University of Georgia

For a warning to be effective, it must motivate the individual to comply with its directives. This chapter argues that perceived threat, or expectations about adverse consequences, is what provides the initial motivation for precautionary behavior. In this regard, most warnings are fear-based communications. Following a brief review of the fear-arousal literature, the importance of consequences are examined in the context of the warnings literature. Two categories of consequences are examined: those associated with the threat itself and those associated with the recommended action. In most respects, a perceived threat is a necessary but not sufficient condition for warning effectiveness. Beliefs about the effectiveness of the recommended actions and the costs associated with compliance also contribute to motivating self-protective behavior. Social influence, source credibility, and other persuasion heuristics also help to explain warnings-related behavior. These heuristics may be most important in the presence of low or ambiguous levels of perceived threat. Finally, the motivation to comply is at least partially determined by the expectations of the message recipient. Three categories of individual difference factors are discussed: familiarity and experience, personal relevance, and risk-taking.

10.1 INTRODUCTION

In the preceding chapter, attitudes and beliefs were portrayed as important ingredients in the formation of expectations. Expectations were defined as the anticipatory outcomes of behavior, and referring back to the basic information-processing model presented in Chapter 2 by Wogalter, DeJoy, and Laughery (Figure 2.1), an effective warning should alter the expectations of the individual concerning the risks posed by the product, object, or activity in question. Precautionary intent is largely a function of the individual's expectations about possible outcomes or consequences. It follows that expectations about adverse consequences and the desire to avoid them is what provides the initial motivation for precautionary behavior. In the most obvious sense, this motivation occurs because the warning message provides the individual with new and relevant information about some hazard. However, warnings may also motivate without providing new information (Hilton, 1993), to the extent that they provide timely prompts or reminders at the point of contact with the hazard.

This chapter begins with a general summary of the literature on fear-arousing communications. To a considerable extent, warnings are fear-based communications. In their most rudimentary form, warnings alert people to hazards. More complete warnings may also indicate the types of injury or other adverse effects that might occur, and recommend actions to avoid or minimize them. The chapter then proceeds to examine the importance of consequences in the context of the warnings literature. Drawing from prominent theories of compliance and self-protective behavior (see reviews by Wilson, Purdon, and Wallston, 1988; Weinstein, 1993), consequences typically take two forms: those associated with the threat itself and those associated with the recommended action or precaution. The first dimension reflects a rather direct link to the likelihood and severity of possible adverse outcomes, while the second focuses on the effectiveness of the recommended precautions and the costs associated with compliance. Both of these dimensions appear to be important to warning effectiveness. Following this, the effects of social influence and other persuasion heuristics are considered. The last section of the chapter looks at the contribution of individual difference factors, including familiarity and experience, personal relevance, and risk-taking.

10.2 FEAR AROUSAL

A large of amount of behavioral research beginning in the 1950s and 1960s was directed at determining whether fear is an effective motivating factor for preventive and self-protective behavior (for reviews see Higbee, 1969; Leventhal, 1970). Fear appeals, as a subclass of emotional persuasion, typically share three characteristics (Averill, 1987). First, the factual content of the message describes or connotes some situation that is socially recognized as emotional (e.g., danger). Second, an emotional appeal relates the situation to the receiver in an immediate and personal way—the message is personally relevant. Third, emotional appeals tend to be non-rational in the strict sense; they make use of imagery, vivid scenes, or other such devices that bolster the emotional impact of the message.

10.2.1 Theoretical Perspectives

According to basic fear-drive theory (e.g., Hovland, Janis, and Kelly, 1953), people follow recommendations because doing so reduces the tension created by the fear message. It was generally assumed that the more fear aroused in the individual, the greater the probability of attitudinal and behavioral change. This linear or 'more the merrier' approach is portrayed in the upper portion of Figure 10.1. This line of thinking would suggest that using explicit or vivid 'bloody fingers' posters in industrial settings should be more effective than less explicit images or warning messages. However, results from early research showed that the relationship between fear arousal and attitude and/or behavior change is neither simple nor direct. Janis and Feshbach (1953), in their classic study, found that a high fear appeal involving explicit, unpleasant photographs was less effective in prompting attitude change than appeals involving milder levels of fear arousal. These findings led to the idea that people may disregard or seek to avoid high fear messages because they are so unpleasant. Too much fear can bring about inappropriate or defensive responses such as denial that diminish the effects of the appeal. According to Leventhal (1970), this type of boomerang effect can occur when the recommended action

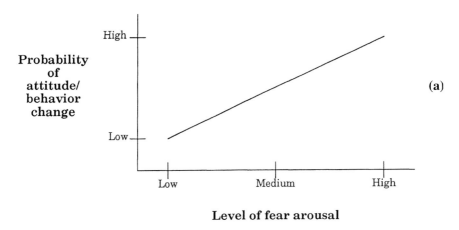

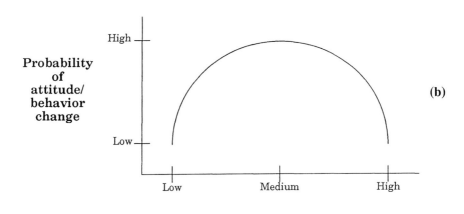

Figure 10.1 Hypothetical curves showing two alternative relationships between fear arousal and attitude/behavior change.

is perceived as falling short of reducing or eliminating the threat at issue. This type of curvilinear or inverted-U relationship is depicted in the lower portion of Figure 10.1.

McGuire's (1980) communication–persuasion model also subscribes to a curvilinear relationship between the level of fear aroused by a message and the probability of compliance. His model is organized around ten stages or phases that depict the successive mediating responses that must be elicited if the risk communication is to be effective. The ten stages are: (1) exposure, (2) attending, (3) reacting affectively, (4) comprehending, (5) yielding, (6) storing and retaining content, (7) information search and retrieval, (8) decision-making, (9) behaving, and (10) post-behavioral consolidating. McGuire argues that the level of threat contained in a warning message tends to enhance some output steps (such as yielding) but tends to reduce other steps (such as reacting affectively and retrieval). McGuire concludes that the net-result of this processing is that an intermediate level of fear arousal is often most effective in producing self-protective behavior.

However, the curvilinear relationship is difficult to test because it is difficult to know whether any set of fear-arousing communications represents the full range of fear arousal. Simply devising three or more levels of fear-arousing messages and not detecting an inverted-U function may mean simply that the point of inversion was not reached or that the intermediate or optimal level was 'skipped over' in selecting the levels of fear arousal. Several authors, in reviewing the work conducted in the 1950s and 1960s, have argued that the bulk of evidence suggests a positive relationship between fear and persuasion, and that inverted-U findings may be more the exception than the rule (e.g., Leventhal, 1970; Sutton, 1982).

Cognitively oriented models

More recent conceptualizations of the fear–persuasion relationship have tended to assign less importance to fear arousal *per se* (Job, 1988; Wilson *et al.*, 1988; Witte, 1992). Leventhal (1970) developed the dual process model which posits that emotional and instrumental behavior are simultaneous, parallel processes. Within this framework, perceived threat initiates two types of response: fear control (guided by internal cues) and danger control (guided by external cues). The purpose of fear control is to reduce arousal and the function of danger control is to reduce the threat. Presumably, these two processes can augment each other, interfere with each other, or be largely independent of each other. Leventhal argues that this model can explain both the facilitating and inhibiting effects of fear. Unfortunately, the difficulty of independently manipulating fear and danger have made this model difficult to test in practice.

Leventhal's model is perhaps most useful because it makes a distinction between emotional and cognitive reactions to fear-based messages. Fear is basically the emotional reaction that occurs when a serious and personally relevant threat is perceived. This emotional arousal may either facilitate or inhibit self-protective action. For example, a highly explicit or graphic warning message may motivate the individual to follow recommended precautions or it may cause him or her to deny or discount the depicted threat. Basically, Leventhal's position is that self-protective behavior is linked more closely to controlling danger or threat than to controlling fear. Danger control involves the selection and execution of responses directed at avoiding or minimizing the threat. Simply arousing a high level of fear through a warning message may not lead to self-protective behavior if the depicted outcome is perceived as being very unlikely or if the recommended precautionary actions are perceived as being difficult to perform or insufficient to counteract the threat.

Models derived from value–expectancy theory further emphasize cognitive processes at the expense of fear arousal. Sutton's application of subjective-expected utility (SEU) theory (e.g., Sutton, 1982) and Rogers' (1983) protection motivation theory assign a major role to perceived threat in the subjective evaluative response elicited by fear messages. In SEU theory, motivation to change is a function of the degree to which the SEU associated with changing is greater than the SEU associated with not changing. The SEU of changing is determined by the utility of the outcome (i.e., the degree to which the outcome is threatening) and the perceived efficacy of the recommended action. In protection motivation theory, the evaluation of threat includes the perception of outcome severity, the perceived probability of occurrence, and the perceived efficacy of the recommended response. Protection motivation theory has been the most widely used model in fear arousal research since about the mid 1970s. Basically, this model focuses almost exclusively on Leventhal's danger-control processes; fear-control processes are essentially ignored (Witte, 1992).

10.2.2 Using Fear Arousal

Fear arousal certainly plays some role in attitude and behavior change (including compliance with warnings), but fear arousal by itself may not always be sufficient to produce these effects. Often the effects of fear are short-lived and behavior motivated by fear tends to be more automatic than voluntary. In some instances, fear may actually inhibit behavior change (Becker and Janz, 1987). In other instances, factual information alone or information in combination with positive appeals can be as or more effective in motivating people to change their behavior. The sheer prevalence of fear-based messages in our society may also help explain why such appeals often produce disappointing results. Still, there is evidence to suggest that reasonable levels of fear arousal can be effective when the recommended actions are perceived to be efficacious and the associated costs are few.

Job (1988) proposes that fear is most likely to be effective when the following five conditions are met: (1) fear onset should occur before the desired or recommended behavior is offered; (2) the fear-arousing event should appear to be likely; (3) the action or actions to offset the fear should be clearly specified; (4) the level of fear aroused should be commensurate with the recommended action's ability to reduce it; and (5) fear offset should occur as a reinforcer for the desired action. The first recommendation deals with the temporal perspective and the idea that the drive state must be created first to activate responding. Generally speaking, this drive state exists to the extent that the message has altered expectations about the severity and/or likelihood of adverse consequences. The second recommendation reflects the observation that people are less likely to take precautions in response to low probability hazards and/or those which might occur sometime in the distant future. The third recommendation is quite straightforward, although it is sometimes ignored by those designing fear-based appeals. The message should feature actions that will lower or eliminate the hazard. Concerning the fourth recommendation, arousing a high level of fear is not likely to be very useful if the recommended precaution is perceived as offering only minimal or limited benefits. High levels of fear arousal can evoke competing responses, such as avoidance or denial, that detract from the message. Finally, the fifth recommendation generally refers to the idea that the recommended precaution or precautions should be perceived as being effective in removing the source of danger (and fear).

10.3 EXPECTATIONS ABOUT CONSEQUENCES

This section reviews warning research pertaining to the two dimensions of consequences outlined earlier: threat related and action or outcome related.

10.3.1 Threat-related Consequences

The emphasis on perceived threat as a motivational factor for warnings is linked directly to fear arousal. Presumably, the expectation of threatening consequences arouses fear which, in turn, increases the likelihood of some response (such as compliance with the warning) that will reduce the fear. Essentially this type of 'fear framing' is a punishment paradigm, in which something bad might happen if you do not behave as instructed. The fear-framed message is negative. The use of punishment to motivate behavior change seems to be valid in common sense terms, but behavioral scientists tend to be reluctant to endorse this approach and argue that punishment often yields unpredictable or disappointing results (e.g., Job, 1988). It is widely acknowledged that punishment works best when it is applied immediately and consistently, and most warnings situations do not fulfill these requirements. For example, failing to wear protective eye wear when using a household hammer might result in immediate eye injury, but it is much more likely that this unsafe behavior will continue over a relatively long period of time before any such injury occurs. Ignoring the admonition to wear protective eye wear does not normally result in immediate injury on any regular or consistent basis.

Perceived hazardousness

While fear arousal has not been investigated directly by warnings researchers, there is considerable evidence that the level of hazard posed by a particular product, object, or activity plays an important role in determining how the individual interacts with warning messages. Virtually without exception, as the level of perceived hazard increases, people are: (1) more willing to look for and read warnings (e.g., Godfrey, Allendar, Laughery, and Smith, 1983; Otsubo, 1988; Silver, Leonard, Ponsi, and Wogalter, 1991; Frantz, 1994); (2) better able to recall the content of warnings (e.g., Friedmann, 1988); (3) more positively inclined to comply (e.g., Young, Brelsford, and Wogalter, 1990; Wogalter, Brelsford, Desaulniers, and Laughery, 1991; Wogalter, Brems, and Martin, 1993a); and (4) more likely to actually comply (e.g., Friedmann, 1988; Otsubo, 1988; Frantz, 1994). The interested reader might also want to refer to Table 9.2 in Chapter 9 by DeJoy, which summarizes these studies.

Injury severity and likelihood

Typically, perceived threat is thought to consist of two primary dimensions: outcome likelihood and outcome severity. Likelihood refers to the perceived probability of experiencing some type of adverse consequence (injury, illness, etc.), and severity pertains to the perceived seriousness of the consequence (e.g., Slovic, Fischhoff, and Lichtenstein, 1979; Lowrance, 1980). A question of considerable theoretical and practical significance to the warnings field concerns the relative importance of these two dimensions in producing precautionary behavior. At this point, research on warnings points to injury severity as the more important dimension, at least for most situations involving consumer products (Young *et al.*, 1990; Wogalter *et al.*, 1991, 1993a). For example, Wogalter *et al.* (1991, Study 1) asked subjects to rate 72 generically named consumer products in terms of hazardousness, injury severity, familiarity, and several other dimensions. They found that perceived likelihood and perceived severity were both rather highly correlated with willingness to read warnings ($r^2 = -0.64$ and 0.89, respectively). However, regression analysis showed that severity by itself accounted for 80% of the variance in willingness to read warnings. The addition of likelihood to the model provided virtually no improvement in explanatory power. Including a severity × likelihood interaction term also failed to boost prediction. A follow-up study (1991, Study 2) found essentially similar effects using perceived hazardousness as the criterion variable. When subjects were asked to generate injury scenarios for various products (1991, Study 3), the severity of the first generated scenario was most predictive of perceived hazard ($r^2 = 0.81$).

Perceived injury severity also may be the mediating factor in the explicitness effect and may also help explain the inverse relationship observed between product familiarity and warning effectiveness. Several studies suggest that the explicitness or concreteness with which consequences are stated in a warning message can increase perceived hazard (Morris and Kanouse, 1981; Loring and Wiklund, 1988; Laughery, Rowe-Hallbert, Young, Vaubel, and Laux, 1991; Laughery, Vaubel, Young, Brelsford, and Rowe, 1993). In particular, the Laughery et al. (1993) findings suggest that providing explicit information about consequences elevates perceived hazard by increasing perceptions of injury severity. Laughery and colleagues also found that non-explicit warnings reduced precautionary intent regardless of the level of injury severity associated with the product. As a general rule, including information about the consequences that might result from failure to obey the warning increases perceived risk and intent to comply (Leonard, Mathews, and Karnes, 1986; Wogalter and Barlow, 1990).

Research also suggests that users who are more familiar or experienced with a product are less likely to look for (e.g., Godfrey et al., 1983), read (e.g., LaRue and Cohen, 1987; Otsubo, 1988; Wogalter et al., 1991), and comply with warnings (e.g., Otsubo, 1988; Wogalter et al., 1993; Wogalter, Barlow, and Murphy, 1995). Specific to the present discussion, several of these studies examined the effects of both perceived hazardousness and familiarity (e.g., LaRue and Cohen, 1987; Otsubo, 1988; Wogalter et al., 1991, 1993). Two of these studies, LaRue and Cohen (1987) and Wogalter and associates (1991, Study 1), provided correlation matrices. In both studies, perceived hazardousness/danger was more strongly correlated with the need for warnings than was familiarity ($r^2 = 0.89$ and 0.95, respectively for hazardousness/danger and -0.63 and -0.62 for familiarity). There was also a fairly substantial intercorrelation between hazardousness and familiarity ($r^2 = -0.59$ and -0.63), indicating that as people become more familiar with a product, they perceive it as being less hazardous. Using regression analysis, Wogalter and colleagues (1991) found that familiarity added very little explanatory power beyond that offered by perceived hazardousness. While the warnings literature indicates that both perceived hazardousness and familiarity influence how people approach and respond to warnings, perceived hazardousness may be the factor that motivates information seeking and precautionary action in most situations. Moreover, perceptions concerning the severity of possible injury may be the key dimension in judgments of perceived hazardousness.

As a final point, care should be exercised before dismissing perceived injury likelihood as an important component of perceived threat. Most risk perception researchers view severity and likelihood as interacting multiplicatively (e.g., Slovic et al., 1979; Lowrance, 1980) and argue that the relative importance of the two dimensions can be expected to vary across different hazards. For example, when severity reaches a certain level, the only uncertainty is the probability of the adverse event. In constructing the perceived threat associated with something like AIDS, which is often fatal, perceived likelihood might very well be the more important dimension of perceived threat. Such also appears to be the case for catastrophic or dreaded societal-level hazards such as nuclear accidents (e.g., Slovic, Fischhoff, and Lichtenstein, 1980).

10.3.2 Action- or Outcome-related Consequences

Outcome-related consequences pertain to the effectiveness of the recommended precaution or precautions in removing or reducing some particular threat. Although perceived threat may create the initial motivation to follow the recommendations contained in a warning message, beliefs about the effectiveness of these actions can also be expected to play a role in motivating precautionary behavior. From a value–expectancy perspective (e.g., Weinstein, 1993), the actual decision to engage in self-protective behavior involves the weighing of expected costs and benefits (see Figure 9.1 in Chapter 9 by DeJoy). With respect to warning messages, the benefits of complying with the indicated precautions are evaluated against the various costs and barriers associated with following the recommendations. The effectiveness of a warning message should increase if either the benefits of taking the action can be heightened or the costs can be reduced.

Perceived effectiveness

Guidelines for the design of warnings generally recommend that at a minimum a warning should contain the following four elements (FMC Corporation, 1980; Peters, 1984; Westinghouse Electric Corporation, 1981): (a) a signal word to convey the seriousness of the hazard or threat, (b) a hazard statement that describes the nature of the threat or hazard, (c) a consequences statement that describes what might happen if the warning is ignored, and (d) instructions about the action or actions that will reduce or eliminate the threat. As discussed earlier in this chapter, information about possible injury or other adverse consequences appears to be especially important in shaping perceived threat and creating the initial motivation for precautionary behavior. The choice of signal word (e.g., Wogalter, Jarrard, and Simpson, 1994) and the way in which the hazard is described (e.g., Laughery *et al.*, 1993; Wogalter and Barlow, 1990) also can enhance the perceived threat. By contrast, the instructions statement is the only part of the standard warning that conveys any information about the action or actions that should be taken to reduce or remove the hazard. The instructions do not usually speak directly to the benefits of the recommended precautions, but they provide a starting point for such thinking. If the warning label on an all-terrain vehicle (ATV) states that the operator should wear a helmet, the reader of this message now has something to evaluate in terms of its potential effectiveness in reducing the threat of serious head injury. Of course, if insufficient threat has been established or if the threat has not been personalized (i.e., this injury could happen to me), little perceived benefit can be realized by taking the recommended precaution and no further processing of the warning is likely.

Although beliefs or expectations about the effectiveness of recommended precautions have received very little direct attention in the warnings literature, several studies do suggest that the instructions portion of a warning message is important to ultimate compliance. Leventhal, Singer, and Jones (1965), in an early fear-arousal study, reported that warnings that provided instructions about reducing or eliminating a hazard were more likely to be complied with. Viscusi and colleagues (Viscusi, Magat, and Huber, 1986) found that messages specifically recommending a particular action had a greater influence than the overall amount of risk information. Wogalter *et al.* (1987) found that removal of any of the four standard warning elements reduced perceived effectiveness, but that removal of either the hazard statement or the instructions statement produced the greatest reductions in rated effectiveness. Frantz (1994) found that including procedurally explicit precautions in the directions for use significantly increased both reading rates and compliance.

Costs of compliance

Warnings researchers have examined costs of compliance from two perspectives. First, several studies suggest that warning effectiveness degrades quite rapidly as the costs of complying with the warning increase. Wogalter *et al.* (1987) placed warnings on a broken exit door in a campus building. The warning sign directed users to an adjacent door (low cost), or another set of doors 15 meters away (moderate cost), or a third set of doors 60 meters away (high cost). The results showed that the warning was almost always obeyed in the low cost condition (94%), but totally ignored in the high cost condition (0%). Similar effects were obtained in a chemistry laboratory simulation (Wogalter, Allison, and McKenna, 1989). When subjects could access the recommended protective equipment (mask and gloves) with little effort, compliance was quite high. By contrast, very few subjects were willing to go even into the next room to the retrieve the needed protective equipment. Dingus and colleagues (Dingus, Wreggit, and Hathaway, 1993) examined costs of compliance in recreational and consumer products contexts. Again, relatively small increments in cost produced large decrements in compliance.

However, these same studies also demonstrate that quite high rates of compliance can be achieved when costs are minimal. Specific to this point, Dingus and colleagues (Dingus et al., 1993, Experiment 2) manipulated cost in terms of whether protective equipment (gloves and respirator masks) were enclosed in the product packaging for an 'industrial strength tile de-scaler.' Providing this equipment along with the product had a dramatic effect on compliance. In fact, slightly over 80% of subjects wore the gloves and 50% used the masks in the experimental condition in which there was no specially crafted warning message urging them to do so. Compliance rates were even higher when the safety equipment was combined with various warning message configurations. Racicot and Wogalter (1995), in a study of video warnings, found that 85% of subjects who reported seeing the safety equipment (gloves and masks) complied with the recommendation to use them.

Certainly compliance costs are not limited to the amount of effort that must be expended to obtain needed safety equipment or to behave in a safe manner. Studies of helmet usage among ATV operators (e.g., Lehto and Foley, 1991), seat belt usage in automobiles (e.g., Fhaner and Hane, 1974), and personal protective equipment usage in industry (e.g, Acton, 1977; Cleveland, 1984), all suggest that virtually any type of discomfort, restriction of movement or freedom, or other encumbrance can serve as a barrier to compliance. The simple fact that behaving unsafely is sometimes more pleasurable or rewarding than behaving safely also qualifies as a cost of compliance (McDowell, 1988). All of these costs or barriers may find their way into the personal weighing of costs and benefits.

Costs of non-compliance

Warning effectiveness can also be affected by manipulating the costs of non-compliance. Following the line of thinking pursued earlier in this chapter, increasing the costs of non-compliance essentially serves to elevate perceived threat. Elevating the costs of non-compliance (perceived threat) has the potential to shift the basic cost–benefit tradeoff in the direction of compliance. Presumably, any warning attribute or design characteristic that increases perceived threat should also alter beliefs about the costs of non-compliance.

As discussed previously, providing explicit and concrete information on the severity of possible injuries appears to increase perceived hazardousness (e.g., Loring and Wiklund, 1988; Morris and Kanouse, 1981; Laughery et al., 1991, 1993). The use of certain 'potent' signal words such as 'lethal' have been shown to increase perceived hazardousness relative to more commonly used words such as 'warning' or 'caution' (Wogalter et al., 1994). Certain colors and shapes in the warning message also can enhance perceived hazardousness. The colors red and yellow have been found to connote greater levels of hazard than the colors green or blue (Bresnahan and Bryk, 1975). Pointed shapes, such as diamonds and triangles, also appear to convey greater hazard than regular rectangles or circles (Collins, 1983; Jones, 1978; Riley, Cochran, and Ballard, 1982). Including the warning information within the instructions for use may increase the hazard level associated with the product under some circumstances (Frantz, 1994); longer and more detailed warnings may also produce the same effect (Silver et al., 1991).

Wogalter et al. (1995) reported that placing a warning so that it actually blocked use of the equipment resulted in better compliance among experienced users. They reasoned that this arrangement might have helped to create the belief in the user that this was a particularly important warning. Interactive warnings increase the noticeability of warnings (e.g, Frantz and Rhoades, 1993; Duffy, Kalsher, and Wogalter, 1995) by disrupting ongoing or 'automatic' behavior, but certain types of interactive warning also may alter people's expectations about the costs of non-compliance.

10.3.3 More on Consequences

Before leaving the topic of consequences, there are two additional points that merit some consideration in trying to optimize the use of perceived threat in warning messages.

Message vividness and persuasiveness

First, there is evidence in the social psychological literature suggesting that the vividness of a message can sometimes undermine its persuasiveness. According to Nisbett and Ross (1980, p. 45), information is vivid 'to the extent that it is (a) emotionally interesting, (b) concrete and imagery-provoking, and (c) proximate in a sensory, temporal, or spatial way.' Communications containing concrete and specific language have not always been found to be more persuasive than those using more abstract and general language (Taylor and Thompson, 1982; Collins, Taylor, Wood, and Thompson, 1988). In fact, the vividness effect may be more the exception than the rule.

In one study, Frey and Eagly (1993) tested the hypothesis that vividness effects are most likely to occur when message recipients are not constrained to pay attention to the information. They reasoned that vividness effects might be suppressed when subjects are forced to pay close attention to the message, which is the case in most persuasion experiments. When people are free to adjust the amount of attention they devote to stimuli, vivid information may gain a memorial and persuasive advantage over more pallid content. However, their results showed that when the message was presented in an incidental manner, vivid messages were less memorable and less persuasive than pallid messages. When subjects were constrained to attend to the message, vividness had no effect on memory or persuasion. Process data suggested that the vivid elements of the message interfered with the reception and processing of the essential meaning of the message and thereby reduced its memorability and persuasiveness.

The results of the Frey and Eagly study are particularly relevant to warnings because typically most warning situations involve incidental exposure. That incidental exposure may actually diminish the impact of vivid information (to the extent that non-vivid information would be more effective) is quite surprising and deserving of some direct examination by warnings researchers. These findings notwithstanding, Frey and Eagly are careful to point out that the interference or distraction associated with vivid information may be confined to complex messages involving abstract and multiple arguments (e.g., newspaper editorials). By contrast, warning messages often are quite brief and do not require the recipient to follow a complex line of thought or to synthesize a large amount of information in reaching a conclusion.

Match between perceived threat and perceived effectiveness

A second reason for why warning effectiveness may not always increase with increased perceived threat involves the relationship between perceived threat and the recommended action. Essentially, creating a high level of perceived threat may be counter-productive if the action is not perceived to be effective in reducing or removing the threat. Extending Leventhal's parallel process model (Witte, 1992), this type of boomerang effect is likely when a significant mismatch exists between perceived threat and perceived efficacy. Witte's perceived efficacy is similar to Bandura's (1977) concept of self-efficacy, in that it consists of two components: (1) beliefs about the outcomes that will result from the action, and (2) beliefs about one's ability to perform the behavior (see also Chapter 9 by DeJoy for a general discussion of self-efficacy). According to Witte (1992, p. 339), 'perceived efficacy is the crucial variable that determines which parallel process [danger control or fear control] will dominate.' In a high perceived threat/low perceived efficacy situation, people may react to the fear and deny the threat or dismiss the message. For example, if an individual does not believe that seat belts are effective in preventing injuries in motor vehicle crashes, messages which focus simply on establishing a high degree of perceived threat are not likely to produce increased usage of occupant restraints. Danger control is most likely to dominate in situations where both perceived threat and perceived efficacy are high. When danger control dominates, individuals are more likely to engage in self-protective behavior. In designing warnings, care should be taken to examine the relative match between perceived threat and the perceived ability of the recommended actions to alleviate the threat. The potential importance of this match further underscores the importance of the instructions portion of the warning.

10.4 SOCIAL INFLUENCE AND OTHER PERSUASION HEURISTICS

Current social psychological theory features two general accounts of message-based persuasion (Tesser and Shaffer, 1990). The first type assumes that the recipient systematically processes and elaborates the persuasive message. This type of processing is most likely to occur when the target individual is both motivated and able to scrutinize the message. Social influence attempts that assume systematic processing emphasize the quality (content) of the message. The second type of message-based persuasion emphasizes the heuristic processing of persuasive messages (e.g., Chaiken, Liberman, and Eagly, 1989), or the idea that people use simple rules of thumb or heuristics to guide their thinking about persuasive messages. Heuristic processing is likely when either the motivation or ability to engage in issues-related thinking is low. Given the number of persuasive messages that the average person is exposed to on a daily basis (mostly from advertising), it seems reasonable to argue that people rely quite heavily on heuristics to guide their decision-making (Cialdini, 1984). The sheer prevalence of warning messages in modern society and the fact that people often possess limited motivation to attend to or read these messages suggests that persuasion heuristics may be important to understanding why people do or do not comply with warnings.

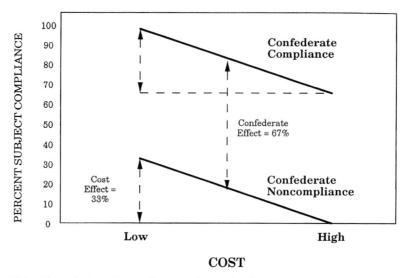

Figure 10.2 The relative effects of cost and social influence (confederate compliance) on warning compliance (adapted from Wogalter et al., 1989). Reprinted with permission. Copyright of the Human Factors and Ergonomics Society.

Eagly and Chaiken (1984) identified five major heuristics that people use when exposed to persuasive messages. First, people may rely on the perceived expertise of the source of the message (see Chapter 5 by Cox for a detailed discussion of source factors). Second, people tend to be more easily persuaded by people they like or admire. Third, people may evaluate the quality of the message by the number of arguments presented. For example, providing people with ten reasons for stopping smoking may be persuasive *not* because each point is valid but because so many apparent reasons for not smoking are presented. Fourth, people are sometimes persuaded by the presence of statistics in support of an argument (e.g., four out of five doctors recommend pain reliever X). Fifth, social influence or consensus cues may influence the persuasion process. Put more simply, people may sometimes use the behavior of others to gauge what is appropriate in a given situation.

Although persuasion heuristics would appear to be quite relevant to warnings, they have received very little attention from warnings researchers. In one interesting series of studies, Wogalter et al. (1989) examined the effects of cost and social influence on warning compliance. Social influence was manipulated by the presence of a confederate who either did or did not comply with the warning. Cost was varied in terms of the level of effort required to comply with the warning. They found that both factors produced significant effects on compliance. However, combining the data from two of the studies showed that although cost and social influence were both important determinants of compliance behavior, the effect of social influence was twice that of cost (see Figure 10.2). More recently, Racicot and Wogalter (1995) found that featuring a role model in a video message also resulted in better compliance when compared to the video message alone.

Returning to the social psychological literature for a moment, research on message-based persuasion (see review by Tesser and Shaffer, 1990) has yielded several findings that deserve to be explored in future research on warnings. Three of these are highlighted below. First, evidence shows that persuasion heuristics or other such 'peripheral' information are likely to be more important when the recipient's motivation and/or ability to undertake systematic processing is low. When motivation and/or ability is high, the impact of peripheral information is attenuated. Using cigarette warnings as the illustration, when perceived threat is high, it probably does not matter much that the warning was issued by the Surgeon General of the United States. It would be interesting to see a warning study that factorially varied both perceived threat and social influence and/or source credibility. Second, there is also social psychological research suggesting that heuristic and systematic processing can co-occur and have additive or interactive effects on the individual's judgments. In particular, people with an intermediate or uncertain level of motivation may scrutinize or systematically process message content when heuristic cues indicate that it would be worthwhile to do so. Under such circumstances, the credibility of the Surgeon General may prompt the recipient to analyze the message carefully, but he or she will be persuaded primarily to the extent that the actual arguments are strong. Third, heuristic cues can lead to biased processing of messages, especially when the individual is not knowledgeable and the message is ambiguous. For example, social influence may cause someone to process and elaborate the majority position but engage in little or no processing of minority views.

10.5 INDIVIDUAL DIFFERENCE FACTORS

Viewing risk communication as an interactive process means that considerable importance is assigned to what the person brings to the situation and to how the person interacts with warnings and related materials. In this chapter, three categories of individual difference factors are emphasized in the context of motivation: familiarity and experience, personal relevance, and risk-taking. These are motivational factors primarily because they have the potential to influence the formation of threat and/or outcome expectations.

10.5.1 Familiarity and Experience

As discussed elsewhere in this book (most notably Chapter 8 by Leonard, Otani and Wogalter and Chapter 9 by DeJoy), research demonstrates quite clearly that users who are more familiar or experienced with a product are less likely to look for (e.g., Godfrey *et al.*, 1983), read (e.g., LaRue and Cohen, 1987; Otsubo, 1988), and comply with warnings (e.g., Otsubo, 1988; Wogalter *et al.*, 1995). Typically, familiarity is defined in terms of the individual's personal knowledge of and/or experience with the product, object, or activity in question. It is worth noting, however, that familiarity and experience are not necessarily identical concepts. People may have some familiarity with products or classes of products that they seldom or never use themselves. In this regard, Wogalter *et al.* (1991) noted that ratings of product familiarity for a variety of consumer products were only correlated 0.66 with frequency of use and 0.28 with time of contact.

Three explanations were offered for the familiarity effect in Chapter 9 by DeJoy: benign experience, habituation, and script theory. The benign experience explanation holds that as people use a product without incurring any safety problems, they become less concerned about its dangers and more confident in using the product. Habituation occurs to the extent that people become accustomed to seeing a particular warning message and essentially it recedes into the background. As such, the ability of even a well designed warning message to attract attention can be expected to decrease as a function of repeated exposure. The sheer prevalence of warning messages in today's environment and the trend to standardize their appearance and content may actually contribute to the problem of habituation. Script theory (e.g., Schank and Abelson, 1977) reasons that experienced users rely on scripts stored in memory and devote little attention to warning labels, instructions, or other materials. A common theme in all these explanations is that familiar or experienced users are less likely to pay much attention to warnings because they have come to perceive the product or situation as relatively benign.

Experienced users might also be less likely to attend to warnings when they switch products within a category of familiar products. This is a potentially serious problem since new versions of familiar products may be more dangerous than older versions, and for some products specific hazards may be detected only after a particular product has been in use for several years or even longer. Two studies bear directly on this issue. Morris, Mazis, and Gordon (1977) found that approximately 78% of women read the patient package insert when they first started using oral contraceptives, but less than 11% read the insert that accompanied subsequent prescription refills. Godfrey and Laughery (1984) surveyed women about tampons and toxic shock syndrome and found that of the women who used tampons, 73% said they noticed a warning on or in the package. By contrast, only 32% of those switching from one tampon product to another noticed a warning on the new product. Furthermore, older women (aged 27 and older) were less likely to have noticed a warning (50%) than younger women (78%).

10.5.2 Personal Relevance

It seems logical to assume that people will not be motivated to comply with warnings that they do not consider to be personally relevant. DeJoy (1991) proposed a dynamic or stage model of the warning process that emphasizes the individual's personal risk appraisal activities (see Figure 10.3). Personal relevance enters into the first two stages of this model in particular.

Awareness

People who are not aware of a hazard may not look for or attend to warning messages. Strawbridge (1986) found that subjects tended to read only the first portion of the warning before going on to the 'uses' section of the label. When queried about this, the majority of subjects indicated that they were not interested in learning about the hazards of product

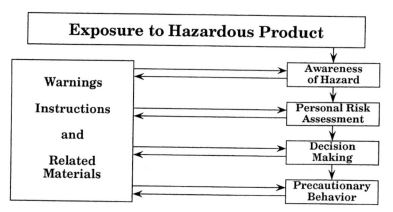

Figure 10.3 Dynamic or stage model of the warning process (adapted from DeJoy, 1991).

use but simply wished to understand how to use the product. In some instances, embedding warning information within instructions may be superior to other locations (Wogalter, Kalsher, and Racicot, 1993b; Frantz, 1994). In these two compliance studies, integrating warning information within usage instructions produced better compliance than placing similar warning information on a different part of the product packaging (Frantz, 1994) or on a warning sign located near where the product was to be used (Wogalter et al., 1993b). These results may have occurred because the experimental task required that the subjects attend to the instructions and the relevant equipment and not the surrounding environment.

Personalizing the warning message may be one way to attract the attention of naive users. Wogalter, Racicot, Kalsher, and Simpson (1993c) used a programmable electronic sign to present subjects with personalized warnings that contained their names. The personalized sign increased compliance (the wearing of safety equipment) compared to the impersonal sign. Racicot and Wogalter (1995) found that videotaped models (depicted as being in the same situation as the subjects) also were effective in improving compliance relative to a more conventional warning sign. These findings suggest that various devices that personalize the warning message may be effective in gaining the attention and subsequent compliance of experimental subjects. It is important to note that this effect is not the same thing as simply increasing the noticeability of the warning. Warning devices intended to enhance noticeability or conspicuity, such as symbols, pictorials, highlighting, and bright colors, may be of limited benefit (see DeJoy, 1989; Lehto and Papastavrou, 1993), in part because they do not increase the personal relevance of the hazard in question (see Chapter 7 by Wogalter and Leonard for a detailed discussion of attention and noticeability).

Personal risk assessment

The second stage of DeJoy's stage model, 'personal risk assessment,' is intended to reflect the idea that the individual must personalize the risk before entering the decision-making or cost–benefit stage. It is entirely possible for an individual to be aware of a particular hazard but conclude that that hazard does not pose a risk to him/her personally. This personalization of risk is important because considerable research suggests that most people tend to be unrealistically optimistic in judging their level of personal risk for a variety of possible adverse outcomes (e.g., Weinstein, 1980, 1987; Svenson, 1981; DeJoy, 1992). Specific to warnings, Rethans (1979) found that generally subjects were optimistic when judging their comparative risk of injury from a variety of consumer products. More recently, Bohannon and Young (1993), in a study of alcoholic beverage advertisements, reported that adolescents (mean age = 13.6 years) were optimistic when judging their risk compared to adolescents in general, and that the inclusion of warnings in the advertisements diminished this optimism. The older subjects (mean age = 23.3 years) in the study did not make optimistic judgements of personal risk. Other warnings-related research indicates that males tend to be more confident than females when judging their ability to avoid hazards and to use products without experiencing adverse consequences (Friedmann, 1988; Young, Martin, and Wogalter, 1989; Vredenburgh and Cohen, 1993).

Direct experience

Direct experience is thought to be a very potent factor in increasing the personal relevancy of a hazard. Weinstein (1989) reviewed research on the effects of motor vehicle crashes on seat belt use, criminal victimization (other than rape) on individual crime prevention efforts, natural hazards experience on both preparedness and compliance with evacuation warnings, and myocardial infarction and smoking. On the basis of this review, he offered five tentative propositions: (1) personal experience with adverse events leads people to perceive the hazards as more frequent and to see themselves as more vulnerable; (2) experience leads people to think more about the risk and with greater clarity; (3) effects of personal experience on perceptions of seriousness and controllability tend to be specific to the type of experience and situation encountered; (4) people take precautions that are relevant to the particular hazard experience encountered in the past; and (5) the duration of increased precautionary intent may be short.

The first two propositions suggest that personal experience tends to enhance perceptions of likelihood and severity. Those having direct experience with a particular hazard can be expected to have more vivid and concrete thoughts about the consequences of that hazard. However, with respect to proposition three, many experiences involve only mild losses, so perceptions of severity may not automatically follow from personal experience. Also, in some cases, the actions demanded to prevent future victimization may actually increase the individual's feelings of uncontrollability. The third and fourth propositions both suggest that personal experience with one particular hazard typically does not produce a generalized feeling of vulnerability. Precautionary intent is more likely to be specific to the particular threat encountered and proportionate to the level of severity experienced and the perceived effectiveness of the available self-protective measures. The last proposition is important because it implies that personal experience does not necessarily produce lasting effects; in some instances, the 'window of opportunity' for increased precautionary action appears to be quite limited indeed.

Although personal experience would seem to predict warning compliance, the small amount of warnings-related research on this topic has yielded mixed results. Martin and Wogalter (1989), in a rating study of 18 consumer products, found that, in general, subjects with greater injury experience with a particular product reported greater precautionary intent with that product. On the other hand, Lehto and Foley (1991) studied ATV riders and found that the proportion of riders wearing helmets did not increase with accident experience, regardless of severity. The authors speculated that this finding might have occurred because the riders did not believe that a helmet would have prevented their injuries. However, they also noted that 98% of riders indicated that wearing a helmet can prevent serious injury or death. This type of attitude–behavior inconsistency, while perplexing, is certainly not unique to this particular example of self-protective behavior.

10.5.3 Risk taking

Two people exposed to the same level of hazard may respond quite differently in terms of precautionary behavior. One possible explanation for this frequent observation is that people differ in their characteristic need for and tolerance of risk. This personality or trait explanation implies that people who are high in sensation-seeking (Zuckerman, 1979) may be less likely to comply with warnings and may even disregard certain warnings intentionally in the interest of experiencing high levels of risk. Sensation-seekers might actually be motivated to ignore warnings. Purswell, Schlegel, and Kejriwal (1986) used discriminant analysis to classify subjects into safe and unsafe groups with respect to their use of four consumer products (saber-saw, electric knife, router, and drain cleaner). Their results indicated that sensation seeking was a useful predictor for hazardous product use.

Unfortunately, the presentation of the results did not provide data on the relative importance of sensation seeking to classification accuracy.

A competing explanation for the observed individual differences in risk taking is that essentially the willingness to take risks is domain- or situation-specific. Basically, this perspective views personal safety as a commodity which people are willing to trade off against other benefits (Denscombe, 1993). The level of perceived threat and the perceived effectiveness of the recommended action are important factors in deciding whether to avoid a hazard, but other factors also enter into the equation. Frequently taking a risk is associated with a number of benefits, such as convenience, pleasure, peer approval, or even economic gain. As a general rule, the particular costs and benefits involved and the importance attached to them can be expected to vary from one situation to the next. Although warnings-related data are scarce, evidence can be found in the psychological and risk perception literature indicating that often people make riskier decisions in some situations than in others (e.g., Kogan and Wallach, 1964; Dion, Baron, and Miller, 1970). For example, people make riskier decisions in activities in which they freely choose to participate (e.g., Starr, 1969), and that people will tolerate substantially different levels of risk depending on the manner in which the risk information is presented or framed (e.g., Kahneman and Tversky, 1982).

One practical implication of this discussion of risk-taking is that the decision to use warning messages should not be made without first analyzing the situation in which they will be applied and without exploring the predominant motives of the target population. For example, in some recreational environments the presence of a warning might actually encourage risk taking or perhaps attract those who are seeking thrills and high levels of risk.

10.6 SUMMARY AND CONCLUSIONS

This chapter began with the proposition that most warnings are fear-based communications: they inform people that something bad might happen if they do not heed the warning. This type of message framing assumes that the expectation of threatening consequences arouses fear, which in turn increases the probability of some type of response that will reduce the fear. A brief review of the fear-arousal literature in psychology and communications, however, showed that the relationship between fear arousal and attitude and behavior change is far from simple. Fear arousal does not always produce the expected outcomes, and in some instances, high levels of fear may actually be less effective than more moderate levels. Also, recent theoretical formulations dealing with fear appeals have tended to de-emphasize emotional arousal *per se* in favor of cognitive processes. Increased attention has been given to the individual's evaluative and decision-making activities and to perceptions related to outcome severity, outcome likelihood, and the perceived efficacy of the recommended actions.

Clearly, perceived threat is an important motivational factor in warning compliance, and warnings studies are quite consistent in showing that warning effectiveness increases as a function of increased perceived hazardousness. Whether this is a linear or curvilinear function essentially is unknown at this point. In addition, perceptions about the severity of possible injury appear to be more important than perceptions about injury likelihood in establishing perceived threat, at least for most consumer products. Providing explicit information about injury consequences is one way to enhance perceived hazardousness and perceived injury severity, but a variety of other warning features and devices have also been shown to enhance perceived threat. It is worth noting that very little attention has been given to examining the positive framing of warning messages. The most straightforward type of positive framing might be to emphasize the benefits of performing the recommended behavior (e.g., wear safety glasses to protect your vision). Results from non-warning studies suggest that positive- or gain-framed appeals may be most beneficial when people are being asked to engage in preventive or health-enhancing behaviors, which is the case in most warning applications (Rothman and Salovey, 1997).

Perhaps the most prudent conclusion about perceived threat is that it is a necessary but not sufficient condition for warning compliance. Beliefs about the effectiveness of the actions called for in the warning and costs associated with complying are also important in motivating precautionary behavior. In short, the effectiveness of a warning message should increase if either the benefits of taking the action can be heightened or if the costs can be reduced. There are two basic ways to increase benefits: (1) increase the perceived effectiveness of the recommended action; or (2) increase the level of perceived threat (or the costs of non-compliance), which has the indirect effect of making the available action seem more necessary or beneficial. Although research indicates that warning effectiveness degrades rapidly as costs increase, impressively high levels of compliance have been achieved in situations where costs can be substantially reduced. While the consequences portion of the standard warning messages is central to establishing perceived threat, the instructions portion provides the basis for assessing the benefits of complying.

Consistent with the decision-making focus of this chapter as well as much current thinking in the fear-arousal literature, care should be taken in the design of warnings to balance perceived threat and perceived effectiveness. A warning that creates a high level of perceived threat may not be effective if the recommended actions are perceived to be inappropriate or inadequate in reducing or eliminating the threat. This type of mismatch may help to explain the boomerang-type effects noted in the fear-arousal literature where message effectiveness sometimes declines when very high levels of fear are produced. Explicit and vivid messages may enhance perceived threat, but they may also force people to deny or discount the warning if effective means to control the threat are not offered. In addition, highly fear-arousing messages may distract the recipient and interfere with the processing of the message itself.

The social psychological literature on persuasion heuristics may provide some clues to understanding how people respond to warning messages. The basic idea is that often people use simple rules or heuristics to guide their thinking about persuasive messages. One persuasion heuristic in particular, social influence, has been investigated by warning researchers. In this research, live and video role models were used to model the actions called for in the warning message. When models were present, compliance rates were higher than when they were absent. In general, persuasion heuristics are likely to be important in the presence of lower or more ambiguous levels of perceived threat. When motivation or perceived threat is high, heuristics, such as the credibility or likeability of the source or social influence, probably will not be very important. This social psychological perspective also emphasizes the potential importance of source factors (credibility, etc.) in motivating self-protective behavior. In general, source factors have not been explored systematically by warning researchers (see also Chapter 5 by Cox).

Finally, the motivation to follow warnings is at least partially determined by the expectations that the message recipient brings to the situation. Three categories of individual difference factors were discussed: familiarity/experience, personal relevance, and risk-taking. First and foremost, a number of studies indicate that those who are familiar and/or experienced with a particular product, object, or activity are less likely to seek out and comply with warning messages. Cognitive reappraisal appears to be the underlying process here; familiar and experienced individuals have come to view the situation as involving limited or minimal personal threat. Still, considerable caution should be exercised in concluding that unfamiliar or novice individuals can be expected to be looking for and alert to warning information. Such individuals usually are in an information-seeking mode, but they may not be seeking warning information *per se*. A higher priority may be to figure out how to assemble or operate the product. They may overlook even a highly conspicuous warning because it simply is not relevant to them at that point in time. The whole issue of personal relevance needs to be explored in more detail by warnings researchers. As a final point, differences in risk taking may cause people to respond very differently to the same potential hazard. This chapter favors the domain- or situation-specific over the personality explanation for these differences; however, direct warning-related evidence is extremely limited on both sides. Since 'human error' or intentional risk taking is implicated frequently in accidents and injury (often by the process of elimination), then there is a clear need for further study of risk taking in the context of warnings.

REFERENCES

ACTON, W.I. (1977) Problems associated with the use of hearing protection. *Annals of Occupational Hygiene*, 29, 387–395.

AVERILL, J.R. (1987) The role of emotion and psychological defense in self-protective behavior. In WEINSTEIN, N.D. (ed.), *Taking Care: Understanding and Encouraging Self-protective Behavior*. New York: Cambridge University Press, pp. 54–78.

BANDURA, A. (1977) *Social Learning Theory*. Englewood Cliffs, NJ: Prentice-Hall.

BECKER, M.H. and JANZ, N.K. (1987) On the effectiveness and utility of health hazard/health risk appraisal in clinical and nonclinical settings. *Health Services Research*, 22, 537–551.

BOHANNON, N.K. and YOUNG, S.L. (1993) Effect of warnings in advertising on adolescents' perceptions of risk for alcohol consumption. In *Proceedings of the Human Factors Society 37th Annual Meeting*. Santa Monica, CA: Human Factors Society, pp. 974–978.

BRESNAHAN, T.F. and BRYK, J. (1975) The hazard association values of accident-prevention signs. *Professional Safety*, 17–25.

CHAIKEN, S., LIBERMAN, A., and EAGLY, A.H. (1989) Heuristic and systematic information processing within and beyond the persuasion context. In ULEMAN, J.S. and BARGH, J.A. (eds), *Unintended Thought: Limits of Awareness, Intention, and Control*. New York: Guilford, pp. 212–252.

CIALDINI, R. (1984) *Influence*. New York: Harcourt Brace Jovanovich.

CLEVELAND, R.J. (1984) Factors that influence safety shoe usage. *Professional Safety*, 29, 26–29.

COLLINS, B.L. (1983) Evaluation of mine-safety symbols. In *Proceedings of the Human Factors Society 27th Annual Meeting*. Santa Monica, CA: Human Factors Society, pp. 947–949.

COLLINS, R.L., TAYLOR, S.E., WOOD, J.V., and THOMPSON, S.C. (1988) The vividness effect: elusive or illusory? *Journal of Experimental Social Psychology*, 24, 1–18.

DEJOY, D.M. (1989) Consumer product warnings: review and analysis of effectiveness research. In *Proceedings of the Human Factors Society 33rd Annual Meeting*. Santa Monica, CA: Human Factors Society, pp. 936–940.

DEJOY, D.M. (1991) A revised model of the warnings process derived from value-expectancy theory. In *Proceedings of the Human Factors Society 35th Annual Meeting*. Santa Monica, CA: Human Factors Society, pp. 1043–1047.

DEJOY, D.M. (1992) An examination of gender differences in traffic accident risk perception. *Accident Analysis and Prevention*, 24, 237–246.

DENSCOMBE, M. (1993) Personal health and the social psychology of risk taking. *Health Education Research*, 8, 505–517.

DINGUS, T.A., WREGGIT, S.S., and HATHAWAY, J.A. (1993) Warning variables affecting personal protective equipment use. *Safety Science*, 16, 655–673.

DION, K., BARON, R., and MILLER, N. (1970) Why do groups make riskier decisions than individuals? *Advances in Experimental Social Psychology*, 5, 305–377.

DUFFY, R.R., KALSHER, M.J., and WOGALTER, M.S. (1995) Interactive warning: An experimental examination of effectiveness. *International Journal of Industrial Ergonomics*, 15, 159–166.

EAGLY, A.H. and CHAIKEN, S. (1984) Cognitive theories of persuasion. *Advances in Experimental Social Psychology*, 17, 267–359.

FHANER, G. and HANE, M. (1974) Seat belts: relations between beliefs, attitude, and use. *Journal of Applied Psychology*, 59, 472–484.

FMC CORPORATION (1980) *Product Safety Signs and Label System*, (3rd Edn). Santa Clara, CA: FMC Corporation.

FRANTZ, J.P. (1994) Effect of location and procedural explicitness on user processing of and compliance with product warnings. *Human Factors*, 36, 532–546.

FRANTZ, J.P. and RHOADES, T.P. (1993) A task analytic approach to the temporal placement of product warnings. *Human Factors*, 35, 719–730.

FREY, K.P. and EAGLY, A.H. (1993) Vividness can undermine the persuasiveness of messages. *Journal of Personality and Social Psychology*, 65, 32–44.

FRIEDMANN, K. (1988) The effect of adding symbols to written warning labels on user behavior and recall. *Human Factors*, 30, 507–516.

GODFREY, S.S., ALLENDAR, L., LAUGHERY, K.R., and SMITH, V.L. (1983) Warning messages: will the consumer bother to look? In *Proceedings of the Human Factors Society 27th Annual Meeting*. Santa Monica, CA: Human Factors Society, pp. 950–954.

GODFREY, S.S. and LAUGHERY, K.R. (1984) The biasing effects of product familiarity on consumers awareness of hazard. In *Proceedings of the Human Factors Society 28th Annual Meeting*. Santa Monica, CA: Human Factors Society, pp. 388–393.

HIGBEE, K.L. (1969) Fifteen years of fear arousal: research on threat appeals 1953–1968. *Psychological Bulletin*, 72, 426–444.

HILTON, M.E. (1993) An overview of recent findings on alcoholic beverage warning labels. *Journal of Public Policy and Marketing*, 12, 1–9.

HOVLAND, C., JANIS, I., and KELLY, H. (1953) *Communication and Persuasion*. New Haven, CT: Yale University Press.

JANIS, I.L. and FESHBACH, S. (1953) Effects of fear-arousing communications. *Journal of Abnormal and Social Psychology*, 48, 78–92.

JOB, R.F.S. (1988) Effective and ineffective use of fear in health promotion campaigns. *American Journal of Public Health*, 78, 163–167.

JONES, S. (1978) Symbolic representation of abstract concepts. *Ergonomics*, 21, 573–577.

KAHNEMAN, D. and TVERSKY, A. (1982) The psychology of preferences. *Scientific American*, 246, 160–173.

KOGAN, N. and WALLACH, M. (1964) *Risk Taking: A Study in Cognition*. New York: Holt, Rinehart, and Winston.

LARUE, C. and COHEN, H.H. (1987) Factors affecting consumers' perceptions of product warnings: an examination of the differences between male and female consumers. In *Proceedings of the Human Factors Society 31st Annual Meeting*. Santa Monica, CA: Human Factors Society, pp. 610–614.

LAUGHERY, K.R., ROWE-HALLBERT, A.L., YOUNG, S.L., VAUBEL, K.P., and LAUX, L.F. (1991) Effects of explicitness in conveying severity information in product warnings. In *Proceedings of the Human Factors Society 35th Annual Meeting*. Santa Monica, CA: Human Factors Society, pp. 481–485.

LAUGHERY, K.R., VAUBEL, K.P., YOUNG, S.L., BRELSFORD, J.W., and ROWE, A.L. (1993) Explicitness of consequence information in warning. *Safety Science*, 16, 597–613.

LEHTO, M.R. and FOLEY, J.P. (1991) Risk-taking, warning labels, training, and regulation: are they associated with the use of helmets by all-terrain vehicle riders? *Journal of Safety Research*, 22, 191–200.

LEHTO, M.R. and PAPASTAVROU, J.D. (1993) Models of the warning process: important implications towards effectiveness. *Safety Science*, 16, 569–595.

LEONARD, S.D., MATHEWS, D., and KARNES, E.W. (1986) How does the population interpret warning signals? In *Proceedings of the Human Factors Society 30th Annual Meeting*. Santa Monica, CA: Human Factors Society, pp. 116–120.

LEVENTHAL, H. (1970) Findings and theory in the study of fear communications. In BERKOWITZ, L. (ed.), *Advances in Experimental Social Psychology*, vol. 5, pp. 119–186.

LEVENTHAL, H., SINGER, R., and JONES, S. (1965) The effects of fear and specificity of recommendations upon attitudes and behavior. *Journal of Personality and Social Psychology*, 2, 20–29.

LORING, B.A. and WIKLUND, M.E. (1988) Improving swimming pool warning signs. In *Proceedings of the Human Factors Society 32nd Annual Meeting*. Santa Monica, CA: Human Factors Society, pp. 910–914.

LOWRANCE, W.W. (1980) *Of Acceptable Risk: Science and the Determination of Safety*. Los Altos, CA: William Kaufman.

MCDOWELL, I. (1988) Man the risk taker. *American Journal of Preventive Medicine*, 4, 172–177.

MCGUIRE, W.J. (1980) The communication-persuasion model and health-risk labeling. In MORRIS, L.A., MAZIS, M.B., and BAROFSKY, I. (eds), *Product Labeling and Health Risks*. Banbury Report No. 6. Coldspring Harbor Laboratory.

MARTIN, E.G. and WOGALTER, M.S. (1989) Risk perception and precautionary intent for common consumer products. In *Proceedings of the Human Factors Society 33rd Annual Meeting*. Santa Monica, CA: Human Factors Society, pp. 931–935.

MORRIS, L.A. and KANOUSE, D.E. (1981) Consumer reaction to the tone of written drug information. *American Journal of Hospital Pharmacy*, 38, 667–671.

MORRIS, L.A., MAZIS, M., and GORDON, E. (1977) A survey of the effects of oral contraceptive patient information. *Journal of the American Medical Association*, 238, 2504–2508.

NISBETT, R.E. and ROSS, L. (1980) *Human Inference: Strategies and Short-comings of Social Judgment*. Englewood Cliffs, NJ: Prentice-Hall.

OTSUBO, S.M. (1988) A behavioral study of warning labels for consumer products: perceived danger and use of pictographs. In *Proceedings of the Human Factors Society 32nd Annual Meeting*. Santa Monica, CA: Human Factors Society, pp. 536–540.

PETERS, G.A. (1984) A challenge to the safety profession. *Professional Safety*, 29, 46–50.

PURSWELL, J.L., SCHLEGEL, R.E., and KEJRIWAL, S.K. (1986) Prediction model for consumer behavior regarding product safety. In *Proceedings of the Human Factors Society 30th Annual Meeting*. Santa Monica, CA: Human Factors Society, pp. 1202–1205.

RACICOT, B.M. and WOGALTER, M.S. (1995) Effects of a video warning sign and social modeling on behavioral compliance. *Accident Analysis and Prevention*, 27, 57–64.

RETHANS, A. (1979) An investigation of consumer perceptions of product hazards. Doctoral dissertation, University of Oregon.

RILEY, M.W., COCHRAN, D.J., and BALLARD, J.L. (1982) An investigation of preferred shapes for warning labels. *Human Factors*, 24, 737–742.

ROGERS, R.W. (1983) Cognitive and physiological processes in fear appeals and attitude change: a revised theory of protection motivation. In CACIOPPO, J. and PETTY, R. (eds), *Social Psychophysiology*. New York: Guilford, pp. 153–176.

ROTHMAN, A.J. and SALOVEY, P. (1997) Shaping perceptions to motivate healthy behavior: the role of message framing. *Psychological Bulletin*, 3–19.

SCHANK, R.C. and ABELSON, R. (1977) *Scripts, Plans, Goals, and Understanding*. Hillsdale, NJ: Lawrence Erlbaum Associates.

SILVER, N.C., LEONARD, D.C., PONSI, K.A., and WOGALTER, M.S. (1991) Warnings and purchase intentions for pest-control products. *Forensic Reports*, 4, 17–33.

SLOVIC, P., FISCHHOFF, B., and LICHTENSTEIN, S. (1979) Rating the risks. *Environment*, 21, 14–20, 36–39.

SLOVIC, P., FISCHHOFF, B., and LICHTENSTEIN, S. (1980) Facts and fears: understanding perceived risk. In SCHWING, R. and ALBERS JR, W.A. (eds), *Societal Risk Assessment: How Safe is Safe Enough*. New York: Plenum.

STARR, C. (1969) Social benefit versus technological risk. *Science*, 165, 1232–1238.

STRAWBRIDGE, J.A. (1986) The influence of position, highlighting, and imbedding on warning effectiveness. In *Proceedings of the Human Factors Society 30th Annual Meeting*. Santa Monica, CA: Human Factors Society, pp. 716–720.

SUTTON, S.R. (1982) Fear-arousing communications: a critical examination of theory and research. In EISER, J.R. (ed.), *Social Psychology and Behavioral Medicine*. London: Wiley, pp. 303–307.

SVENSON, O. (1981) Are we all less risky and more skillful than our fellow drivers? *Acta Psychologica*, 47, 143–148.

TAYLOR, S.E. and THOMPSON, S.C. (1982) Stalking the elusive 'vividness' effect. *Psychological Review*, 89, 155–181.

TESSER, A. and SHAFFER, D.R. (1990) Attitudes and attitude change. *Annual Review of Psychology*, 41, 479–523.

VISCUSI, W.K., MAGAT, W.A., and HUBER, J. (1986) Informational regulation of consumer health risks: an empirical evaluation of hazard warnings. *Rand Journal of Economics*, 17, 351–365.

VREDENBURGH, A.G. and COHEN, H.H. (1993) Compliance with warnings in high risk recreational activities: skiing and scuba. In *Proceedings of the Human Factors Society 37th Annual Meeting*. Santa Monica, CA: Human Factors Society, pp. 945–949.

WEINSTEIN, N.D. (1980) Unrealistic optimism about future life events. *Journal of Personality and Social Psychology*, 39, 806–820.

WEINSTEIN, N.D. (1987) Unrealistic optimism about illness susceptibility: conclusions from a community-wide sample. *Journal of Behavioral Medicine*, 10, 481–500.

WEINSTEIN, N.D. (1989) Effects of personal experience on self-protective behavior. *Psychological Bulletin*, 105, 31–50.

WEINSTEIN, N.D. (1993) Testing four competing theories of health-protective behavior. *Health Psychology*, 12, 324–333.

WESTINGHOUSE ELECTRIC CORPORATION (1981) *Product Safety Label Handbook*. Trafford, PA: Westinghouse Printing Division.

WILSON, D.K., PURDON, S.E., and WALLSTON, K.A. (1988) Compliance to health recommendations: a theoretical overview of message framing. *Health Education Research*, 3, 161–171.

WITTE, K. (1992) Putting the fear back into fear appeals: the extended parallel process model. *Communication Monographs*, 59, 329–349.

WOGALTER, M.S., ALLISON, S.T., and MCKENNA, N.A. (1989) Effects of cost and social influence on warning compliance. *Human Factors*, 31, 133–140.

WOGALTER, M.S. and BARLOW, T. (1990) Injury likelihood and severity in warnings. In *Proceedings of the Human Factors Society 34th Annual Meeting*. Santa Monica, CA: Human Factors Society, pp. 580–583.

WOGALTER, M.S., BARLOW, T., and MURPHY, S. (1995) Compliance to owner's manual warnings: influence of familiarity and the task-relevant placement of a supplemental directive. *Ergonomics*, 38, 1081–1091.

WOGALTER, M.S., BRELSFORD, J.W., DESAULNIERS, D.R., and LAUGHERY, K.R. (1991) Consumer products warnings: the role of hazard perception. *Journal of Safety Research*, 22, 71–82.

WOGALTER, M.S., BREMS, D.J., and MARTIN, E.G. (1993a) Risk perception of common consumer products: judgments of accident frequency and precautionary intent. *Journal of Safety Research*, 24, 97–106.

WOGALTER, M.S., GODFREY, S.S., FONTENELLE, G.A., DESAULNIERS, D.R., ROTHSTEIN, P.R., and LAUGHERY, K.R. (1987) Effectiveness of warnings. *Human Factors*, 29, 599–622.

WOGALTER, M.S., JARRARD, S.W., and SIMPSON, S.W. (1994) Influence of signal words on perceived label of product hazard. *Human Factors*, 36, 547–556.

WOGALTER, M.S., KALSHER, M.J., and RACICOT, B.M. (1993b) Behavioral compliance with warnings: effects of voice, context, and location. *Safety Science*, 16, 637–654.

WOGALTER, M.S., RACICOT, B.M., KALSHER, M.J., and SIMPSON, S.N. (1993c) Behavioral compliance to personalized warning signs and the role of perceived relevance. In *Proceedings of the Human Factors and Ergonomics Society 37th Annual Meeting*. Santa Monica, CA: Human Factors Society, pp. 950–954.

YOUNG, S.L., BRELSFORD, J.W., and WOGALTER, M.S. (1990) Judgments of hazard, risk, and danger: do they differ? In *Proceedings of the Human Factors Society 34th Annual Meeting*. Santa Monica, CA: Human Factors Society, pp. 503–507.

YOUNG, S.L., MARTIN, E.G., and WOGALTER, M.S. (1989) Gender differences in consumer product hazard perceptions. In *Proceedings of Interface 89*. pp. 73–78.

ZUCKERMAN, M. (1979) *Sensation Seeking: Beyond the Optimal Level of Arousal*. Hillsdale, NJ: Lawrence Erlbaum Associates.

CHAPTER ELEVEN

Behavior

N. CLAYTON SILVER
University of Nevada at Las Vegas

CURT C. BRAUN
University of Idaho

This chapter explores the effects of warnings on behavior. Research employing behavioral intention and/or behavior measures is reviewed. Studies manipulating the presence or absence of warnings show that warnings can influence behavior. Also, both physical and content characteristics of warnings influence compliance. In addition to the characteristics of warnings, properties of the target audience are factors in compliance. Familiarity, perceived risk, gender, and locus of control are examples of personal factors that play a role in warning effects on behavior. Three situational variables, time pressure, cost of compliance, and modeling, have been shown to substantially affect the extent to which warnings influence behavior. Finally, some of the work on mass media efforts to influence safety behavior is reviewed.

11.1 INTRODUCTION

At the most basic level, a warning is a communication that conveys the existence of a hazard that can be avoided or minimized through appropriate behavior. The previous four chapters have focused on the sequential stages that warning information goes through (attention/noticeability, comprehension/memory, attitudes and beliefs, and motivation) culminating in behavior. By definition, behavior is the collection of observable, overt acts that can be measured (Lefton, 1997). In this case, the behavior is compliance or non-compliance with the warning. In Chapter 4, Wogalter and Dingus explained how behavior can be evaluated. In some cases, 'behavior' is evaluated through measures of behavioral intentions, that is, what people report they will or would do in response to a warning. This chapter will examine factors that influence behavioral intentions and behavioral compliance.

Two meta-analyses by Kim and Hunter (1993a,b) suggest the relationship between intention and behavior might be stronger than previously believed. Using data collected from over 90 000 individuals across 138 different attitude and behavior situations and adjusting for measurement reliabilities (i.e., attenuated correlations), Kim and Hunter (1993a) reported the corrected correlation between attitudes and behavior to be strong at $r = 0.79$.

A similar analysis (Kim and Hunter, 1993b) of the relationship between behavioral intentions and behavior produced similar results with $r = 0.82$.

Kim and Hunter's (1993b) findings suggest three criteria for evaluating the utility of subjective measures in warnings research. First, subjective measures must display a high degree of correspondence and relevance to the desired behavior. For example, assessing an individual's willingness to comply with a warning might be a better predictor of actual compliance than their perception of the hazards associated with a particular activity or product. Second, the validity of intention scores must take into consideration the level of volitional control needed to actually perform the act. An individual's willingness to comply, for example, must be weighted by his/her ability to comply. As volitional control decreases, so too does the predictive utility of these subjective measures. Finally, consideration must be given to the reliability of the scores. Scores from subjective measures demonstrates decreasing predictive validity as reliability decreases. Given these three criteria, there is evidence suggesting that one can infer behavior from behavioral intentions and that such measures are useful in understanding warning effectiveness.

In the following sections of this chapter we review research that has addressed the effects of warnings on behavioral intentions and behavior. The sections are organized on the basis of factors that may influence such effects. They included presence or absence of a warning, physical characteristics (location, color, and multiple modalities), content (language and pictorial symbols), individual difference characteristics (familiarity, perceived risk, prior injury, sex differences, locus of control and time stress), cost of compliance, and modeling. The last section of the chapter reviews some related findings, namely, in the area of mass communications.

11.2 PRESENCE OR ABSENCE OF WARNINGS

Research has been reported that examined effects of warnings by manipulating whether warnings were present or absent. Wogalter, Jarrard, and Simpson (1994) found that ratings of perceived hazard and precautionary intentions were lower for products with labels containing no warnings or warning signal words than for products with either a warning or a warning and an icon. Behavioral data are somewhat analogous. Wogalter et al. (1987) showed that a higher proportion of individuals completing a chemistry task donned protective equipment when a warning was present than when it was absent. Similar findings were noted for tasks involving a photocopier, pay telephone, and a broken door. Moreover, in a chemistry task similar to Wogalter et al. (1987), Bouhatab (1991) found that providing safety pictorials on the wall of the laboratory facilitated greater compliance (using goggles) than when no pictorials were shown. Ferrari and Chan (1991) introduced warning signs placed around elevator doors in a dwelling, signifying that the sound volume of personal stereos should be reduced. The volume went down after the warning was posted. When the warning was removed, the publicly audible stereos returned back to baseline.

A field study by Thyer and Geller (1987) explored the effectiveness of a dashboard sticker warning that read 'SAFETY BELT USE REQUIRED IN THIS VEHICLE.' The use of seat belts by front seat passengers was the dependent measure. During an initial two-week baseline phase, belt use was 34%. The sticker was then placed on the dashboard, and the usage increased to 70%. Two weeks later the stickers were withdrawn and passenger belt use dropped to 41%. Finally replacement of the stickers for two final weeks resulted in 78% belt use.

Another field study by Laughery, Laughery, and Lovvoll (1998) explored the effect of a warning on whether or not service station attendants would mount a 16-inch tire on a 16.5-inch rim, a hazardous mismatch. The tire and rim were taken to service stations in the back of a pickup truck, and the station attendants were asked to mount them. In one condition of the study, a warning label was placed on the rim, whereas in another condition there was no label. At 10 of 18 stations when the label was present the attendant refused to mount the tire, whereas 9 of 10 started to mount the tire at stations when no label was present.

The above laboratory and field studies indicate that the presence of a warning resulted in changes in behavioral intentions and behavior. A good deal of the research addressing warnings effects on behavior has manipulated various characteristics of the warning as independent variables. These characteristics include both physical and content factors.

11.3 PHYSICAL CHARACTERISTICS

Several physical characteristics of warnings have been investigated with regard to behavioral effects, including location (placement), color and modality.

11.3.1 Location

In the real estate industry it is stated that three factors affect the value of a piece of property: location, location, location. Although location is not the only factor that accounts for product warning effectiveness, it plays an important role. For example, Frantz and Rhoades (1993) found increased compliance for warnings that were placed so they temporarily interfered with the task at hand, namely file cabinet installation and loading. Furthermore, Duffy, Kalsher, and Wogalter (1993) reported that when individuals were plugging items into an extension cord, an interactive warning decreased the number of individuals who performed the task in an unsafe manner.

There are other warning location decisions that influence compliance. For instance, Strawbridge (1986) investigated how a product warning's position, highlighting, and embeddedness influenced effectiveness. The warning information mentioned that the product contained acid and must be shaken to avoid severe burns. A control condition was used that instructed the person to shake the product. When the critical warning message was at the beginning of the warning section (not embedded) more individuals shook the product (47%) than when it was embedded in the middle of the section (27%). Under certain non-embedded conditions, behavioral compliance reached as high as 60%. The reduced compliance resulting from placing a warning in a cluttered environment has been replicated (e.g., Wogalter, Kalsher, and Racicot, 1993a). In some circumstances keeping warning information separate and distinct enhances compliance because it is potentially more attention-capturing and easier to read.

But manufacturers do not always separate instructions from precautions. Would compliance be enhanced if the warning were placed in a precautions section rather than in the directions? Frantz (1993) had individuals use a drain cleaner to unclog a kitchen sink drain. The warning messages were either placed in the Directions for Use section or the Precautions section. Surprisingly, the compliance rate for placing the warning in the Directions for Use section was 83%, whereas the compliance rate was only 48% for the Precautions section. In a separate study, Frantz (1994) found similar results. These results are contradictory to the suggestions that precautions should be separated from usage information (Ryan, 1991). When queried about their behavior, most of the participants indicated that they were searching for information as to how to use the product when they encountered the warning.

Lending some credence to these findings, Zlotnik (1982) reported that individuals receiving warnings embedded within instructions committed fewer errors in assembling hobby kits than those who received instructions without embedded warnings. Moreover, Wogalter, Kalsher, and Racicot (1992) and Wogalter, Magurno, Rashid, and Klein (1998) reported that individuals performing a chemistry laboratory task were more likely to comply with the warning when it was within a task instruction sheet than on a sign posted on a wall in front of them.

Evaluating where to put the warning within the instructions, Wogalter *et al.* (1987) found that individuals were about twice as likely to comply with warnings placed at the beginning of instructions than with those at the end. Hence, these results indicate that the warning has a greater chance of being complied with when it is in a place that will be most often seen by individuals who are seeking information about the product's use.

Thus far, we have examined warning placement as a function of embeddedness. Yet another concern is: where does one physically place the warning on the product in order to enhance its noticeability? Wogalter, Barlow, and Murphy (1995) asked individuals to connect an external disk drive to a microcomputer. They had to ground a plug, eject the disk, and turn off the computer before connecting the drive. An additional safety directive ('read the second page of the manual') was placed in various locations (e.g., front of the drive, cover page of the manual, shipping box). Greater compliance was observed when the directive was placed on the front of the disk drive. Similarly, Wogalter and Young (1994) had participants glue together a model plane. The warning information was provided as a tag attached to the mouth of the bottle, wings surrounding the bottle, or a warning on the label. Over 80% complied (i.e., put on gloves) for the tag, 35% for the wings, and only 13% for the standard warning label. Moreover, Wogalter, Glover, Magurno, and Kalsher (1999) found that using the tag label to convey instructions about connecting automobile battery jumper cables also led to greater compliance.

The majority of research focusing on the placement of warnings has used tasks that require the use of instructions. Clearly, research participants are unlikely to be familiar with tasks involving chemistry experiments, installing computer equipment, or cleaning drains. The low familiarity with these tasks dictated the use of the instructions. In situations where the tasks were more familiar, however, results concerning the placement of the warning have been less optimistic. Gill, Barbera, and Precht (1987) evaluated the effect of three different warning placements on the use of an electric heater. Regardless of whether a traditional tag-based warning or an interactive warning was used, there was no compliance in all conditions. Similarly, Hunn and Dingus (1992) tested the effect of three different warning placements on compliance. Using a general liquid cleaning product, warnings were presented on the product label, on a card glued to the nozzle of the sprayer, or affixed to the handle. Compliance rates for the three warnings did not differ. Thus, although most of the research indicates that location has a substantial effect on compliance, the literature is not entirely consistent.

11.3.2 Color

One of the physical characteristics of a warning that can also influence its effectiveness is color. Wogalter *et al.* (1987) found that a multi-feature chromatic (color) warning placed on a drinking fountain signifying unpotable water was more effective than a simpler achromatic (black and white) warning in deterring drinking. Expanding upon these results, Braun and Silver (1995) found that a higher proportion of participants complied (e.g., donned protective gloves) when the warning was printed in red rather than in green or black. Red, which is associated with terms such as hot and danger, has a greater hazard association value than either green or black. Furthermore, Wogalter *et al.* (1999) found that individuals were more likely to connect automobile-battery jumper cables correctly when the colors yellow and red were added to a manufacturer's original achromatic warning label. Hence, because a color such as red can convey hazard, it can also help produce higher compliance rates.

11.3.3 Multiple Modalities

Another possibility for enhancing warning effectiveness is the use of multiple modalities. Wogalter *et al.* (1993a) reported that compliance increased when a warning sign was paired with a voice warning compared to the sign or voice warning alone. Moreover, Wogalter and Young (1991) reported a field study simulating a slippery floor in a shopping center. Compared to a voice warning alone, a printed warning alone, and no warning, maximum compliance occurred when voice and printed warnings were provided together. Clearly, physical characteristics of warnings such as location, color and multiple modalities can influence behavioral compliance with warnings.

11.4 CONTENT CHARACTERISTICS

Two dimensions of warnings that have received attention in studying behavioral compliance are characteristics of the language used in text and pictorials.

11.4.1 Language

A variety of studies have evaluated the effects of different language characteristics on intentions and behavior. These studies have looked at factors such as message valence (also called 'framing,' which can range from very negative to very positive), the use of fear, and explicitness. These manipulations are all probably related in some fashion, but it is difficult to synthesize the findings because of differences in terminology and definitions. In the case of valence, deTurck and Goldhaber (1989) found that individuals reported greater awareness of a negatively stated warning (that outlined behavior to avoid) than a positive warning (that outlined behavior to adopt). A statement from the negative warning, for example, read: 'Never remove lamphouse from column unless column is in uppermost, fully extended position.' The same statement, in the positive condition, read: 'Always be sure column is in uppermost fully extended position before removing lamphouse.' In addition to differences in reported awareness, the authors also noted that participants exposed to the negatively stated warning reported that they were more likely to see, read, and comply (behavioral intention) with the warning.

The use of fear-inducing and explicit messages has produced similar findings. Laughery, Rowe-Hallbert, Young, Vaubel, and Laux (1991) found that explicit warnings produced an increase in perceived severity of injury and intent to comply with the warning. Wogalter and Barlow (1990) found that individuals working on a chemistry laboratory task complied more often (81%) with a low likelihood high severity warning message that read: 'Contact with skin *can* cause *intense* skin irritation' than with a low likelihood, low severity warning message (44%) that stated: 'Contact with skin *can* cause *mild* skin irritation.' Moreover, in a study in which individuals used a water-repellent sealer, Frantz (1994) reported that procedural explicitness (e.g., wear rubber gloves) enhanced behavioral compliance as compared to procedural non-explicitness (e.g., avoid contact with skin). The effects of fear arousal, both linear and nonlinear, can be found in Chapter 9 by DeJoy.

11.4.2 Pictorial symbols

Although pictorial symbols are used widely in warnings for increasing their noticeability and as a 'culture-free' or nonverbal method of conveying hazard information, their effects, as measured by changes in behavior, are still somewhat unclear. Although there is evidence that pictorials increase behavioral compliance, evidence also exists showing the opposite. On the positive side, Schneider (1977) reported that Mr. Yuk and skull and crossbones symbols added to a warning reduced, although not significantly, the percentage of preliterate children opening a presumably hazardous container. Similarly, Jaynes and Boles (1990) found that the presence of a pictorial in a warning significantly increased compliance over the no pictograph warnings. Finally, Wogalter, Begley, Scancorelli, and Brelsford (1997) reported that behavioral compliance increased after a pictorial was added to an elevator usage warning.

Not all studies evaluating the effect of pictorials, however, have reported positive results. Friedmann (1988), for example, found that showing either a proactive symbol (i.e., showing the safety precaution to take) or a reactive symbol (i.e., showing the potential consequence of not following the precaution) did not increase compliance to a written warning. Although a variety of explanations might account for the failure of pictorials to affect behavior consistently, one possibility is that any accompanying language warnings provided enough information that could not be improved upon dramatically by the pictorial symbol. Another possibility is that some pictorial symbols were not understood by the target audience. Hence, it is important that, when developing pictorial warnings, the issue of comprehension be addressed through pilot testing and prototyping development techniques.

11.5 INDIVIDUAL DIFFERENCE CHARACTERISTICS

In addition to physical and content design characteristics, there are other factors that influence warning effects on intentions and behavior. Individual difference characteristics such as familiarity, perceived risk, prior injuries and other personal and demographic variables are examined in this section.

11.5.1 Familiarity

Familiarity refers to the extent to which one is acquainted/experienced with a particular activity or product. To examine this issue, Venema (1989) asked individuals to use either methylated spirits to refill a burner for fondue or to use paint remover on a piece of furniture. Each product had one of three different types of label: a neutral label with no safety information, the current label for each product, and an improved label which conformed more to recommended design standards. For the methylated spirits, closing the bottle and extinguishing the flame before refilling had compliance rates of over 80% for all label conditions. For the paint remover, however, behavioral compliance (i.e., putting on gloves) for all warning label conditions was less than 40%. Many individuals indicated that they did not read the label because they were familiar with the product. Hence, as an individual becomes increasingly familiar with a product or activity, the likelihood that one will engage in precautionary behavior decreases (Godfrey, Allender, Laughery, and Smith, 1983; Godfrey and Laughery, 1984; Wogalter, Desaulniers, and Brelsford, 1986; Zeitlin, 1994). Moreover, Otsubo (1988) placed participants in situations in which there were various warning conditions for either a jig saw (low level of danger) or a circular saw (high level of danger). She discovered that those who complied with the warning had less experience with tools or saws and were less confident in using the saw. Johnson (1992) also found that workers who were new to working on a scaffold were more likely to comply with a warning associated with its use.

Familiarity might account for many reports of ineffective warnings. For instance, McGrath and Downs (1992) found that warning compliance was lowest for hammers and plastic bags. Both products were quite familiar to the participants. Furthermore, Lehto and Foley (1991) reported that helmet use among all-terrain vehicle riders decreased as the frequency of riding increased.

11.5.2 Perceived Risk

Slovic, Fischhoff, and Lichtenstein (1979, 1980) suggested that perceptions of hazard result from the combination of severity and likelihood of injury information. In other work, however, hazard perception appears to be more strongly a function of severity of injury (Wogalter et al., 1986). Similarly, Young, Wogalter, and Brelsford (1992) suggested that hazard perceptions are based primarily on severity in situations where the outcomes (i.e., the possible injuries) are less than catastrophic. When the outcomes are extremely severe, however, perceptions of hazards are more a function of likelihood of injury. In general, greater levels of behavioral compliance are associated with warnings that describe higher levels of injury severity (Otsubo, 1988; Wogalter and Barlow, 1990; Vredenburgh and Cohen, 1993). Obviously, if an individual believes that the likelihood or severity of injury is minimal, then the probability of warning compliance also is minimal.

To illustrate this issue, Hatem (1993) examined the effects of warning labels describing the hazards and precautions when using glue. The warning message and instructions directed the user to open all windows and doors before using and to turn on a fan if available. The fan was within four feet of the working area. Only one out of 59 participants complied either by opening the window or turning on the fan. When queried about why they did not comply, about 50% said that the task was too short so compliance was unnecessary; 20% mentioned that because it was a research study they assumed that there was no danger; 15% felt that the room was large enough to dissipate the fumes; and 10% thought that they were not instructed to do so or that it was not necessary to adopt the precautionary behavior for other reasons. Many of these individuals might have perceived minimal personal risk, and thus saw little reason to comply with the warning.

Although the general trend is that explicit warnings describing negative outcomes are associated with higher levels of precautionary behavior, this is not always the case. For instance, Chy-Dejoras (1992) found that participants who viewed a slightly aversive behavior (spilling tile remover and verbally expressing pain) had significantly higher glove use than either the control group (spilling tile remover without any pain) or the highly aversive condition (spilling tile remover and showing pictures of a burned hand).

Why was there less self-protective behavior in the above study when there was evidence of a more serious injury? One suggestion stems from the protection motivation model (Rogers, 1984). Rogers proposes that motivation to protect one's self is based on one's assessment of four beliefs: (a) the threat is severe (threat magnitude is high); (b) the probability of the threat occurring is likely; (c) being able to respond to reduce the threat; and (d) the response will be effective in diminishing the threat. When there is strong belief in all four factors, there is a greater probability that preventive behavior will occur. In this protection motivation model, fear is not an important component. Therefore, in assessing the Chy-Dejoras (1992) findings, it is possible that individuals viewed either the magnitude of the threat as lower, or the probability of spilling again as less likely. Slovic (1978) has also pointed out that higher levels of risk are acceptable if the risk is controllable, voluntary, familiar, or known.

11.5.3 Prior Injuries

Of the different attitudes an individual might hold, those resulting from experience are most likely to be predictive of behavior. Such is the case with respect to experience with prior injuries. Individuals who reported they have been injured by a particular activity or product (or knew others close to them who had) reported higher levels of perceived hazard for those products than their non-injured counterparts. For example, Smither, Watzke, and Braun (1997) presented older adults with a list of 20 different products and activities. In addition to ratings of hazardousness, participants indicated if they had been injured while using the product. The authors found that perceptions of product hazardousness were significantly higher for individuals who had been injured. A similar pattern of differences was noted also between individuals who had knowledge of someone being injured and those who did not.

Wogalter, Brems, and Martin (1993b) reported similar findings with respect to injury experience. In their study, research participants were instructed to provide ratings of precautionary intent for items ranging from vacuum cleaners to hammers to bicycles. Participants also indicated whether they had been injured by any of the products. When comparing ratings of precautionary intent, the authors noted that injured individuals reported significantly more precautionary intent than non-injured participants. However, not all data concerning prior injury and warning-related behavior are consistent. For instance, Lehto and Foley (1991) found that helmet use when riding all-terrain vehicles was lower if individuals had been moderately or seriously injured in all-terrain vehicle accidents. A more detailed description of responses after prior injury experience is given in Chapter 9 by DeJoy.

11.5.4 Sex Differences

Surprisingly, there are few studies that have examined sex differences in behavioral compliance. Glover and Wogalter (1997) reported that in a computer simulation of a coal-mine environment, females were more likely than males to comply with warning signs. Similarly, Vredenburgh and Cohen (1993) indicated that females reported complying with skiing and scuba diving warnings more than males did.

11.5.5 Locus of Control

Locus of control refers to the extent to which an individual believes that events are under his/her control. For example, an internal safety locus of control would indicate that an accident was under personal control and thus one's own fault. Conversely, an external safety locus of control would indicate that an accident was beyond personal control and thus not one's own fault. Donner (1991) administered a safety locus of control scale (Laux and Brelsford, 1989) to individuals who had to complete a task using either a bench grinder or a fabric protector. External safety locus of control individuals perceived the products as more hazardous than those who had an internal safety locus of control and more often complied with the warning.

11.6 SITUATIONAL VARIABLES

11.6.1 Time Stress

Wogalter *et al.* (1998) examined the effects of time pressure on behavioral compliance. In a chemistry lab task, participants either were given as much time as they needed (low time stress) or were severely limited in the amount of time they had to complete the task (high time stress). Compliance (the wearing of protective equipment) was higher for individuals in the lower time stress condition than those in the higher time stress condition. It is possible that this result might be due to the time stress narrowing perceptions and impeding judgment and decision-making ability, or that the individuals simply might not have taken the time to comply.

11.6.2 Cost of Compliance

One of the factors that has been shown to affect warning-related behavior is cost of compliance. Costs might include expenditure of time, money, or effort. Dingus, Hunn, and Wreggit (1991) had individuals use a spray bottle filled with cleaning liquid under various warning label content conditions. Gloves were either provided with the product package (low cost) or not provided with the package (high cost). Compliance was greater in the low cost condition (88%) than in the high cost condition (25%). Moreover, in separate experiments, Dingus, Wreggit, and Hathaway (1993) examined racquetball players using personal protective equipment. In all cases, providing eye protection on the court (low cost) resulted in greater compliance than either providing eye protection at a checkout center 60 feet away (medium cost) or not providing eye protection (high cost). Interestingly, no individuals in the medium cost condition wore eye protection. Wogalter, Allison, and McKenna (1989) found that when individuals were performing a chemistry lab task and gloves were placed on a work table (low cost) there was 73% compliance, whereas only 17% compliance was exhibited when the gloves were placed in another room. Finally, Godfrey, Rothstein, and Laughery (1985) placed a sign on an exit door stating that it was broken and that another exit should be used. More people obeyed the sign when the alternative exit was adjacent to the broken door than when the sign told them to use an alternative exit 50 feet away.

Aside from cost of compliance being an expenditure of energy to obtain the necessary protection, Lehto and Foley (1991) found that helmet use for individuals using all-terrain vehicles was substantially lower when helmets were perceived as uncomfortable. Hence, discomfort can also be considered a cost of compliance, which will reduce adoption of the appropriate behavior.

Concerning money versus time cost effects, Godfrey et al. (1985) placed a warning on the telephone explaining that one would lose money, and a warning on a copy machine that stated that a delay was expected due to technical problems. As anticipated, the warning concerning money had a greater impact on behavior than the warning concerning time.

11.7 MODELING

In many cases, people do not need to learn through direct experience. Sometimes we learn what to do or not to do by observing other people. This observational learning and imitating others' behavior is known as modeling (Myers, 1995). Although much of the research in this area concerns children and whether they will imitate an adult model, the concept has been applied to whether an individual will adopt some precautionary behavior if he/she observes another individual behaving that way. Wogalter et al. (1989) performed a series of experiments to evaluate this issue. First, gloves were placed on a work table, and when a confederate put them on, 100% of the participants also put them on. When the confederate did not put on the gloves, the compliance rate for the participants dropped to 33%. In a field study, a sign was placed on the door of an elevator in a women's dormitory stating, in essence, that the elevator might stick between floors and to use the stairs. In the control condition (no confederate present), 31.5% of the participants used the stairs. When the confederate used the elevator, 27.7% of the participants used the stairs. However, when the confederate used the stairs, participants' behavioral compliance increased to 88.9%.

Similarly, Chy-Dejoras (1992) had individuals watch a videotape of an individual wearing gloves while performing a floor tiling task in which an adhesive remover was used. Those who watched the individual wearing gloves while performing the task were significantly more likely to use gloves (87%) than the control group (50%). Racicot and Wogalter (1995) also demonstrated that when a participant was shown a video image of a warning sign and a male person putting on gloves and a mask there was greater compliance than when seeing the video image of a warning sign alone. These studies show that modeling can affect behavioral compliance dramatically.

11.8 MASS MEDIA AND CAMPAIGNS

The research described thus far in this chapter has focused primarily on the behavioral effects of warnings that are presented on signs, labels, or other types of written document plus warnings presented auditorily. There is another research domain concerned with safety communications that could influence on people's behavior, namely, the mass media.

In an attempt to enhance precautionary behavior, campaigns and programs are often used to educate the public or to facilitate attention to information that might not be considered ordinarily when using the product. The mass media can provide a number of important benefits. First, mass media can alert individuals to health and safety risks that they ordinarily might not know (Lau, Kane, Berry, Ware, and Roy, 1980). Second, the mass media can provide safety information repetitively. The constant presence of this information might promote its retention in long term memory and ultimately affect behavior.

Mass media campaigns differ from the types of warning discussed thus far in that they target specific segments of the population or the population as a whole, whereas many of the discussed warnings target individuals (i.e., product warnings). Flora, Maibach, and Maccoby (1989) make a similar distinction, contrasting massed versus targeted communications. This difference in the target audience is associated with a variety of other differences that increase the dichotomy between studies of massed and targeted warnings. For example, mass media research has relied on communication methods that reach large numbers of people such as television, radio, newspaper, and magazines. By contrast, research on targeting individuals has relied on communication methods such as labels, signs, booklets, newsletters, videos, package inserts, etc. In this section, we will examine the mass media as a means of influencing people's compliance with warnings.

Although the mass media are a natural choice to promote behavioral changes within a variety of contexts, most research has focused on changes in behavioral intentions rather than behavior. For instance, mass media campaigns have shown significant increases for individuals intending to exercise (Owen, Bauman, Booth, Oldenburg, and Magnus, 1995) and intending to use condoms (Middlestadt *et al.*, 1995).

The mass media efforts have obtained a moderate degree of success with regard to modifying behavior. For example, the Stanford Three Community Study (Maccoby, Farquhar, Wood, and Alexander, 1977) is probably the most frequently cited example of a successful mass media campaign. Targeting coronary heart disease, the researchers evaluated the effectiveness of an education program in two different California communities. A third community was used as a control. In both experimental communities, television, radio, newspaper, cookbooks, and health information pamphlets were used. In one of the experimental communities, face-to-face contacts were added. The initial intervention lasted nine months following the collection of baseline data. Using measures of knowledge and plasma cholesterol levels, the researchers reported significant increases in knowledge and reductions in cholesterol levels for the experimental communities but not for the control community.

Some efforts aimed at reducing smoking also have shown positive effects. Warner (1977), for example, addressed smoking through the use of commercials and advertisements that included celebrity endorsements explaining the risks and hazards of smoking. He concluded that the anti-smoking campaign helped reduce Americans' smoking behavior by 20–30%. Similarly, Lewitt, Coate, and Grossman (1981) reported that an anti-smoking broadcast over radio and television reduced the rate of adolescent smoking. These reports lend support to the use of mass media to promote behavior change.

However, the literature on the efficacy of the mass media is mixed. Some mass media efforts have shown limited effectiveness in modifying behavior. For example, Flay, Hansen, and Johnson (1987) reported that a coordinated television and school-based education program failed to alter the smoking intentions and behavior of school aged children. Bauman, LaPrelle, Brown, Koch, and Padgett (1991) found similar results following the introduction of three different radio and television messages in six southeastern communities. Using cigarette smoking as a dependent variable, the researchers found no significant reduction in smoking when contrasting the six experimental communities with the four control communities. In fact, the authors noted that the rates of smoking increased over the intervention period. Similar negative findings were reported by Barber and Grichting (1990), who evaluated Australia's drug abuse campaign. Following a three-year mass media effort, no evidence for behavioral change was found. Other studies aimed at promoting birth control (Udry, Clark, Chase, and Levy, 1972), increasing seatbelt use (Robertson et al., 1974), improving nutrition (Patterson et al., 1992), and changing driving behavior (Lourens, Van der Molen, and Oude Egberink, 1991) also failed to produce significant changes in behavior.

Efforts designed to assess the effectiveness of fear-arousing messages have also produced mixed results. A meta-analysis of fear-arousing campaigns (Boster and Mongeau, 1984) reported that the relationship between fear manipulations and perceived fear, attitudes, and behavior were relatively low (correlations ranged from 0.10 to 0.36). The authors concluded that these low correlations were attributable largely to weak manipulations of fear that were not adequate in power to elicit behavior in the desired direction. In studies where fear was more successfully manipulated, the magnitude of the fear was related positively to the effectiveness of the campaign.

According to Taylor (1991), there are a number of other reasons why the effectiveness of mass media is limited in modifying behavior. First, people might not pay attention to the message, or they may never experience the medium that contains the message. Second, mass media messages might not focus on the specific experiences that would be important to the individuals. Third, the messages might not include specific actions that the audience could adopt as behavior. Adler and Pittle (1984) noted that many mass media campaigns promote abstract concepts rather than concrete behavior. Fourth, messages are often dictated by advertisers' hypotheses rather than having a focus on empirical research. Finally, it is difficult to extract the specific factors that influence health or safety behavior; in short, there may be confounds. One way to view the mass media is that they set the stage for behavioral change to occur rather than producing it *per se*. Mass media in conjunction with methods affecting individuals on a more personal level may have the greatest potential to elicit behavior change.

11.9 SUMMARY AND CONCLUSIONS

In this chapter we have examined various factors that influence a warning's effects on behavioral intentions and behavior. We have shown that, in general, the presence of warnings, as compared to their absence, enhances behavioral compliance. Moreover, we examined a number of message characteristics and showed how they were related (or not related) to behavioral compliance. We summarize these findings briefly below.

First, integrating information from various works, it appears that consideration should be given to placing warning messages at the beginning of the directions for use section (e.g., Wogalter et al., 1987; Frantz, 1993). Individuals learning how to use the product inevitably will read this information. Moreover, the memory literature shows that by placing information at the beginning (primacy effect) it is more likely to be remembered than if placed either in the middle or end (Glanzer and Cunitz, 1966). However, there has been some research to suggest that when the warning messages are not embedded (i.e., separate and distinct from instructions), compliance is increased. The implication is that placing warning information in a cluttered environment reduces compliance. Possibilities for this finding might be that either the individual will not take the time or effort to sift through the information or that the information simply is not salient.

Color and pictorial symbols in warnings can benefit behavioral compliance, partly because of added noticeability and understandability. For example, some colors have distinct connotations (e.g., red implies danger, cf. Braun and Silver, 1995). Indeed, the use of color and pictorial symbols is analogous to using another redundant modality; hence, the same message is conveyed in two or more different ways.

We also addressed a number of individual difference characteristics. Individuals who have been injured (or who know of others close to them who have been injured) by a particular product will provide higher perceived hazard ratings to that product than those who were not injured. Usually this means that behavioral compliance will increase. Once again, as we indicated earlier, familiarity might play a role. Another factor is cost of compliance. If precautions are quick, easy, cheap and/or convenient, the prospects of behavioral compliance increase. On the other hand, if cost is high, compliance decreases dramatically.

Although the majority of studies in the warning field are based predominantly on intentions, attitudes, or beliefs, there is evidence to suggest that these measures are related to behavior. For example, as nicely illustrated in Chapter 9 by DeJoy, as perceived hazardousness (a belief strength) increases, warning compliance increases (e.g., Otsubo, 1988). Although this is not always the case (e.g., Lehto and Foley, 1991), many studies seem to support this conclusion (e.g., Friedmann, 1988). Moreover, familiarity not only decreases the likelihood of reading the warning but also it decreases compliance (e.g., Otsubo, 1988). In our estimation, many of these concepts (e.g., perceived hazardousness, carefulness, likelihood of injury, severity of injury) are highly intercorrelated (e.g., Wogalter and Silver, 1990); therefore, it seems only logical that the behavioral measures follow the same lines. We qualify this, however, by saying that a variable, familiarity, for example, could moderate the effects of behavioral compliance; that is, although individuals might perceive that a particular situation (or product) could result in a highly severe injury, those who are familiar with the situation might not comply with the warning, whereas those who are less familiar will be more likely to comply with the warning.

In order to foster precautionary behavior, mass media campaigns are often used. The available evidence indicates that warnings presented via mass media channels are more likely to affect pre-behavioral attitudes and intentions than behavior. However, it is important to note that only a few of these studies actually measured behavior. This reliance on measures of attitudes and intentions is promoted in part by the variable temporal relationship between the presentation of the warning and the behavior in question. In many studies of product warnings, compliance or the lack of is readily observable over a short period of time. The same cannot be said of studies concerning changes in much health-related behavior. The temporal variability between the presentation of the warning and the opportunity to display compliant behavior creates unique demands on the selection of criterion measures, the timing of these measures, and the interpretation of the results. It is difficult to determine, without additional research, if the lack of observable behavioral changes is the result of an ineffective warning, inappropriate dependent measures, or just poor timing. The mass media, however, are not without their merits.

Given that attitudes, beliefs, knowledge, and intentions are the precursors of behavior, mass media efforts can play a vital role in the promotion of behavioral change. One important facet of the mass media is to provide the same message in a number of ways. Therefore, novelty (e.g., using different celebrity endorsements) of the campaign will not only target many individuals, but also cognitively instill the message into memory. Although we have pointed out numerous drawbacks to mass media campaigns, we believe that if the campaign is credible (e.g., using expert sources), if individuals are targeted over an extended period of time, and if individual feedback is provided (which is almost impossible from a mass media campaign), then the behavioral effects might be more long term.

On a more individualized basis, modeling seems to be quite effective in facilitating behavioral compliance. There are a number of reasons why this is effective. First, it is possible that the participants perceive models as experts or individuals who are quite familiar with the task. Second, it could be simply a matter of conformity; that is, there might be some social pressure to conform and put on protective gear. Third, it enhances self-efficacy.

When asked if warning labels were effective tools for preventing injuries, David Pittle (1991, p. 110), director for Consumer Union, stated 'They do no harm, and may do some small amount of good.' Potentially, they can change behavior if the conditions are right. In general, warnings can be extremely important in increasing safe behavior (Cox, Wogalter, Stokes, and Murff, 1997). Compliance is surely an interactive function of the warning, product, situational factors, and characteristics of users (Stewart and Martin, 1994). Warnings need to be noticeable, understandable, convey the appropriate message, and take into account various characteristics of the target audience (such as their beliefs, attitudes, and motivation) to have a chance at influencing safety-related behavior.

REFERENCES

ADLER, R. and PITTLE, R. (1984) Cajolery or command: are education campaigns an adequate substitute for regulation? *Yale Journal on Regulation*, 1, 159–193.

BARBER, J.J. and GRICHTING, W.L. (1990) Australia's media campaign against drug abuse. *The International Journal of the Addictions*, 25, 693–708.

BAUMAN, K.E., LAPRELLE, J., BROWN, J.D., KOCH, G.G., and PADGETT, C.A. (1991) The influence of three mass media campaigns on variables related to adolescent cigarette smoking: results of a field experiment. *American Journal of Public Health*, 81, 597–604.

BOSTER, F.J. and MONGEAU, P. (1984) Fear-arousing persuasive messages. In BOSTROM, R.N. and WESTLEY, B.H. (eds), *Communication Yearbook 8*. Beverly Hills, CA: Sage Publications, pp. 330–375.

BOUHATAB, A. (1991) Effects of individual characteristics, context and posted warnings used in industrial work places on hazard awareness and safety compliance. Unpublished doctoral dissertation, University of Oklahoma.

BRAUN, C.C. and SILVER, N.C. (1995) Interaction of signal word and colour on warning labels: differences in perceived hazard and behavioural compliance. *Ergonomics*, 38, 2207–2220.

CHY-DEJORAS, E.A. (1992) Effects of an aversive vicarious experience and modeling on perceived risk and self-protective behavior. In *Proceedings of the Human Factors Society 36th Annual Meeting*. Santa Monica, CA: Human Factors Society, pp. 603–607.

COX, E.P., WOGALTER, M.S., STOKES, S.L., and MURFF, E.J.T. (1997) Do product warnings increase safe behavior?: a meta-analysis. *Journal of Public Policy and Marketing*, 16, 195–204.

DETURCK, M.A. and GOLDHABER, G.M. (1989) Effectiveness of product warnings: Effects of language valence, redundancy, and color. *Journal of Products Liability*, 12, 93–102.

DINGUS, T.A., HUNN, B.P., and WREGGIT, S.S. (1991) Two reasons for providing protective equipment as part of hazardous consumer product packaging. In *Proceedings of the Human Factors Society 35th Annual Meeting*. Santa Monica, CA: Human Factors Society, pp. 1039–1042.

DINGUS, T.A., WREGGIT, S.S., and HATHAWAY, J.A. (1993) Warning variables affecting personal protective equipment use. *Safety Science*, 16, 655–674.

DONNER, K.A. (1991) Prediction of safety behaviors from locus of control statements. In *Proceedings of Interface 91*. Santa Monica, CA: Human Factors Society, pp. 94–98.

DUFFY, R.R., KALSHER, M.J., and WOGALTER, M.S. (1993) The effectiveness of an interactive warning in a realistic product-use situation. In *Proceedings of the Human Factors Society 37th Annual Meeting*. Santa Monica, CA: Human Factors Society, pp. 935–939.

FERRARI, J.R. and CHAN, L.M. (1991) Interventions to reduce high-volume portable headsets: 'turn down the sound.' *Journal of Applied Behavior Analysis*, 24, 695–704.

FLAY, B.R., HANSEN W.B., and JOHNSON, C.A. (1987) Implementation effectiveness trial of a social influences smoking prevention program using schools and television. *Health Education Research*, 2, 385–400.

FLORA, J.A., MAIBACH, E.W., and MACCOBY, N. (1989) The role of media across four levels of health promotion intervention. *Annual Review of Public Health*, 10, 181–201.

FRANTZ, J.P. (1993) Effect of location and presentation format on attention to and compliance with product warnings and instructions. *Journal of Safety Research*, 24, 131–154.

FRANTZ, J.P. (1994) Effect of location and procedural explicitness on user processing of and compliance with product warnings. *Human Factors*, 36, 532–546.

FRANTZ, J.P. and RHOADES, T.P. (1993) A task-analytic approach to the temporal and spatial placement of product warnings. *Human Factors*, 35, 719–730.

FRIEDMANN, K. (1988) The effect of adding symbols to written warning labels on user behavior and recall. *Human Factors*, 30, 507–515.

GILL, R.T., BARBERA, C., and PRECHT, T. (1987) A comparative evaluation of warning label designs. In *Proceedings of the Human Factors Society 31st Annual Meeting*. Santa Monica, CA: Human Factors Society, pp. 476–478.

GLANZER, M. and CUNITZ, A.R. (1966) Two storage mechanisms in free recall. *Journal of Verbal Learning and Verbal Behavior*, 5, 351–360.

GLOVER, B.L. and WOGALTER, M.S. (1997) Using a computer simulated world to study behavioral compliance with warnings: effects of salience and gender. In *Proceedings of the Human Factors and Ergonomics Society 41st Annual Meeting*. Santa Monica, CA: Human Factors and Ergonomics Society, pp. 1283–1287.

GODFREY, S.S., ALLENDER, L., LAUGHERY, K.R., and SMITH, V.L. (1983) Warning messages: will the consumer bother to look? In *Proceedings of the Human Factors Society 27th Annual Meeting*. Santa Monica, CA: Human Factors Society, pp. 950–954.

GODFREY, S.S. and LAUGHERY, K.R. (1984) The biasing effects of product familiarity on consumers' awareness of hazard. In *Proceedings of the Human Factors Society 28th Annual Meeting*. Santa Monica, CA: Human Factors Society, pp. 483–486.

GODFREY, S.S., ROTHSTEIN, P.R., and LAUGHERY, K.R. (1985) Warnings: do they make a difference? In *Proceedings of the Human Factors Society 29th Annual Meeting*. Santa Monica, CA: Human Factors Society, pp. 669–673.

HATEM, A.T. (1993) The effect of performance level on warning compliance. Unpublished master's thesis, School of Industrial Engineering, Purdue University.

HUNN, B.P. and DINGUS, T.A. (1992) Interactivity, information, and compliance cost in a consumer product warning scenario. *Accident Analysis and Prevention*, 24, 497–505.

JAYNES, L.S. and BOLES, D.B. (1990) The effect of symbols on warning compliance. In *Proceedings of the Human Factors Society 34th Annual Meeting*. Santa Monica, CA: Human Factors Society, pp. 984–987.

JOHNSON, D. (1992) A warning label for scaffold users. In *Proceedings of the Human Factors Society 36th Annual Meeting*. Santa Monica, CA: Human Factors Society, pp. 611–615.

KIM, M.S. and HUNTER, J.E. (1993a) Attitude–behavior relations: a meta-analysis of attitudinal relevance and topic. *Journal of Communications*, 43, 101–142.

KIM, M.S. and HUNTER, J.E. (1993b) Relationships among attitudes, behavioral intentions, and behavior: a meta-analysis of past research: II. *Communication Research*, 20, 331–364.

LAU, R.R., KANE, R., BERRY, S., WARE JR, J.E., and ROY, D. (1980) Channeling health: a review of televised health campaigns. *Health Education Quarterly*, 7, 56–89.

LAUGHERY, K.R., LAUGHERY, K.A., and LOVVOLL, D.R. (1998) Tire rim mismatch explosions: the role of on-product warnings. In *Proceedings of the Human Factors Society 35th Annual Meeting*. Santa Monica, CA: Human Factors Society, pp. 481–485.

LAUGHERY, K.R., ROWE-HALLBERT, A.L., YOUNG, S.L., VAUBEL, K.P., and LAUX, L.F. (1991) Effects of explicitness in conveying severity information in product warnings. In *Proceedings of the Human Factors Society 42nd Annual Meeting*. Santa Monica, CA: Human Factors Society, pp. 1088–1092.

LAUX, L. and BRELSFORD, J.W. (1989) Locus of control, risk perception, and precautionary behavior. In *Proceedings of Interface 89*. Santa Monica, CA: Human Factors Society, pp. 121–124.

LEFTON, L.A. (1997) *Psychology*, 6th Edn. Needham Heights, MA: Allyn & Bacon.

LEHTO, M.R. and FOLEY, J.P. (1991) Risk-taking, warning labels, training, and regulation: are they associated with the use of helmets by all-terrain vehicle riders? *Journal of Safety Research*, 22, 191–200.

LEWITT, E.M., COATE, D., and GROSSMAN, M. (1981) The effects of government regulation on teenage smoking. *Journal of Law Economics*, 24, 545–569.

LOURENS, P.F., VAN DER MOLEN, H.H., and OUDE EGBERINK, H.J.H. (1991) Drivers and children: a matter of education? *Journal of Safety Research*, 22, 105–115.

MCGRATH, J.M. and DOWNS, C.W. (1992) The effectiveness of on-product warning labels: a communication perspective. *For the Defense*, October, 19–24.

MACCOBY, N., FARQUHAR, J.W., WOOD, P., and ALEXANDER, J. (1977) Reducing the risk of cardiovascular disease: effects of a community-based campaign on knowledge and behavior. *Journal of Community Health*, 3, 100–144.

MIDDLESTADT, S.E., FISHBEIN, M., ALBARRACIN, D., FRANCIS, C., EUSTACE, M.A., HELQUIST, M., and SCHNEIDER, A. (1995) Evaluating the impact of a national AIDS prevention radio campaign in St. Vincent's and the Grenadines. *Journal of Applied Social Psychology*, 25, 21–34.

MYERS, D.G. (1995) *Psychology*, 4th Edn. New York: Worth.

OTSUBO, S. (1988) A behavioral study of warning labels for consumer products: perceived danger and use of pictographs. In *Proceedings of the Human Factors Society 32nd Annual Meeting*. Santa Monica, CA: Human Factors Society, pp. 536–540.

OWEN, N., BAUMAN, A., BOOTH, M., OLDENBURG, B., and MAGNUS, P. (1995) Serial mass-media campaigns to promote physical activity: reinforcing or redundant? *American Journal of Public Health*, 85, 244–248.

PATTERSON, B.H., KESSLER, L.G., WAX, Y., BERSTEIN, A., LIGHT, L., MIDTHUNE, D.N., PORTNOY, B., TENNEY, J., and TUCKERMANTY, E. (1992) Evaluation of a supermarket intervention: the NCI-giant food eat for health study. *Evaluation Review*, 16, 464–490.

PITTLE, D. (1991) Product safety: there's no substitute for safer design. *Trial*, 27, 110–114.

RACICOT, B.M. and WOGALTER, M.S. (1995) Effects of a video warning sign and social modeling on behavioral compliance. *Accident Analysis and Prevention*, 27, 57–64.

ROBERTSON, E.K., KELLEY, A., O'NEILL, B., WIXOM, C., EISWORTH, R., and HADDON JR, W. (1974) A controlled study of the effect of television messages on safety belt use. *American Journal of Public Health*, 64, 1071–1080.

ROGERS, W. (1984) Changing health-related attitudes and behavior: the role of preventative health psychology. In HARVEY, J.H., MADDUX, E., MCGLYNN, R.P., and STOLTENBERG, C.D. (eds), *Social Perception in Clinical and Counseling Psychology*, Vol. 2. Lubbock, TX: Texas Tech University Press, pp. 91–112.

RYAN, J.P. (1991) *Design of Warning Labels and Instructions*. New York: Van Nostrand Reinhold.

SCHNEIDER, K.C. (1977) Prevention of accidental poisoning through package and label design. *Journal of Consumer Research*, 4, 67–74.

SLOVIC, P. (1978) The psychology of protective behavior. *Journal of Safety Research*, 10, 58–68.

SLOVIC, P., FISCHHOFF, B., and LICHTENSTEIN, S. (1979) Rating the risks. *Environment*, 21, 14–39.

SLOVIC, P., FISCHHOFF, B., and LICHTENSTEIN, S. (1980) Facts and fears: understanding perceived risk. In SCHWING, R.C. and ALBERS JR, W.A. (eds), *Societal Risk Assessment*. New York: Plenum.

SMITHER, J.A., WATZKE, J.R., and BRAUN, C. (1997) Perceptions of consumer product dangers among older adults. Unpublished manuscript.

STEWART, D.W. and MARTIN, I.M. (1994) Intended and unintended consequences of warning messages: a review and synthesis of empirical research. *Journal of Public Policy and Marketing*, 13, 1–19.

STRAWBRIDGE, J.A. (1986) The influence of position, highlighting, and embedding on warning effectiveness. In *Proceedings of the Human Factors Society 30th Annual Meeting*. Santa Monica, CA: Human Factors Society, pp. 716–720.

TAYLOR, S.E. (1991) *Health Psychology*. New York: McGraw-Hill.

THYER, B.A. and GELLER, E.S. (1987) The 'buckle-up' dashboard sticker: an effective environmental intervention for safety belt promotion. *Environment and Behavior*, 19, 484–494.

UDRY, J., CLARK, L., CHASE, C., and LEVY, M. (1972) Can mass media advertising increase contraceptive use? *Family Planning Perspectives*, 4, 37–44.

VENEMA, A. (1989) Research Report 69, *Product Information for the Prevention of Accidents in the Home and during Leisure Activities: Hazard and Safety Information on Non-durable Products*. The Netherlands: Institute for Consumer Research, SWOKA.

VREDENBURGH, A.G. and COHEN, H.H. (1993) Compliance with warnings in high risk recreational activities. In *Proceedings of the Human Factors Society 37th Annual Meeting*. Santa Monica, CA: Human Factors Society, pp. 945–949.

WARNER, K.E. (1977) The effects of the anti-smoking campaign on cigarette consumption. *American Journal of Public Health*, 67, 645–650.

WOGALTER, M.S., ALLISON, S.T., and MCKENNA, N.N. (1989) Effects and cost and social influence on warning compliance. *Human Factors*, 31, 133–140.

WOGALTER, M.S. and BARLOW, T. (1990) Injury severity and likelihood in warnings. In *Proceedings of the Human Factors Society 34th Annual Meeting*. Santa Monica, CA: Human Factors Society, pp. 580–583.

WOGALTER, M.S., BARLOW, T., and MURPHY, S. (1995) Compliance to owner's manual warnings: influence of familiarity and the task-relevant placement of a supplemental directive. *Ergonomics*, 38, 1081–1091.

WOGALTER, M.S., BEGLEY, P.B., SCANCORELLI, L.F., and BRELSFORD, J.W. (1997) Effectiveness of elevator service signs: measurement of perceived understandability, willingness to comply, and behaviour. *Applied Ergonomics*, 28, 181–187.

WOGALTER, M.S., BREMS, D.J., and MARTIN, E.G. (1993b) Risk perception of common consumer products: judgments of accident frequency and precautionary intent. *Journal of Safety Research*, 24, 97–106.

WOGALTER, M.S., DESAULNIERS, D.R., and BRELSFORD JR, J.W. (1986) Perceptions of consumer product hazardousness and warning expectations. In *Proceedings of the Human Factors Society 30th Annual Meeting*. Santa Monica, CA: Human Factors Society, pp. 1197–1201.

WOGALTER, M.S., GODFREY, S.S., FONTENELLE, G.A., DESAULNIERS, D.R., ROTHSTEIN, P.R., and LAUGHERY, K.R. (1987) Effectiveness of warnings. *Human Factors*, 29, 599–612.

WOGALTER, M.S., JARRARD, S.W., and SIMPSON, S.W. (1994) Influence of signal words on perceived level of product hazard. *Human Factors*, 36, 547–556.

WOGALTER, M.S., GLOVER, B.L., MAGURNO, A.B., and KALSHER, M.J. (1999) Safer jump-starting procedures with an instructional tag warning. In ZWAGA, H.J.G. BOERSEMA, T., and HOONHOUT, H.C.M. (eds), *Visual Information for Everyday Use: Design and Research Perspectives*. London: Taylor & Francis, pp. 127–132.

WOGALTER, M.S., KALSHER, M.J., and RACICOT, B.M. (1992) The influence of location and pictorials on behavioral compliance to warnings. In *Proceedings of the Human Factors Society 36th Annual Meeting*. Santa Monica, CA: Human Factors Society, pp. 1029–1033.

WOGALTER, M.S., KALSHER, M.J., and RACICOT, B.M. (1993a) Behavioral compliance with warnings: effects of voice, context, and location. *Safety Science*, 16, 637–654.

WOGALTER, M.S., MAGURNO, A.B., RASHID, R., and KLEIN, K.W. (1998) The influence of time stress and location on behavioral compliance. *Safety Science*, 29, 143–158.

WOGALTER, M.S. and SILVER, N.C. (1990) Arousal strength of signal words. *Forensic Reports*, 3, 407–420.

WOGALTER, M.S. and YOUNG, S.L. (1991) Behavioral compliance to voice and print warnings. *Ergonomics*, 34, 79–89.

WOGALTER, M.S. and YOUNG, S.L. (1994) Enhancing warning compliance through alternative product label designs. *Applied Ergonomics*, 25, 53–57.

YOUNG, S.L., WOGALTER, M.S., and BRELSFORD JR, J.W. (1992) Relative contribution of likelihood and severity of injury to risk perceptions. In *Proceedings of the Human Factors Society 36th Annual Meeting*. Santa Monica, CA: Human Factors Society, pp. 1014–1018.

ZEITLIN, L.R. (1994) Failure to follow safety instructions: faulty communication or risky decisions. *Human Factors*, 36, 172–181.

ZLOTNIK, M.A. (1982) The effects of warning message highlighting on novel assembly task performance. In *Proceedings of the Human Factors Society 26th Annual Meeting*. Santa Monica, CA: Human Factors Society, pp. 93–97.

PART FOUR

Practical Issues of Warning Design

This section provides practical guidance on warnings development. Government regulations, industry standards, and general guidelines can serve as a basis for the design of warnings, but these rules may not be adequate by themselves. Procedures for developing warnings and testing them on appropriate target populations are described.

CHAPTER TWELVE

Standards and Government Regulations in the USA

BELINDA L. COLLINS

National Institute of Standards and Technology

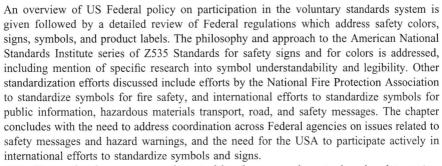

An overview of US Federal policy on participation in the voluntary standards system is given followed by a detailed review of Federal regulations which address safety colors, signs, symbols, and product labels. The philosophy and approach to the American National Standards Institute series of Z535 Standards for safety signs and for colors is addressed, including mention of specific research into symbol understandability and legibility. Other standardization efforts discussed include efforts by the National Fire Protection Association to standardize symbols for fire safety, and international efforts to standardize symbols for public information, hazardous materials transport, road, and safety messages. The chapter concludes with the need to address coordination across Federal agencies on issues related to safety messages and hazard warnings, and the need for the USA to participate actively in international efforts to standardize symbols and signs.

Note: Figures that do not appear in the text of this chapter are shown in the color plate section.

12.1 BACKGROUND FOR WARNING AND SYMBOL ACTIVITIES

12.1.1 Government as Participant in Voluntary Standards

In the USA, standards are developed through a private, voluntary process, led by industry with active participation by government. As such, there are many private sector standards developing organizations (SDOs) which develop and publish standards using the consensus process. Typically these organizations represent different sectors of the economy, such as automotive, electronics, fire protection, textiles, and the like. The US approach to standards is different from that of most nations which usually have a single national standards body, that very often is government backed and financed. In the USA, a non-governmental organization, the American National Standards Institute (ANSI) performs an umbrella function, coordinating US positions internationally, and accrediting SDOs as following consensus procedures. However, ANSI does not make standards itself, rather it publishes standards developed by others as American National Standards.

Voluntary standards are documents which have been established by consensus and/or approved by a voluntary consensus standards body. As implied by the term 'consensus,' these standards represent mutually agreed upon common and repeated use of rules, guidelines or characteristics for products or related processes and production methods. The World Trade Organization (WTO) defines standards as documents approved by a recognized body that provides for common and repeated use of rules, guidelines or characteristics for products or related processes and production methods with which compliance is *not* mandatory. By contrast, technical regulations are defined as documents which provide for product characteristics or related processes and production methods with which compliance is *mandatory*. Both standards and technical regulations can include terminology, symbols, and packaging, plus requirements for marking or labeling, as these apply to product process and production methods. The distinction between standards and regulations, while useful, is often blurred in reality.

In the USA, oversight of regulations and standards is a very complex process with extensive government and private sector involvement. At the Federal level, a number of agencies are responsible for developing regulations pertaining to safety requirements. Each agency typically has sector specific authority. For example, the Food and Drug Administration (FDA) produces regulations for drugs and medical devices, the Consumer Product Safety Commission (CPSC) regulates consumer products, while the Department of Transportation (DOT) is responsible for the safe movement of traffic by air, sea, or land, and also for the safety of the products for transporting goods and people. In the Department of Labor, the Occupational Safety and Health Administration (OSHA) oversees workplace safety, while the Mine Safety and Health Administration (MSHA) oversees mine safety.

These and other federal agencies notwithstanding, much of the overseeing of human safety, including products and services, occurs in the private sector by voluntary standards developers, such as National Fire Protection Association (NFPA), Underwriters Laboratory (UL), ASTM (American Society for Testing and Materials), and others. For example, NFPA develops standards for life and fire safety, including the life safety code (NFPA 101) which is then incorporated by most code authorities as an integral element of a local building code. UL develops standards and certifies (or lists) products for electrical safety, and has been doing so for more than 100 years. Along with other laboratories, UL is recognized by OSHA as a Nationally Recognized Test Laboratory (NRTL) for testing products for workplace safety. Many voluntary standards are used by Federal agencies as a basis or supplement for their rules and regulations. Because the legal community and marketplace may also consider them as *de facto* requirements and use them in legal actions, the importance of voluntary standards to public safety should not be underestimated. The field of safety is thus a complex mix of federal regulations and voluntary standards. This mix is backed up by the US court system, which can litigate alleged violations of both voluntary safety standards and regulations.

12.1.2 Overview of Federal Activities

In addition to developing regulations, personnel from many US Federal agencies participate in the voluntary standards process as technical experts and in support of their agency's fundamental mission. Federal agencies are increasingly moving away from the development of unique in-house regulations to greater reliance on voluntary standards, particularly in procurement, but also as an extension of their regulatory mission. This is true of most Federal agencies including the Department of Defense (DOD), OSHA, and CPSC. Such reliance has long been supported by the Office of Management and Budget

(OMB, 1993) through its Circular A-119 which encourages Federal use of voluntary standards, and participation in their development. Public law 104-113 (1996), the National Technology Transfer and Advancement Act, recently codified the Circular's requirements by directing Federal agencies to use voluntary standards to the extent possible.

In the next sections, Federal regulations which address the use of safety signs, labels, colors or symbols for safety applications are discussed. The focus is on the graphic layout of safety messages, rather than the wording and procedures for selecting wording. A recurring theme is the need to evaluate symbols and safety messages for their understandability and effectiveness. The role of the National Institute of Standards and Technology (NIST) in working with the American National Standards Institute (ANSI) Z535 Standards for Safety Signs and Colors (particularly the Z535.3 Standard for Safety Symbols) is discussed, followed by discussion of other voluntary efforts for safety messages. Finally, suggestions and concerns about the need for better systems for safety signs throughout both the public and private sectors are presented.

12.2 US GOVERNMENT REGULATIONS

12.2.1 Transportation

The *Manual on Uniform Traffic Control Devices* (MUTCD) sponsored by the Federal Highway Administration (FHWA, 1988) (cited as Code of Federal Regulations 23 CFR 1204.4 and 49 CFR 1.48) is the principal resource document for the transportation engineer. Published as a voluntary standard (ANSI D6.1e, 1989), it has been adopted also as a regulation by all state and local jurisdictions seeking Federal funds. Consequently, it is almost universally used by jurisdictions across the USA. The MUTCD contains extensive information about highway signage and symbols, including a set of symbols to indicate curves, deer (and other animal) crossings, yield, and the familiar set of symbols used on roads throughout the USA (and Canada). The MUTCD provides not only symbols, but also color codes to indicate information. Thus, information signs are blue (e.g., hospital); prohibition signs are red (e.g., stop); caution signs are yellow (e.g., road narrows or deer crossing); highway information signs are green (e.g., directions and speed limits); temporary work zone signs are orange (e.g., detour or watch out for construction worker); and recreation signs are brown (e.g., campsite location). Use of familiar colors and symbols was shown in a series of research projects sponsored (and/or conducted) by DOT to speed reaction time at interstate speeds, and thus contribute to traffic safety. Many of the DOT symbols were chosen on the basis of comprehension testing with samples of highway users (King, 1971, 1975).

12.2.2 Hazardous Materials Transport

DOT regulates the use of signs and symbols for the transport of hazardous materials (in 49 CFR 172.300). DOT has largely adopted the protocols set forth by the United Nations, so that essentially symbols and signs indicating hazardous materials are standardized throughout the world. According to the United Nations protocol, the warning placard is to be a diamond, with the symbol indicating the type of hazard in the upper half, effectively appearing as a triangular symbol sign. The lower half of the diamond provides supplementary word information. These signs are used in all forms of transportation, including trucking, railroads, ships and containerized shipments, and provide uniform information to those handling hazardous materials regardless of origin or final destination. Specifications for the different types of sign are given in CFR 49 Part 100. Unfortunately, these symbols were never evaluated for understandability, legibility or even good graphic layout. Yet, their widespread use over many decades has provided the opportunity for them to be learned by a wide variety of users. Consequently, these symbols generally can meet the goal of communicating safety information effectively and rapidly.

12.2.3 Drugs and Medical Devices

Unlike DOT which mainly uses symbols for communicating safety information, the Food and Drug Administration (FDA) relies on word labeling to regulate both drugs and medical devices. The FDA's extensive labeling requirements (21 CFR Part 201 [4-1-95]) pertain primarily to the type of drug or device to be labeled, the type of information that must be contained on the label, and general guidance for avoiding misrepresentation or fraud on the label. Adequate directions for use also must be given in a form suitable for a layperson to use the drug safely and for the purposes for which it was intended. Ingredients must be listed accurately, along with an expiration date if required. Graphic requirements typically relate to size and prominence of the wording on the label. Labeling in Spanish is authorized only for medications available by prescription and distributed in the Commonwealth of Puerto Rico. Requirements for supplementary information are given also if the total package and label are too small to accommodate all the required information. The FDA's specific guidance for labeling prescription drugs includes section headings for labels, which must be displayed in the following order: description; clinical pharmacology; indications and usage; contraindications; warnings, precautions, adverse reactions; drug use and dependence; over dosage; dosage and administration; how supplied; and additional information on animal pharmacology and clinical studies if appropriate. The FDA defines each of these terms, and specifies the type of information to be included on the label. Although the Code specifies the information that must be on the label, it does not include any use of symbols or colors.

The FDA gives specific guidance for labeling over-the-counter drugs (CFR 201.60). These design considerations include the use of conspicuous and easily legible boldface print or type in distinct contrast (by typography, layout, color, embossing or molding) to other matter on the package; the ratio of height to width of letters must not exceed a differential of 3 units to one unit; and letter heights pertain to upper case or capital letters. When upper and lower case or all lower case letters are used, the lower case letter 'o' or its equivalent must meet the minimum standards. Letters and numerals must be in 'a type size established in relationship to the area of the principal display panel of the package and shall be uniform for all packages of substantially the same size.' Guidance is provided also for minimum height requirements for labels used on different size packages (CFR 201.60, p. 34). The FDA specifies detailed information to be included on the label, and particular drugs for which warnings must be given (including estrogenic hormone preparations, wintergreen oil, tannic acid, potassium salts, ipecac syrup, among others). As noted earlier, all of the guidance focuses on the kind of information to be supplied by the drug maker, with no emphasis on the understandability or effectiveness of the label itself.

The FDA also furnishes general guidance for warning and caution statements used with 'drugs' and 'devices' (21 CFR Part 369). According to this portion of the CFR, warning statements 'should appear in the labeling prominently and conspicuously as compared to other words, statements, designs, and devices, and in bold type on clearly contrasting background' (p. 291). Numerous specific warning and caution statements are given for drugs ranging from acetanilid and acetophenetidin to vibesate preparations. Specific wording for insulin drug labels can be found in Part 429.11. Unlike other sections, this one provides color specifications that relate to the amount of insulin in the preparation. FDA specifies red for indicating 40 U.S.P. units of insulin per milliliter, white for 100 U.S.P. units, and brown and white diagonal stripes for 500 U.S.P units and so forth, depending on the original form of the insulin.

12.2.4 Food and Vitamins

The FDA also gives general provisions for food labeling, including identity, number of servings, nutrition, and warnings (FDA, 1995 21 CFR part 101). An example is shown in Figure 12.1. Although they are not warnings in the usual sense, the format has a standardized look providing information useful in determining personal risk. For example, the label indicates the amount of sodium in a product, which is useful for persons on sodium-reduced medical diets or with high blood pressure.

Nutrition Facts
Serving Size 1 cup (228g)
Servings Per Container 2

Amount Per Serving
Calories 260   Calories from Fat 120

	% Daily Value*
Total Fat 13g	**20%**
Saturated Fat 5g	**25%**
Cholesterol 30mg	**10%**
Sodium 660mg	**28%**
Total Carbohydrate 31g	**10%**
Dietary Fiber 0g	**0%**
Sugars 5g	
Protein 5g	

Vitamin A 4%	•	Vitamin C 2%
Calcium 15%	•	Iron 4%

* Percent Daily Values are based on a 2,000 calorie diet. Your daily values may be higher or lower depending on your calorie needs:

	Calories:	2,000	2,500
Total Fat	Less than	65g	80g
Sat Fat	Less than	20g	25g
Cholesterol	Less than	300mg	300mg
Sodium	Less than	2,400mg	2,400mg
Total Carbohydrate		300g	375g
Dietary Fiber		25g	30g

Calories per gram:
Fat 9 • Carbohydrate 4 • Protein 4

Figure 12.1 Nutrition labeling format required by the Food and Drug Administration (FDA), showing type face, size, and typical content information.

The FDA also lists specific food labeling requirements for spices, flavorings, colorings and chemical preservatives, kosher, percentage of juice in fruit or vegetable juice beverages, and D-erythroascorbic acid and monosodium glutamate. For these labels, the FDA stipulates that the 'principal display panel' for nutritional labeling must 'be large enough to accommodate all the mandatory label information required to be placed thereon by this part with clarity and conspicuousness and without obscuring design, vignettes, or crowding' (CFR, Part 101.1, p. 14). It also specifies the area of the display panel necessary relative to the size and geometry (rectangular, cylindrical, etc.) of the package. The information panel is located immediately to the right of the principal display panel and must allow information to be displayed prominently and conspicuously. A minimum letter height of 1/16th inch is specified, again with the requirement of legibility and conspicuity. All information required by the FDA on a label must be located on either of these two panels.

The FDA also specifies the graphic format for daily recommended vitamins in food labels (CFR, Part 101.9, d). The guidelines state that nutrition information must be set off in a box by hairlines which are black or one color type on a white or neutral contrasting background. They require the use of a single, easy to read type style (with a type face not less than 8 point for headings and 6 point for other information) using both upper and lower case letters, along with specific spacing between lines. Furthermore, letters should never touch. Highlighting or bolding is specified for information such as nutrition facts, amount per serving and percentage daily value. Finally, the headline 'Nutrition Facts' must be in a type size larger than any other print on the label, and, if possible, be the full width of all the information provided. Further specific format instructions relate to text presentation within the label for particular types of information such as 'calories from fat' or percentage of vitamin nutrient. A sample label is given in the CFR for guidance on layout and information type. Use of a second language is permitted, either within the English label, or adjacent to it, provided that all nutritional information is given in both languages.

Similar guidance is given for labels for multi-vitamins, with graphic examples of appropriate labels in Part 101.36. Finally, in Appendix B to Part 101, the FDA provides examples of appropriate 'graphic enhancements' permitted by the FDA. This information pertains primarily to type face and size, separation of groupings, and rules for box markings. Thus, the label should be boxed with all black or one color type on a white or neutral background, with an example of overall label layout given in Part 101 (see Figure 12.2). These guidelines result in a readily recognizable label with information consistently displayed and located so that a lay person (who reads English) can easily determine the nutritional content of a particular item. In Part 101.15 the FDA suggests reasons why a word, statement or other information may lack the required prominence and conspicuity, and directs attention to label size, space, typeface, etc. Although the FDA gives specific wording for food labeling, graphic advice for labels for products such as self-pressurized containers is limited to the need to provide warnings that are prominent and conspicuous.

In a departure from the labeling requirements discussed above, the FDA, in 21 CFR 101.11, specifies warning sign graphics to be used in retail establishments that sell products containing saccharin. These signs must be printed in red and black ink on white cardboard and be at least 11 by 14 inches (28×36 cm). 'The background of the bold heading "Saccharin Notice" and the boxed warning statement shall be bright red and the lettering, white. The remaining background shall be white with black ink. All lettering shall be in gothic typeface' (Part 101.11 a, p. 59). This notice is to be displayed in three locations: establishment entrance, central area selling soft drinks, and area selling

Nutrition Facts
Serving Size 1 tablet

Amount Per Serving	% Daily Value
Vitamin A 5000 I.U.	100%
50 % as Beta Carotene	
Vitamin C 60 mg	100%
Vitamin D 400 I.U.	100%
Vitamin E 30 I.U.	100%
Thiamin 1.5 mg	100%
Riboflavin 1.7 mg	100%
Niacin 20 mg	100%
Vitamin B_6 2.0 mg	100%
Folate 0.4 mg	100%
Vitamin B_{12} 6 mcg	100%
Biotin 0.03 mg	10%
Pantothenic Acid 10 mg	100%

Figure 12.2 Nutrition Fact format for vitamins required by the Food and Drug Administration (FDA) showing layout, vitamin type, and percentage daily requirement for children and for adults.

packaged goods with saccharin with specific exceptions for smaller establishments. The above review of the FDA requirements suggests that because each application is mandated separately, the agency has not been empowered to take a single, comprehensive agency approach to graphic design.

12.2.5 Controlled Substances

The Drug Enforcement Administration (DEA) of the Department of Justice in 21 CFR Part 1302 lists labeling and packaging requirements for controlled substances. It requires all commercial containers to have a symbol printed on the label designating the schedule in which the controlled substance is listed. The 'symbols' consist of the alphanumeric characters ranging from CI to CV, and covering Schedules I through V. The DEA requires that the symbol be located on the upper right corner of the main panel of the label and be at least twice the size of the largest type otherwise printed on the label. If a symbol is overprinted on the label (rather than located in the upper right-hand corner), it must be at least one-half the total height of the label, and be 'in a contrasting color providing clear visibility against the background color of the label' (Part 1302, p. 33). Furthermore, the symbol must be clear and large enough to be read without being removed from the shelf.

12.2.6 Tobacco Products and Alcoholic Beverages

In the USA, warning labels have been required on cigarette packages since the mid 1960s, and more recently similar requirements have been extended to smokeless tobacco products and alcoholic beverages. The emphasis in these warnings has been, almost exclusively, on words rather than symbols or other graphic displays.

The Federal Trade Commission (FTC) has responsibility under various Federal laws to insure the proper display of health warnings in advertising and on packaging of tobacco products sold in the USA. Soon after the Surgeon General released the 1964 report of the Advisory Committee on Smoking and Health (Public Health Service, PHS, 1964), the FTC proposed several administrative rules that would have required health warnings on cigarette packages and advertising. These proposed rules were preempted by the Federal Cigarette Labeling and Advertising Act of 1965 (Public Law 89-92) which required that all cigarette packages contain the following health warning: 'CAUTION: Cigarette Smoking May Be Hazardous to Your Health.' The 1965 legislation specifically prohibited states or other jurisdictions from using any other types of warning statement or devices and included a three-year prohibition of warning labels in cigarette advertisements. Section 4 of the law provided general guidance regarding format and typographical requirements for the mandated warnings: 'Such [warning] statement shall be located in a conspicuous place on every cigarette package and shall appear in conspicuous and legible type in contrast by typography, layout, or color with other printed matter on the package.'

Subsequent legislation, the Public Health Cigarette Smoking Act of 1969 (Public Law 91-222), banned cigarette advertising on television and radio and strengthened the package warning label to read: 'WARNING: The Surgeon General Has Determined That Cigarette Smoking is Hazardous to Your Health.' The statutory language of the Act, however, continued to omit specific reference to the risks and consequences of smoking and extended the preemption on requiring any additional warnings on cigarette packaging. A consent order issued in 1972 required that all cigarette advertising 'clearly and conspicuously' display the same warning required by Congress for cigarette packages (Federal Trade Commission, FTC, 1972).

The warnings seen today on cigarette packages and advertisements in the USA are the result of the Comprehensive Smoking Education Act of 1984 (Public Law 98-474). This law required cigarette companies to rotate the following four warnings on all cigarette packages and advertising.

> SURGEON GENERAL'S WARNING: Smoking Causes Cancer, Heart Disease, Emphysema, and May Complicate Pregnancy
>
> SURGEON GENERAL'S WARNING: Quitting Smoking Now Greatly Reduces Serious Risks to Your Health
>
> SURGEON GENERAL'S WARNING: Smoking by Pregnant Women May Result in Fetal Injury, Premature Birth, and Low Birth Weight
>
> SURGEON GENERAL'S WARNING: Cigarette Smoke Contains Carbon Monoxide

The current warnings are noteworthy in at least three respects: (1) they include specific health consequences associated with smoking, (2) they feature four rotating warnings intended to help offset the effects of habituation, and (3) they call attention to the message source by using uppercase letters. Still, the current warnings retain the simple rectangular shape and the black lettering on white background visual format of previous warnings. They are also relatively small and inconspicuous in terms of size and location on the cigarette package. Warnings on outdoor billboard advertisements are allowed to be somewhat abbreviated from those appearing in newspaper, magazines, and product packaging. Also, the 1984 legislation did not require warning labels on speciality advertising items, such as pens and pencils, clothing, and sporting goods, that carry cigarette company logos, brand names, or other promotional messages.

In 1986, Congress extended requirements for warning labels to smokeless tobacco products by passing the Comprehensive Smokeless Tobacco Health Education Act of

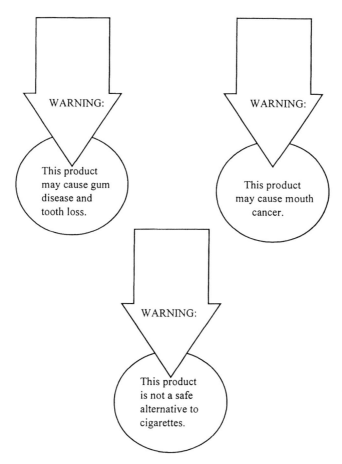

Figure 12.3 Health warnings required by the Federal Trade Commission for smokeless tobacco advertisements, describing the possible adverse health consequences of the product.

1986 (Public Law 99-252). This legislation requires tobacco manufacturers to display and regularly rotate the following three warnings on all smokeless tobacco packages and advertising (except billboards).

WARNING: This product may cause mouth cancer

WARNING: This product may cause gum disease and tooth loss

WARNING: This product is not a safe alternative to cigarettes

The smokeless tobacco Act stipulates that the warnings displayed in advertisements appear in a circle-and-arrow format (see Figure 12.3). Interestingly, this format was recommended unsuccessfully by the Federal Trade Commission (FTC, 1981) for use with cigarette warnings.

Recent proposals for tobacco-related legislation and liability settlement (Broder, 1997) include specifications for more prominent and explicit warning messages on cigarette packages and for stricter limits on tobacco marketing and advertising, including bans on vending machines and outdoor billboards. The proposed warnings would cover 25% of the front label and would be in bold white-on-black lettering. The messages would be strengthened to indicate that smoking is addictive, causes fatal lung disease, causes cancer, and can be deadly. As of this writing, these proposals are still being debated in Congress and elsewhere. Meanwhile, other countries, notably Australia, have recently instituted a new set of warnings on cigarettes that are more prominent, informative, and explicit in communicating the health risk posed by cigarettes.

With respect to warning labels on alcoholic beverages, the Alcoholic Beverage Labeling Act of 1988 (Public Law 100-690) requires that all alcoholic beverage containers have the following warning on the label: 'GOVERNMENT WARNING: (1) According to the Surgeon General, women should not drink alcoholic beverages during pregnancy because of the risk of birth defects. (2) Consumption of alcoholic beverages impairs your ability to drive a car or operate machinery, and may cause health problems.'

The label must be designed so that the required statement is readily legible under ordinary conditions. Minimum type sizes are specified for different size beverage containers. For example, for containers between 237 ml (8 fluid ounce) and 3 l (101 fluid ounce), the mandatory statement 'shall be in script, type or printing not smaller than 2 millimeters' (CFR Sec.16.22). This warning message can be part of the brand label, or be located separately from the brand label on a back or side label. As with the tobacco warnings, this legislation prohibits the use of any other warning statements on alcoholic beverages (and their associated packaging) sold in the USA. Although warnings have also been proposed for magazine and television advertisements, such requirements have not yet been enacted.

12.3 SAFETY SIGNS AND SYMBOLS

The Occupational Safety and Health Administration (OSHA) has issued requirements for safety signs and symbols for industrial safety. In its regulations, OSHA specifies signage for means of egress, ionizing and non-ionizing radiation, personal protective equipment, and general environmental conditions (OSHA, 1994, 29 CFR Part 1910). OSHA has treated each of these requirements differently, with no systematic approach for the design and content of warning signs and labels. For example, under means of egress, OSHA defines characteristics of means of egress, including routes and general specifications, as well as exit marking, specifying the word 'Exit.' OSHA states that these signs must be 'readily visible,' and 'located and of such size , color and design as to be readily visible' (29 CFR, Part 1910, p. 130). Furthermore, exit signs must consist of the letters 'Exit' at least 6 inches (15 cm) high with stroke widths no less than 0.75 inches (2 cm) wide. Exit signs are to be 'distinctive in color and provide contrast with decorations, interior finish, or other signs' (29 CFR, Part 1910, p. 130). OSHA also states that exit signs must be illuminated by a 'reliable light source giving a value of not less than 5 fc on the illuminated surface.' (The abbreviation fc is footcandle, a non-SI unit of illuminance; the SI unit is lux and there are 10.76 lux in 1 fc.) The CFR stipulates that this entire 'subpart is promulgated from NFPA 101-1970, Life Safety Code' (29 CFR, Part 1910, p. 131), making it clear that the voluntary standard and the Federal regulation are identical, at least for the 1970 NFPA Code.

In the same Part, OSHA specifies design requirements for warning signs for ionizing radiation (such as x-rays, gamma rays and the like, but excluding sound or light waves). In 29 CFR Part 1910.96, OSHA defines ionizing radiation explicitly, and prescribes a specific symbol (see Figure 12.4). This symbol has particular design characteristics, including three fan-like blades surrounding a central disk with an outer angular separation of 60% between each blade. The regulation specifies that the color of the cross-hatched area of the central disk and each blade be magenta or purple, with a yellow background.

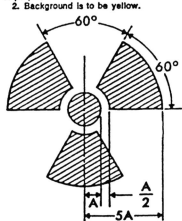

Figure 12.4 Guidelines for the design of ionizing radiation hazard signs as prescribed by the US Occupational Health and Safety Administration (OSHA).

Those familiar with the Z535.1 (ANSI, 1991) or the ISO 3860 approach, will recognize a conflict here. The latter two standards specify that the symbol color should be black, rather than magenta. This change was made for Z535.1 because the magenta color fades very rapidly on exposure to visible light, particularly daylight, leaving a sign with a yellow and grayish symbol. Similarly, at the US Federal level, the DOT specifies the yellow and black symbol, while OSHA and the Nuclear Regulatory Commission (NRC) specify yellow and magenta. The conflict between the two approaches has yet to be resolved. Furthermore, a more durable magenta pigment has not yet been identified.

In Part 1910, OSHA also specifies locations for using the radiation hazard sign, and states that both the symbol and the words 'CAUTION, HIGH RADIATION AREA or AIRBORNE RADIOACTIVITY AREA' or other hazard identification must be used. The NRC requires that this sign be used for containers used to transport radioactive materials (10 CFR Part 20).

OSHA specifies a symbol for warnings for non-ionizing radiation (e.g., radio frequency or RF radiation) in Part 1910.97. Again, 'symbol' refers to the 'overall design, shape, and coloring of the RF radiation sign' (Part 1910.97, p. 221) which consists of a 'red isosceles triangle above and inverted black isosceles triangle, separated and outlined by an aluminum color border,' with the words 'Warning-Radio-Frequency Radiation Hazard' appearing in the upper triangle (Part 1910.97, p. 222). Although the word 'symbol' is used, it refers to the whole layout of the sign, including the word message, not just the graphic image. Additional warning or precautionary instructions are allowed at the discretion of the user. Although Part 1910 was updated in 1994, it continues to refer the reader to the out-of-date, ANSI Z53.1(1953), edition of the *Safety Color Code* (not the 1979 or 1991 editions). In Part 1910, OSHA gives lengthy specifications for signs and tags indicating biological and other hazards. For biological hazards, OSHA specifies the biological hazard symbol (**see Figure 12.5 in color section**) and adds the term WARNING to the existing two terms DANGER and CAUTION as an acceptable signal word.

OSHA specifies safety color codes in Part 1910.144 for marking physical hazards. When such hazards occur in general workplace areas, specific locations or on particular equipment, they must be color coded to signify the most important hazard. Red identifies 'Danger' or 'Stop.' Use of red for fire protection equipment and apparatus was 'reserved' from the regulation even though it is present in the voluntary standard referenced in this section. Yellow designates 'Caution' and marks physical hazards. In the next section, OSHA (1994, 1910.145) provides requirements for colors to be used on accident prevention signs and tags. 'Danger' signs must be red, black, and white 'as specified in Table 1 of Fundamental Specification of Safety Colors for CIE Standard Source C, American National Standard Z53.1 (1967).' As noted earlier, this rule does not reference the 1991 edition of the ANSI Z535 *Safety Color Code* which replaced the Z53.1 standard. (It is important to note that within two consecutive pages of the Code of Federal Regulations different editions of the ANSI Standards for *Safety Colors and Safety Signs* are specified, making it very difficult for users, who must check with ANSI to obtain out-of-date versions of the standard.) Part 1910.144 states that 'Caution' signs must have a yellow background, with a panel of black with yellow letters, and use the color specifications given in ANSI Z53.1, 1967. Safety instruction signs must be white, with a panel in green with white letters. Signage for protective gear (such as hard hats or safety shoes) is not specified explicitly by Part 1910, although requirements for use are given.

In an appendix to Part 1910, OSHA (1994) recommends that the entire 'Danger' sign be red with lettering or symbols in a contrasting color, 'Caution' signs should be yellow, with lettering or symbols in a contrasting color, 'Warning' signs should be orange, again with letters or symbols in a contrasting color; and 'Biological Hazard' signs should be fluorescent orange, or orange-red with letters or symbols in a contrasting color, and use any of the three signal words. References in Part 1910.145 cite an earlier edition of the *Safety Color Code* (Z53.1, 1979) for determining the colorimetric specifications for safety colors. Nowhere in Part 1910 is reference given to its successor, the more recent ANSI (1991) series of Standards for safety signs, colors, symbols, labels, and tags, which give more current colorimetric specifications and color codes.

In Part 1910.145 (p. 433) OSHA (1994) states that 'The wording of any sign should be easily read and concise. The sign should contain sufficient information to be easily understood. The wording should make a positive, rather than negative suggestion, and should be accurate in fact.' Tags (with no particular graphic design) are specified for lockout tag-out applications. They must be legible, understandable, and attached securely. Finally, OSHA requires that the 'slow moving vehicle' emblem should 'consist of a fluorescent yellow-orange triangle with a dark red reflective border' (Part 1910.144, p. 432). Detailed design specifications, including specific dimensions for the symbol are given in the regulations, in accordance with the American Society of Agricultural Engineers emblem for identifying slow-moving vehicles.

In another section of the CFR (Part 1910.134), OSHA uses explicit color coding and word labels to identify gas mask canisters. In this Part, OSHA (1994) notes that 'each gas mask canister shall be painted a distinctive color or combination of colors' (Part 1910.134, p. 417). The colors must be clearly identifiable by the user and distinguishable from one another, while the coating itself should resist peeling, blistering, fading, etc. This Part assigns the following color schemes: acid gases are to be shown by white, with an additional 0.5 inch (1.3 cm) stripe around the bottom of the canister to indicate the presence of hydrocyanic acid gas, and a 0.5 inch (1.3 cm) yellow stripe at the bottom to indicate chloric gas. Organic vapors are to be identified by black, and ammonia gas by green, while acid gases plus ammonia should have a 0.5 inch (1.3 cm) white stripe around the bottom of the green canister. Carbon monoxide is to be identified with blue, and acid gases and organic vapors with yellow, while the presence of hydrocyanic acid gas and chloropicrin vapor are to be indicated by yellow with a 0.5 inch (1.3 cm) blue stripe around the bottom of the canister. Acid gases, organic vapors and ammonia gases are to be brown, while radioactive materials are to be indicated by purple (magenta). Particulates in combination with any of the gases are to be identified by the canister color for the gas contaminant plus a 0.5 inch (1.3 cm) gray stripe at the bottom. Finally, if all the atmospheric contaminants require protection, the gas mask canister must be red with a 0.5 inch (1.3 cm) gray stripe, which is located at the top of the canister. No guidance is given for teaching the user the meanings of these complex color codes, however.

12.3.1 Consumer Product Labels

The Consumer Product Safety Commission (CPSC), provides guidance for labeling safety hazards likely to be encountered by consumers. 16 CFR Part 1205.6 (CPSC, 1994) specifies a warning label for use on reel-type and rotary-type power mowers. This label must contain a signal word (Danger), specific wording (Keep hands and feet away), a symbol (showing a hand being severed), specific colors (white, black and yellow), and dimensions, as well as an overall shape for the sign (a rectangle in the upper quadrant with a triangle in the lower quadrant). This graphic approach is different from all others covered in the present chapter! More typically, CPSC references appropriate voluntary standards, as required by the CPSC operating statutes.

12.3.2 Railroad

Finally, the Federal Railroad Administration (FRA) of the Department of Transportation requires a rear-end marking device for passenger, commuter and freight trains (49 CFR

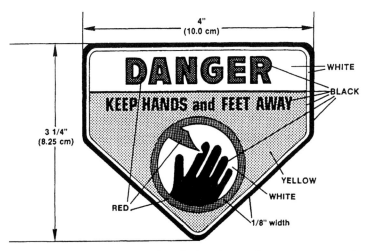

Figure 12.6 Guidelines for the design of symbols to be used on reel-type mowing machines as prescribed by the US Consumer Product Safety Commission (CPSC).

Part 221). In Part 221.13, FRA (1994) defines the marking devices as a 'display on the trailing end of the rear car of that train.' This marking is neither a sign nor a symbol but a 'lighted device.' In addition, each train must 'continuously illuminate or flash a marking device' as prescribed in this subpart. This device must be illuminated or flash 'during the period between one hour before sunset and one hour after sunrise, and during all other hours when weather conditions so restrict visibility that the end silhouette of a standard box car cannot be seen from 0.5 mile (0.8 km) on a tangent track by a person having 20/20 corrected vision' (Part 221.13, p. 216). The marking devices may be operated by a photocell which is operational when there is less than 1 cd/m^2 of ambient light. Further, the device must be located at least 48 inches above the top of the rail. Device intensity (or effective intensity for flashing lights) is restricted to the range of 100–1000 candela, and the color range of red-orange-yellow as defined by the International Lighting Commission or CIE (with specific chromaticity boundaries given for the particular color range). Flash rate for flashing lights must be not less than 1.3 seconds nor more than 0.7 seconds. The FRA provides guidance for testing the warning device and indicates the specific data required for test record acceptance.

12.3.3 Review of Specifications in Federal Regulations

The preceding review of specifications in US Federal regulations for safety signs, labels, symbols, and color codes, reveals a lack of consistency in approach to warning graphics among, or even within, agencies. Color is not used consistently; symbols are developed and used on an *ad hoc* basis to fulfill a specific need; overall sign communicability is not addressed; and no research basis for decisions about warning graphics is given. Only DOT relied on research into symbol effectiveness in selecting symbols for highway use. The result is an inconsistent and unsatisfactory situation for warning of hazards and for communicating sometimes life-threatening information. As the section on tobacco labeling demonstrates, typically Federal requirements for labeling come from very specific laws for different, individual situations. Perhaps as a result, in less heavily regulated areas, voluntary standards have been developed to provide consistent graphic formats and approaches for warnings in product categories. These voluntary, private sector approaches for a more coherent system of warning graphics for a wide range of products and situations will be described in the next section.

12.4 VOLUNTARY STANDARDS FOR SAFETY SIGNAGE

12.4.1 Federal Role in Voluntary Standards

OMB (1993) Circular A-119 states that Federal agencies should, to the extent possible, use voluntary standards in their regulatory and procurement processes. Congress supplemented these recommendations in 1996 when it passed the National Technology Transfer and Advancement Act (PL 104-113), which charges agencies with using voluntary standards where appropriate, and with informing Congress when such standards are not used. This Act also requires the National Institute of Standards and Technology (NIST) to coordinate activities in standards (and conformity assessment) with other government agencies and with the private sector.

NIST is a non-regulatory agency of the Department of Commerce that works closely with industry to meet infrastructural needs for commerce and trade. Since its founding in 1901, NIST has worked with industry and the private voluntary standards system to provide technical expertise, standard test methods, and the overall technical infrastructure. NIST also maintains the physical standards required for fundamental measurements (such as those for length, mass, and time). As PL 104-113 asserts, NIST's role is to work with others to develop written standards; despite its name, it does *not* develop, publish or maintain documentary standards. Many NIST staff along with those from other Federal agencies, serve as technical experts on voluntary, consensus standards-writing committees, however, and often perform technical research to support these activities (in response to requests from the private sector and other Federal agencies). Many other Federal agencies also participate actively in the development of voluntary standards. This participation is expected to increase in the future as a result of the 1996 law which also directs agencies to participate in the development of voluntary standards.

12.4.2 US Voluntary Safety Sign Standards

An example of NIST's support for voluntary standards activities can be seen in its work with the ANSI Accredited Standards Committee (ASC) Z535. In 1978, industry came to the National Bureau of Standards (NBS), now NIST, and expressed the need for a comprehensive system for safety signs and symbols. For many years, the committees for workplace safety signs and for workplace colors had operated independently of each other with no liaison. Industry wanted to develop a system that could encompass all warning messages, including product labels, workplace signs and placards, symbols, temporary tags, and colors. Under ANSI guidance, NBS formed the Z535 committee. At the time, its active research programs in human factors and color made it a logical home for the secretariat for the Z535 committee. The ANSI Z535 Committee on Safety Signs and Colors was later rechartered as an Accredited Standards Committee by ANSI when ANSI gave up direct sponsorship of standards committees. Throughout the long development of the initial five standards, NIST held the secretariat, at first by itself, and then jointly with the National Electrical Manufacturers Association (NEMA) which now holds the secretariat by itself.

The combined Z535 committee on Safety Signs and Colors had initial members from companies such as IBM, Caterpillar, Deere, from government agencies such as OSHA and CPSC, from societies such as the American Welding Society, from standards developers such as UL, from insurance interests such as the Alliance of American Insurers, and from private experts in color, graphics and human factors. Although there were several 'industry' standards, industry expressed the need for a national standard for safety signs, labels, and symbols to remove inconsistencies among the various individual approaches. For example, FMC Corporation (1978) had developed a manual for safety signs and symbols for its products, closely followed by Westinghouse Electric Corporation (1981), which developed a somewhat different manual for its facilities and products. Since their products were both used in still other facilities, users were exposed to different signs, symbols and colors for the same message, and often did not understand what was intended.

The Z535 committee had the following scope: 'To develop standards for the design, application, and use of signs, colors, and symbols intended to identify and warn against specific hazards and for other accident prevention purposes'. Five subcommittees were created and their work resulted in the publication of five standards in 1991.

Z535.1 *Safety Color Code*, which updated Z53.1 (1979)

Z535.2 *Environmental and Facility Safety Signs*, which updated Z35.1 (1972)

Z535.3 *Criteria for Safety Symbols*, a new standard

Z535.4 *Product Safety Signs and Labels*, a new standard

Z535.5 *Accident Prevention Tags*, which updated Z35.2 (1974)

With the passage of the five standards in 1991, the committee met its goal of designing a coherent approach for a system of safety alerting messages, using specified colors, words and symbols in standard, recognizable formats across different products and applications—both industrial and consumer. The standards covered both workplace safety and consumer products, and were somewhat compatible with the colors and symbols found in the international standard ISO 3864, *Safety Signs and Colours*. Following much debate among committee members, the five standards were finally adopted in 1991 as American National Standards (ANSI, 1991) have been revised in 1998 as part of the normal ANSI cycle.

The Z535.3 Standard *Criteria for Safety Symbols* was the first US standard to provide image content and suggested graphic art for symbols, and suggested surround shapes and colors for symbols used as stand-alone signs. In Z535.3 (1991, p.2) image is defined as 'that portion of the symbol which is a graphic rendering, either abstract or representational, of the safety message' and symbol as 'a configuration, consisting of an image with or without a surround shape, which conveys a message without the use of words. It also includes graphic art, such as pictograms, pictorials and glyphs' (Z535.3, 1991, p. 2). Finally, the standard also defines 'panel,' 'signal words,' and 'surround shape' (for use in the immediate vicinity of a safety symbol).

The Z535 Standard does not mandate use of a particular symbol design; rather, it standardizes image content, and suggests artwork to represent this content. Image content and graphical artwork for symbols are given for the following categories: hazard alerting (such as 'Flammable'); mandatory or permitted actions (such as 'Wear your hard hat'); general safety messages (such as 'Safety shower'); and prohibited actions (such as 'Do not smoke'). The standard also gives graphic advice for the design of new symbols for applications not contained in the standard. Central to the philosophy of the Z535.3 Standard is the concept that a symbol must be understandable and communicate its message effectively without the use of words, even when it is used on a sign with word messages. To aid in achieving this goal, an appendix to the standard provides a procedure for evaluating new symbol images, including testing protocols for evaluating their effectiveness in reaching the intended target audience.

In keeping with the concept of a *system* for visual alerting, Z535.3 designates a surround shape for additional information about the safety message (usually for use only when a symbol is the complete sign) as shown in **Figure 12.7 (see color section)**. This standard suggests either an equilateral triangle or a diamond to indicate a hazard, with the severity of the hazard being conveyed by the signal word and color. (The triangle is in conformance with ISO 3864; the diamond conforms to the UN/DOT system for hazardous materials transport discussed earlier in this chapter). A solid color circle conveys mandatory action, while a square or rectangle indicates general safety information, such as equipment location, egress, and permitted actions. Finally, a circular band with a diagonal slash at 45° from upper left to lower right indicates prohibition. The standard also specifies the following colors for use on symbol-only signs.

Hazard Alerting Symbols

Danger	White image on a red background
Warning	Black image on an orange background
Caution	Black image on a yellow background (consistent with ISO Standard 3864)

Mandatory Action Symbols

General	White image on a blue background

General Safety Information Symbols

General	White image on a green background, or black image on a white background with a green border
Egress	White image on a green background, or black image on a white background with a green border
Fire	White image on a red background, or black image on a white background with a red border

Prohibition Symbols

Black image, red circular band with slash, and white background, is preferred. (Consistent with ISO 3864)

Because the prevailing approach to safety information in the USA has long relied exclusively on word messages, the Z535 standard also provides guidance for using symbols as elements of word signs, as well as on stand-alone signs. Consequently, the Z535 standard distinguishes between single-panel (or symbol only) signs and multi-panel safety signs which use word messages (with or without an accompanying safety symbol). The symbol standard specifically states that, for multi-panel signs, the color and format given in the other Z535 standards for environmental signs, product safety signs and labels, and accident prevention tags should be used. The single-panel only signs are in relative conformance with the international provisions for symbol colors and surround shapes given in ISO Standard 3864, and for use when supplementary word messages are not needed.

As noted above, the Z535.3 standard provides symbol artwork as examples for communicating hazard information and safety messages. Importantly, many of these symbols, as shown in Figure 12.8, were chosen based on data obtained from procedures developed and implemented at NBS/NIST (Lerner and Collins, 1980; Collins, 1982, 1983; Collins, Lerner, and Pierman, 1982).

As a result of this research base, the Z535.3 standard also reiterates the need to demonstrate understandability for any safety symbol, particularly a new one, through acceptable test procedures. It states that any symbol that does not demonstrate understandability also must have a word message explaining the intent of the symbol. For example, a number of common symbols, such as those for radiation and laser hazards, were not well understood by the NBS/NIST test subjects (Collins, 1983). Finally, the Z535.3 standard recommends using training/recognition procedures to familiarize users with the symbols and their meanings, particularly by employers or product manufacturers who plan to use the symbol in workplaces or on products.

Going well beyond the scope of then-available safety symbol standards, including the ISO 3864 Standard, the Z535.3 standard provides graphic considerations for designing new symbols. It stresses that individual symbols should be designed, wherever possible,

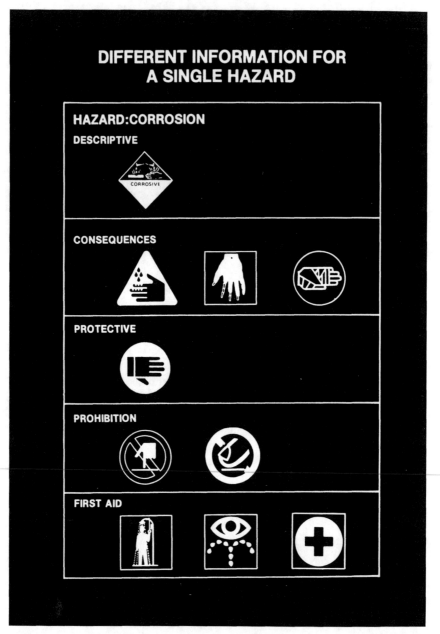

Figure 12.8 Example of symbols tested in NBS/NIST research on safety symbol understandability—many of which subsequently were adopted by the Z535.3 *Standard for Safety Symbols* showing how a symbol might be used to communicate different types of safety messages.

as elements of a consistent visual system, and that visual consistency within a system can be strengthened by attention to the graphics guidelines. These guidelines include use of representational rather than abstract symbols whenever possible, and the use of full human figures in the form of the figure used by the US Department of Transportation (AIGA, 1979). Furthermore, it recommends that when objects, faces, or the full human body are shown, the view (front or side) that is most easily recognized should be used. In addition, symbols should depict action and, where possible, both the human element and the hazard. Z535.3 also recommends graphic elements that contribute to good symbol design, including proportion, symmetry, direction of elements, solid versus outline forms, and simple versus complex detail. Since symbols can be used in widely varying applications, the standard recommends that the designer consider the intended viewing distance when sizing symbols for products or workplaces, as well as likely environmental factors such as dirt, light level, and light quality, which can degrade the effectiveness of a symbol.

An appendix to the Z535.3 standard contains a suggested procedure for evaluating new safety symbols. This test and evaluation procedure stresses the understandability of a symbol as the primary criterion for determining its effectiveness, rather than legibility or attractiveness. The recommended procedure consists of asking a selected target audience to indicate the comprehensibility of a proposed symbol or symbols so that the relative performance of several different symbols representing a particular idea can be compared. The appendix outlines procedures for determining the intended meaning for the symbol, selecting a reasonable subset of a population of target users to evaluate it, and testing it. Once testing is complete, one procedure recommends scoring the data in four categories: correct, wrong, critical confusion (that is, the opposite of correct), and no answer. Knowledge of critical confusion is particularly important for symbols that could be lethally misinterpreted, such as 'exit' for 'no exit' or 'campfire permitted' for 'flammability.'

The criterion for selection of a symbol for a sign is at least 85% correct answers, with fewer than 5% critical confusion. Symbols which do not meet these criteria should be redesigned and retested (with a different group of people). This suggested procedure asks test subjects to write down definitions for the symbols, although an alternative procedure allows use of multiple-choice testing (after reasonable choice alternatives have been determined). Lerner and Collins (1980) found relatively good agreement between definitions and multiple-choice procedures when the multiple choice answers were based on reasonable choice alternatives (derived from definition-style answers). These authors noted the importance of including 'critical confusion' or 'opposite-to-correct' answers among the choices to weed out symbols that convey exactly the wrong meaning. Because Lerner and Collins reported difficulty in interpreting subjects' handwriting and understanding some responses, as well as much greater use of experimenters' time when using a definition-style procedure, they recommended the multiple-choice approach. Of course, newer computer-based technologies might eliminate some of these problems, and provide less constraint on the subjects' responses. See also Wolff and Wogalter (1998) on testing procedures.

Prior to the Z535.3 standard, the USA did not have a comprehensive standard for specifying, selecting, designing or testing symbols for use on safety signs and labels. An underlying rationale for the Z535.3 standard was the need for a set of symbols for use with a 'US population which is multi-ethnic, highly mobile, and derived from a multiplicity of social and educational backgrounds, with different reading skills and word comprehension' (ANSI, 1991, Z535.3, p. 1). Because such factors are likely to reduce the effectiveness of the word-only signs specified in other US standards, symbols and pictorial signs are likely to communicate warnings better and should provide greater safety for the general population. To meet these challenges, the Z535.3 Standard provides general criteria for the design, evaluation, and use of safety symbols to identify and warn against specific hazards and avoid personal injury. Use of the standard should promote the adoption and use of uniform and effective safety symbols for warning people of potential dangers.

12.4.3 Fire Safety Symbols

About the same time that the Z535 standards were being drafted, the National Fire Protection Association (NFPA) charged a committee with developing standards for fire protection symbols. This committee, also chaired by NBS/NIST staff, developed several fire-safety symbol standards, including NFPA 171, *Life Safety Symbols*, and NFPA 178, *Standard Symbols for Fire Fighting Operations* (NFPA, 1980a,b). (The committee also developed a standard for symbols used to indicate fire-safety equipment on architectural and engineering drawing: NFPA 172, 1980.) NFPA 171 and 178, adopted in 1980, contained symbols for familiar fire-safety applications, such as 'extinguisher,' as well as symbols for 'exit' and 'no exit.' These two standards were based on research conducted at NBS/NIST. In particular, this research (Collins and Lerner, 1983) found that several commonly suggested symbols for 'exit,' shown in **Figure 12.9 (see color section)**, were not understood clearly.

For example, one exit symbol recommended for use in transportation facilities was graphically very simple but not well understood by naive subjects. Another much more representational symbol, used in Japan and shown in **Figure 12.10 (see color section)** was found to be much more effective.

When this symbol was studied, it quickly became apparent that not only was it more understandable on first encounter than almost all of its competitors, but also it was more 'visible' in simulated smoke (Collins and Lerner, 1983). The NFPA committee selected this symbol as the standard for 'exit' and recommended its use to ISO for an international standard. Although this symbol was standardized for use in the USA, it is rarely used in the USA. Nonetheless, its selection provides a useful illustration of the use of both understandability and visibility as criteria for selecting symbols for use in both normal and emergency conditions.

Unlike the Z535 committee, the NFPA committee chose not to try to develop signs with symbols and word messages. As a result, the NFPA *Life Safety Code*, NFPA 101, continues to use 'EXIT' as the designation for 'exit' in the USA. Because this standard is referenced by most local building codes, the symbol designated in NFPA 171 remains purely optional. In Canada, however, dual language messages are required, so that both 'EXIT' and 'SORTIE' must appear on an exit sign. UL 924 (UL, 1989a), referenced by the NFPA, provides explicit guidance for the design of 'exit' signs which are acceptable in the USA, and specifies letter height and shape, but avoids specifying color. Unlike the ISO 3864 standard, which specifies green as the color for indicating 'exit,' the NFPA (and UL) standards leave decisions about exit sign color to the local authority. This results in red, green or other color exit signs depending on the locality; this may be very confusing to those accustomed to the international use of green for exit (and red for fire or prohibition).

UL 924 also contains a specification for exit directional indicators, again based on research on visibility and comprehensibility done at NIST (see Collins, 1991). This specification is for directional indicators used on either side of the word 'EXIT,' and sized to meet UL visibility requirements for exit signs. Subsequently, NIST researchers examined the visibility of exit signs in smoke using various types of internal illuminants in the sign, including fluorescent, incandescent, and tritium sources (Collins, 1992).

Again this research was provided to both UL and NFPA for their deliberations about the specifications for exit design characteristics. In addition to exit sign specifications, UL publishes a *Standard for Marking and Labeling Systems*, UL 969 (UL, 1989b), which provides additional information about label durability and placement.

12.4.4 Public Information Symbols

The area of public information symbols for travelers has proved quite amenable to both research and application of graphic design principles at both the national and international level. Thus, the US Department of Transportation (DOT) commissioned the American Institute of Graphic Artists (AIGA) to design a set of 34 public information symbols after a thorough review of existing designs. The result was a set of internally consistent graphic images (AIGA, 1979). The Franklin Research Institute evaluated the understandability of each of the 34 initial images (Freedman and Berkowitz, 1977; Freedman, 1978). These researchers suggested that symbols which were recognized by 60% or fewer of the participants clearly are unacceptable; those recognized by more than 85% are acceptable; and those in the mid-range need improvement. These criteria were then used in the selection of symbols for public information. These researchers used a variety of tests to determine effectiveness, including multiple choice, matching, and behavioral effectiveness (namely ability to walk through a space using the signs correctly). They found strong correlations between the results from each type of approach. These symbols have been widely adopted in the USA for airports and other public transportation facilities, and they appear to satisfy the need for rapidly understandable, well designed symbols for the traveling public.

12.5 INTERNATIONAL STANDARDS ACTIVITIES

12.5.1 Transportation Symbol Standards

During the 1970s and 1980s, a great deal of activity was aimed at developing an international graphic language for communicating messages without the use of words. This effort may have been spurred by the need for symbols for use in transportation systems in Europe which faced the very real challenge of communicating effectively without using multiple languages on signs. In fact, attempts to standardize symbols for road use began with a convention in Paris in 1909 on the 'International Circulation of Motor Vehicles' which adopted symbols for curve, bump, intersection, and railroad grade crossings (King, 1971). In 1931, the League of Nations expanded this set to 26 symbols, with uniform specifications for shape and color. Finally, in 1949, the United Nations adopted a protocol for traffic signs and symbols; it continues to be used to this day in Europe. Only in 1971 did the USA adopt some of the UN protocol and symbols. Until then, the USA had relied on a system of word signs developed initially by the American Association of State Highway and Traffic Officials in 1925. While many of the international symbols were evaluated for understandability in the USA (King, 1971, 1975), they were not originally designed nor selected on the basis of sound graphic criteria, unlike those for transportation facilities. Nonetheless, the USA is now in reasonably good compliance with most, but not all, of the international symbols for traffic control. Slowly, then, international acceptance of common symbols for use on highways has evolved to provide information to motorists rapidly and accurately, as well as to facilitate the safe movement of travelers, goods and hazardous materials between different nations. Their familiarity is considered to be essential for communicating highway information rapidly and accurately.

12.5.2 Safety Symbol Standards

As noted earlier in this chapter, a system of safety symbols was adopted under UN sponsorship for Hazardous Materials Transport. This standard was intended to facilitate safe passage of materials from one locality to another, and is in widespread use today. Because it was a relatively early attempt at standardization, the symbols and graphic format were never evaluated for understandability, visibility, or graphic consistency. However, its long use has led to familiarity with many of the symbols and messages.

The Treasury Board of Canada (1980) sponsored the design and evaluation of a comprehensive system for safety symbols, but its efforts have not met with broad international acceptance. Still another international standardization effort, centered in Europe, focused on symbols and colors for workplace safety, and was comparable in scope to the ANSI (1991) Z535.3 Standard. In the late 1970s, both ISO and the European Commission (EC) issued standards and directives for safety symbols and signs. These two approaches focused on the use of symbols and colors to provide safety information, and not on provisions for using word messages for safety signs or product labels, as ANSI Z535 did. When the EC Directive appeared in 1977, it generated some controversy among member nations, largely because the selected symbols were not based on any research into understandability, and because they were not graphically 'modern,' 'pleasing,' or designed using a consistent format. As a result, when the members of ISO Technical Committee (TC) 80 promulgated the ISO 3864 Standard for *Safety Colours and Signs* (ISO, 1984), they provided symbols only as examples of the graphic concept, not as mandatory graphic art for use in all cases (ISO 3864, 1984). Both international efforts relied on the use of color and graphic imagery to provide meaning, and did not allow for the use of word supplements (because of the difficulty of standardizing one language for Europe and the world). **Figure 12.11 (see color section)** demonstrates a possible bilingual approach based on Z535.

The international and regional efforts (by ISO and the EC) differ significantly from the US approach to warning signs by using only one level of hazard alerting, namely 'CAUTION,' although they all agree on coding it in black and yellow. Unlike the USA, these standards do not provide for degrees of alerting or the use of signal words, such as 'DANGER,' 'WARNING,' and 'CAUTION,' or even the use of supplementary word messages. Because the ISO standard is almost identical with those used by various member states in the European Union, safety symbols and colors effectively are standardized throughout Europe, a situation reinforced by the regulatory nature of the EC Directive. Unfortunately, the regulatory nature of the Directive provided very little opportunity for change, even though ISO members expressed dissatisfaction with its artwork. A number of proposed revisions to ISO 3864 have circulated to the ISO Committee, now TC 145, including several demonstrating the ANSI Z535 three-level approach to hazard warning; but as of this writing this standard has not been revised to incorporate them.

12.6 SUMMARY AND CONCLUSIONS

Under OMB Circular A119 and PL 104-113, Federal agencies are directed to move away from reliance on development of agency-specific standards toward use of voluntary standards, including performance-based standards. During the 1990s, US agencies with major procurement missions, such as DOD, NASA and GSA, began major efforts to replace their in-house specifications (typically military or Federal specifications) with applicable voluntary standards for procuring items for the US Federal Government. This recent (PL104-113) legislation should strengthen the movement toward similar use of voluntary standards by other agencies of the Federal Government. Yet, there are still numerous instances where federal agencies do not make effective use of standards, particularly those for the design of warning labels and signs, developed by the private sector. Such instances include numerous references to out-of-date voluntary standards; and outright failure to reference applicable, relevant standards.

Clearly there is no consistent policy for specification, design or use of symbols or labels throughout the US Federal Government (other than the occasional suggestion that labels be conspicuous and legible). While agencies such as the FDA or FTC provide a consistent and clear approach for labels for a particular topic, they appear to cover other topics in a somewhat idiosyncratic fashion. Thus, the FDA developed a singular approach for warning signs for saccharin, while CPSC developed an entirely different approach for signs for lawn mower safety, and neither agency was able to use these approaches for other warnings. These unique graphic approaches reflect little coordination within or between agencies on their regulations for safety labels and signs. Again, PL 104-113 provides some means for cross-agency coordination so this situation may change in the future, unless a specific law provides for a unique set of graphics and warnings. Nonetheless, currently there is little reliance (at least, as of the 1994 editions of the Code of Federal Regulations) on any one set of voluntary standards, including those of ANSI, NFPA, or ISO, or on any single Federal approach to graphics for warning messages. Neither the OSHA nor CPSC regulations reference the Z535 series of standards, although they do not appear to preclude their use. The present review reinforces the need to coordinate Federal activities with regard to safety signs and labels, and to heighten agency awareness of existing voluntary standards such as the Z535 series.

In addition, very little of the regulatory material specifying the design of labels, signs, and symbols in the Code of Federal Regulations appears to have benefited from research on its likely effectiveness. Thus, while 'conspicuity,' 'legibility,' and 'prominence' are called for as important elements of warning, the only guidance for achieving them is by use of black and white written information in a particular type size and font. Furthermore, while these are reasonable goals, neither procedures nor guidance are given for predicting or evaluating their likely effectiveness for the intended user. There is also no guidance for determining effectiveness of colors or symbols, used either alone or as part of word labels or signs. As a result, a product manufacturer is forced to guess how to design and apply a label, without clear guidance on the 'correct' approach. Additionally, there is no consistency across various Federal agencies for using color, symbols, and graphics as a system for providing effective warnings, nor has there been any evaluation of the effectiveness of existing warnings. While these ideas were of concern to the transportation community when the DOT was developing its systems for highway safety and public information symbols, this philosophy appears to have evaporated in subsequent hit-or-miss attempts to provide safety information.

As a result, the US Federal Government is not well positioned to deal with new issues, such as use of multi-lingual signs and the need to warn those who may not read English well. Policies for the use and design of multi-lingual signs and labels are likely to become an increasingly important topic within the voluntary and regulatory standards community in the USA. Unlike the current voluntary standards (such as NFPA or Z535), however, some Federal regulations do provide for foreign or dual language labels. For example, the FDA allows use of Spanish language labels for materials intended for Puerto Rico or other Spanish-speaking territories. The FDA goes even further in the section on nutritional labeling by giving examples of a bilingual label with numerical information given once, and identification information given in both Spanish and English (rather than require the whole label to be repeated in each language). Other agencies and voluntary standards do not address the issue of multi-lingual signs or the use of symbols to communicate to non-English-literate populations. Clearly, this topic deserves much more consideration, particularly when one considers the variety of 'second' languages common in different areas of the USA. The need for a sensible approach, such as effective symbol imagery, to avoid extremely long and illegible signs, is readily apparent, as is the need to warn all likely viewers of a potential hazard or provide a safety message.

In conclusion, there is a need for much greater coordination within the whole affected community, both voluntary and mandatory, on issues related to effective labeling and warning. This involves such simple activities as selecting the most recent edition of a voluntary standard and relying on the most appropriate voluntary standard, to more complex activities such as resolution among Federal agencies on conflicting requirements (such as the color for radiation hazard symbols), consideration of effective multi-lingual signs, and awareness, support for, and participation in standards activities related to safety signs (both national and international). Finally, there are particular issues related to the use of international safety signs for the USA. In this arena, the USA relies heavily on the use of word signs, reinforced by many years of litigation related to the 'duty to warn,' unlike the European and ISO approaches which use symbols (to avoid multiple language signs and labels). As a result the USA must be increasingly active in domestic and international deliberations which could create standards which conflict with our long established approaches for safety signs and labels, or serve as barriers to entry for our products. Active participation in international standards for safety and warning messages will also provide an avenue for incorporation of US approaches based on sound research in these standards, to the benefit of all.

REFERENCES

AIGA (1979) (US Department of Transportation Publication No. DOT-RSPA-DPB-40-79), *The Development of Passenger/Pedestrian Oriented Symbols for Use on Transportation-Related Facilities*. American Institute of Graphic Arts. Washington, DC: US Department of Transportation.

ANSI (1991) Z535.1, *Safety Color Code*, which updated Z53.1 (1979); Z535.2, *Environmental and Facility Safety Signs*, which updated Z35.1 (1972); Z535.3, *Criteria for Safety Symbols*; Z535.4, *Product Safety Signs and Labels*; Z535.5, *Accident Prevention Tags*, which updated Z35.2 (1974). New York American National Standards Institute.

BRODER, J.M. (1997, June 21) Cigarette makers reach $368 billion accord to curb lawsuits and curtail marketing. *New York Times*, pp. A1, A8.

COLLINS, B.L. (1982) NBS Publication No. BSS 141, *The Development and Evaluation of Effective Symbol Signs*. Gaithersburg, MD: National Bureau of Standards.

COLLINS, B.L. (1983) NBS Publication No. NBSIR 83-2732, *Use of Hazard Pictorials/Symbols in the Minerals Industry*. Gaithersburg, MD: National Bureau of Standards.

COLLINS, B.L. (1991) Visibility of exit signs and directional indicators. *Journal of the Illuminating Engineering Society*, 20, 117–133.

COLLINS, B.L. (1992) Visibility of exit signs in clear and smokey conditions. *Journal of the Illuminating Engineering Society*, 21, 69–84.

COLLINS, B.L. and LERNER, N.D. (1983) NBS Publication No. NBSIR 83-2675, *An Evaluation of Exit Symbol Visibility*. Gaithersburg, MD: National Bureau of Standards.

COLLINS, B.L., LERNER, N.D., and PIERMAN, B.C. (1982) NBS Publication No. NBSIR 82-2485, *Symbols for Industrial Safety*. Gaithersburg, MD: National Bureau of Standards.

CPSC (1994) *Safety Standard for Reel-type and Rotary Power Mowers* (Code of Federal Regulations 16 CFR Part 1205, Sections .6-.35, pp. 250–258). Consumer Product Safety Commission. Washington, DC: US Government Printing Office.

EUROPEAN ECONOMIC COMMUNITY (1977) Council Directive: Safety Signs at Places of Work. *Journal of the European Communities*, 25, L229/12–L229/20.

FDA (1995) (a) *Labeling* (Code of Federal Regulations 21 CFR Ch. 1, Part 201, Sections .1-.39, pp. 8–55); (b) *Packaging and Labeling* (Code of Federal Regulations 21 CFR Ch. 1, Part 429, Subpart B, Sections .10-.12, pp. 300–302); (c) *Interpretative Statements Re Warnings on Drugs and Devices for Over-the-Counter Sale*, and Subpart B *Warning and Caution Statements for Drugs* (Code of Federal Regulations 21 CFR Ch. 1, Part 369, Sections .1-.21, pp. 290–299); (d) *Animal Food Labeling* (Code of Federal Regulations 21 CFR Ch. 1, Part 501, Sections .1-.18, pp. 14–32); (e) *Cosmetic Labeling* (Code of Federal Regulations 21 CFR Ch. 1, Part 701, Sections .1-.13, pp. 181–192); (f) *Food Labeling* (Code of Federal Regulations 21 CFR Part 101, Sections .1-.108, Appendix A & B, pp. 12–164); (g) *Labeling and Packaging Requirements for Controlled Substances* (Code of Federal Regulations 21 CFR Part 1302, Sections .1-.8, pp. 31–34). Food and Drug Administration. Washington, DC: US Government Printing Office.

FHWA (1988) *Manual on Uniform Traffic Control Devices* (Code of Federal Regulations 23 CFR Part 1204, Section .4 and 49 CFR Part 1, Section .48). Federal Highway Administration. Washington, DC: US Government Printing Office. Also published as an American National Standard (1989), ANSI D6.1e, New York, NY: American National Standards Institute.

FMC CORPORATION (1978) *Product Safety Signs and Labels*, 2nd Edn. Santa Clara, CA: FMC Corporation.

FRA (1994) *Rear End Marking Device – Passenger, Commuter and Freight Trains* (Code of Federal Regulations 49 CFR Ch. II Part 221, Sections .1-.17 Federal Railroad Administration. Washington, DC: US Government Printing Office, pp. 214–218.

FREEDMAN, M. (1978) DOT Publication No. DOT-OS-6071, *Symbol Signs – The Testing of Passenger/Pedestrian Oriented Symbols for Use in Transportation Related Facilities*. Washington, DC: Department of Transportation.

FREEDMAN, M. and BERKOWITZ, M.J. (1977) DOT Publication No. DOT-OS-6071, FIRL No. C4448, *Preliminary Report on Laboratory and Pilot Field Testing: Testing Criteria and Techniques of Evaluation for Passenger/Pedestrian Oriented Symbols for Use in Transportation Related Facilities*. Washington DC: Department of Transportation.

FTC (1972) Complaint in the matter of Lorillard; Phillip Morris, Inc.; American Brands, Inc.; Brown and Williamson Tobacco Corporation; R.J. Reynolds Tobacco Company; Liggett & Meyers, Inc., consent orders, etc., in response to the alleged violation of the Federal Trade Commission Act, Complaints, March 30, 1972-Decisions, March 30, 1972, Federal Trade Commission Decisions 80:455–65. Washington, DC: Federal Trade Commission.

FTC (1981) *Staff Report on the Cigarette Advertising Investigation*. Washington, DC: Federal Trade Commission.

ISO (1984) Publication No. ISO 3864, *Safety Colours and Safety Signs*. Geneva, Switzerland: International Organization for Standardization.

KING, L.E. (1971) A laboratory comparison of symbol and word roadway signs. *Traffic Engineering and Control*, 12, 518–520.

KING, L.E. (1975) Recognition of symbol and word traffic signs. *Journal of Safety Research*, 7, 80–84.

LERNER, N.D. and COLLINS, B.L. (1980) NBS Publication No. NBSIR 80-2003, *Workplace Safety Symbols: Current Status and Research Needs*. Gaithersburg, MD: National Bureau of Standards.

NFPA Publication No. NFPA 101 (1970) *Life Safety Code*. Quincy, MA: National Fire Protection Association.

NFPA (1980a) Publication No. NFPA 171, *Life Safety Symbols*. Quincy, MA: National Fire Protection Association.

NFPA (1980b) Publication No. NFPA 178, *Standard Symbols for Fire Fighting Operations*. Quincy, MA: National Fire Protection Association.

OMB (1993) OMB Circular No. A119, *Federal Participation in the Development and Use of Voluntary Standards*. Washington, DC: Office of Management and Budget.

OSHA (1994) *Occupational Safety and Health Standards* (a) *Means of Egress* (Code of Federal Regulations 29 CFR XVII Part 1910, Subpart E, Sections .35-.40, pp. 125–133); (b) *Occupational Health and Environmental Control* (Code of Federal Regulations 29 CFR XVII Part 1910, Subpart G, Sections .94-.100.29, pp. 213–224); (c) *Personal Protective Equipment* (Code of Federal Regulations 29 CFR XVII Part 1910, Subpart I, Sections .132-.140, pp. 410–423); (d) *General Environmental Controls* (Code of Federal Regulations 29 CFR XVII Part 1910, Subpart J, Sections .144-.145, pp. 431–435). Occupational Safety and Health Administration. Washington, DC: US Government Printing Office.

PHS (1964) PHS Publication No. 1103, *Smoking and Health. Report of the Advisory Committee to the Surgeon General of the Public Health Service*. Public Health Service. Washington, DC: US Department of Health, Education, and Welfare.

PUBLIC LAW 89-92 (July 27, 1965) *Federal Cigarette Labeling and Advertising Act*. 79 Stat. 282 (Title 15, Sec. 1331 et seq.).

PUBLIC LAW 91-222 (April 1, 1970) *Public Health Cigarette Smoking Act of 1969*. 84 Stat. 87 (Title 15, Sec 1331 et seq.).

PUBLIC LAW 98-474 (October 12, 1984) *Comprehensive Smoking Education Act*. 98 Stat. 2200 (Title 15, Sec. 1331 et seq.).

PUBLIC LAW 99-252 (February 27, 1986) *Comprehensive Smokeless Tobacco Health Education Act of 1986*. 100 Stat. 30 (Title 15, Sec. 4401 et seq.).

PUBLIC LAW 100-690 (November 18, 1988) *Alcoholic Beverage Labeling Act of 1988*. 102 Stat. 41518 (Title 27, Sec. 213 et seq.).

PUBLIC LAW 104-113 (March 7, 1996) *National Technology Transfer and Advancement Act of 1995*. *15* USC 3701.

TREASURY BOARD OF CANADA (1980) *Graphic Symbols for Public Areas and Occupational Environments*. Ottawa, Canada: Treasury Board of Cauada.

UL (1989a) Publication No. UL 924, *Standard for Exit Markings*. Northbrook, IL: Underwriters Laboratories, Inc.

UL (1989b) Publication No. UL 969, *Standard for Marking and Labeling Systems*. Northbrook, IL: Underwriters Laboratories, Inc.

WESTINGHOUSE ELECTRIC CORPORATION (1981) *Product Safety Label Handbook*. Trafford, PA: Westinghouse Printing Division.

WOLFF, J.S. and WOGALTER, M.S. (1998) Comprehension of pictorial symbols. *Human Factors*, 40, 173–186.

CHAPTER THIRTEEN

Practical Considerations Regarding the Design and Evaluation of Product Warnings

J. PAUL FRANTZ, TIMOTHY P. RHOADES

Applied Safety and Ergonomics, Inc.

MARK R. LEHTO

Purdue University

This chapter provides practical advice and guidance for the development of product warnings. Topics covered include project planning, identifying and understanding product hazards, developing warning prototypes, and evaluating product warnings. Example techniques and strategies for identifying and analyzing product hazards and potential accident scenarios are provided. In addition, methods for determining warning content and placement are described. Finally, the topic of warning evaluation is discussed.

13.1 INTRODUCTION

This chapter provides practical advice and guidance for the development of product warnings. While primarily the chapter addresses the design of graphic (written and pictorial) warnings attached to or accompanying products, many of the topics covered also apply to nongraphic warning stimuli such as auditory and tactile signals. The suggestions and guidelines found in the chapter stem from experience participating in the design and evaluation of warnings for a wide variety of products and situations, published and unpublished research related to warnings, and evaluation of many product warnings at issue in product or premises liability lawsuits. Because of the generic nature of this chapter, it is important to remember that the suggestions and guidelines provided will not be applicable in all situations.

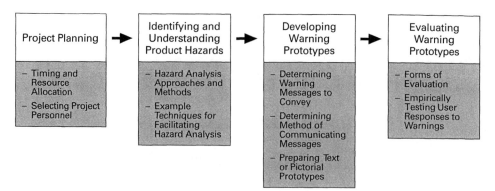

Figure 13.1 Basic model of activities in the warnings design process along with selected topics addressed in this chapter.

The development of warnings and safety-related instructions is a much more complex process than simply sitting down and writing a warning. As Figure 13.1 illustrates, the process of developing warnings can be organized into at least four broad categories: project planning; identifying and understanding product hazards; developing warning prototypes; and evaluating warning prototypes. It is this more complete process, the design of warnings, that this chapter attempts to describe.

This chapter does not provide a checklist or complete 'how-to' procedure for any given product or type of product. The diversity of products and product use scenarios does not lend itself to a specific procedure that would be applicable in all situations. Also, only a brief discussion of the role of standards and regulations related to warnings design will be provided, as this topic is covered in Chapter 12 by Collins.

13.2 PROJECT PLANNING

13.2.1 Timing and Resource Allocation

Frequently the cost and success of a warnings design project is related to the extent to which the effort is effectively planned. Project timing and resource allocation are particularly important. Resisting the tendency to view the preparation of warnings and instructions as a relatively minor last step in the overall product design effort can be beneficial. While this approach can still produce sufficient warnings, considering warnings and instructions much earlier in the design process and involving product designers is likely to be more efficient. Early development allows the warning design process to interact with and overlap the product design process. Such integration can serve to enhance user performance and satisfaction, and help manufacturers avoid costly design modification required to eliminate hazards that might otherwise be discovered late in the product development process.

To facilitate the allocation of necessary resources often it is important to receive management support for the warning design process. Often the key to this support is to persuade management that product warnings and instructions should be *designed*. As a design process, the development of product warnings and instructions requires resources, support, lead time, and expertise like that of other product design functions.

13.2.2 Selecting Project Personnel

Selecting the right people can have a tremendous impact on the efficiency and effectiveness of the warning design process. It is helpful to think of the project as multidisciplinary in nature. At some point in the development process, it may be helpful to receive input from personnel in areas such as marketing, product engineering, product safety, product service/field service engineering, human factors/ergonomics, regulatory affairs, standards and trade association affairs, packaging, and legal counsel. Often a key person to include is one who, by training, education, or experience, has a 'user-centered' orientation. This is a role that human factors engineers specializing in warnings design and evaluation can fill. Another frequently encountered personnel issue is selecting a person to lead or coordinate the inputs from various people. Useful qualities of a project leader include the technical and interpersonal skills necessary to interface with product designers, scientists, and engineers. Meeting these objectives can help minimize design time and iterations, increase the likelihood that information gained during the warning design process can have maximum impact on the ultimate design, packaging, and sale of the product, and increase the likelihood of accurate, complete and consistent warnings. With larger companies, there may be individuals from these different areas that are called on to assist at various times in the design process, whereas with smaller companies, one or two people may represent multiple disciplines.

13.3 IDENTIFYING AND ANALYZING PRODUCT HAZARDS

There are a variety of methods and strategies for identifying and analyzing product hazards (Clemens, 1982). In fact, the topic of hazard analysis is too great to cover in one chapter or even one book. The objective of this present discussion is to provide an overview of hazard analysis techniques available, references to more detailed information, and some specific suggestions and information that may be difficult to find elsewhere.

It is important to note that *identifying* potential hazards may only be the first step in the consideration of product hazards. Often the initial recognition of a potential hazard requires a deeper *analysis* to determine both the conditions under which the hazard might produce injury and the likelihood and severity of such injuries. Such an analysis can be of great assistance in determining the need for a warning and selecting appropriate and meaningful content for a warning.

It is also worth noting that, for purposes of warnings design, typically the identification and analysis of product hazards is different for new products as opposed to existing products. Compared to an existing product, a new product tends to involve a more comprehensive investigation into product hazards. With existing products, a number of hazards may have been identified already, and the warning project may be limited to revising existing warnings in response to product redesign, recent standards activities, or other circumstances.

13.3.1 Hazard Analysis Approaches and Methods

Hazard analysis techniques vary in terms of their formality; the extent to which they are quantitative; the type of logic used (inductive vs. deductive); the cost, expertise, time, and information required to conduct the analysis; the focus of the analysis (e.g., product centered vs. user centered); and their ability to provide clear direction for the design of a product and its warnings (see Hammer, 1972, 1993; Firenze, 1978; Chapanis, 1980; Clemens, 1982; Roland and Moriarty, 1983; Seiden, 1984; Lehto and Salvendy, 1991). Since there is no single method that is best in all situations and because often it is useful to examine a product from a variety of perspectives, what follows is a sample of hazard identification and analysis methods and approaches that may be useful in the development of product warnings.

Analysis of warnings and instructions accompanying similar products

In terms of product design, this approach is similar to product benchmarking. In the warning design process, one reason to examine warnings and instructions accompanying similar products is to discover potential hazards or accident scenarios not previously contemplated. A by-product of this analysis can be the discovery of alternative methods of product use, assembly, installation, or other product-related activities. This technique is less likely to lead to the discovery of a basic product hazard (e.g., exposure to electricity, toxic substance, pinch point, fall potential, etc.) than it is to reveal potential product misuses or specific manifestations of a product hazard that might suggest the need for a more specific warning. In addition, warning messages on similar products may be the result of actual experiences of other companies that may be virtually impossible to foresee prior to introduction of a product in the marketplace. Although there can be benefits of examining warnings accompanying similar products, caution should be exercised in readily adopting elements of a competitor's warning due to differences in product design and differences in the expected use and expected users of products made by different manufacturers.

Product/system centered analyses are directed toward the conditions under which components or systems of a product might fail and the effects of such failures. Failure Modes and Effects Analysis is an example of this type of analysis. The potential limitation of this type of analysis is that it may not be designed to reveal hazards that arise from use or misuse of products which do not actually fail. Consider, for example, the hazards associated with step ladders or extension ladders that are not related to structural failure of the ladder (e.g., loss of balance, ladders tipping over, etc.).

Catastrophic event deductive analyses are those in which possible failures or adverse outcomes are postulated and then examined to determine how or if they might occur. Fault Tree Analysis is an example of such a method (see Hammer, 1993). Obviously, this type of analysis requires initial awareness of an unwanted event.

Product use/task centered analyses involve breaking down the intended use of the product into small elements which are then scrutinized for potential hazards and accident scenarios. Such an analysis may be particularly useful for products used in manufacturing or industrial environments where the procedure for product use is fairly well established and relatively invariable. Job Safety Analysis or Job Hazard Analysis is an example technique used in occupational settings that involves breaking a job or task into basic elements and examining each element for potential hazards. A simple example of a Job Hazard Analysis is provided in Table 13.1. This type of analysis can be useful also for consumer products that are used in a less routine manner and can facilitate the discovery of hazards associated with unintended uses of a product.

User/operator centered analyses are similar to both job/task and product use centered approaches, except that the focal point is the user or operator. One example is the Critical Incident Technique in which experienced operators or users are queried regarding hazards

Table 13.1 Sample job hazard analysis for grinding castings (US Department of Labor, Occupational Safety and Health Administration, USDL/OSHA, 1994).

Step	Hazard	Cause	Preventive measure
1. Reach into right box and select casting	Strike hand on wheel	Box is located beneath wheel	Relocate box to side of wheel
	Tear hand on corner of casters	Corners of casters are sharp	Require wearing of leather gloves
2. Grasp casting, lift and position	Strain shoulder/elbow by lifting with elbow extended	Box too low	Place box on pallet
	Drop casting on toe during positioning	Slips from hand	Require wearing of safety shoes
3. Push casting against wheel and grind burr	Strike hand against wheel	Wheel guard is too small	Provide larger guard
	Wheel explodes	Incorrect wheel installed	Check rpm rating of wheel
	Respirable dust	Dust from caster metal and wheel material	Provide local exhaust system
	Sleeves caught in machinery	Loose sleeves	Provide bands to retain sleeves
4. Place finished casting into box	Strike hand on castings	Build-up of completed stock	Remove completed stock routinely

they have encountered, the 'near misses' or 'close calls' they have experienced, and the mistakes that they have made (Clemens, 1982).

Aside from feedback regarding compatibility between humans and systems, structured or unstructured in-depth interviews with people experienced in using the same or similar products can reveal unforeseen accident scenarios, potential misuses of a product, various styles or patterns of product use among different user groups, and some information regarding the root cause of various accidents (e.g., users lacking knowledge regarding a hazard or users choosing to act in an unsafe manner despite knowledge of a risk). The questions raised during these types of interview can vary across products; however, often it is useful to step through the various stages of product use with the interviewee to facilitate a more thorough examination of possible hazards and to prompt recall of potential accident scenarios. Depending on the situation, these interviews can be conducted in person or over the telephone. In our experience, such interviews are particularly useful if the product is used somewhat frequently, as often is the case with products used by tradespeople. It should be noted that often the amount of information gathered and value associated with in-depth interviews is dependent on the interview skills of the investigator.

Hazard centered analyses involve the consideration of generic types of hazard or source of energy that might be associated with a product. Examples of this type of analysis include Energy Analysis and checklists addressing generic potential hazards. Hammer (1993) provides an extensive discussion of generic hazards and a series of checklists related to hazards such as acceleration, pressure, chemical reactions, electrical energy, flammability, heat, mechanical systems, vibration, and radiation.

Product life stages analyses consider the hazards of a product that may arise during the different stages of its life. In practice, this activity may be less of a formal analysis and more of a shift in perspective that helps to avoid exclusively attending to product use, as opposed to other situations where people interact with a product and where hazards might arise (e.g., assembly or maintenance). For the interested reader, Seiden (1984) provides a detailed description of possible stages in a product's life.

Simulating product use and/or product usability testing can be used also to identify potential hazards and accident scenarios. This is an example of how testing related to marketing or customer satisfaction can be used to enhance product safety. From a procedural standpoint, information can be gained by actually observing product use (unobtrusively or otherwise, in controlled or natural settings) or by gathering reports from people who use the product at their leisure. In situations where people are using a product in a controlled setting, often it is useful to have the person verbally report their impressions, intentions, and thought processes and then to follow the session with either a structured or informal interview. One method of eliciting verbal responses is to provide the user with an assistant who asks questions of the subject during the session. This method can be useful for evaluating prototype warnings and instructions.

Analysis of product affordances. Although other hazard identification approaches address the physical design of a product and the interaction between the product and its users, another approach is to focus on the apparent features of a product and attempt to identify potential accident scenarios. Particular features of a product can provide users with a vast amount of non-verbal information from which they can make judgments about a product's structural composition, its method of operation, its basic functional and structural limitations, and its assembly or operational procedures (see Baggett and Ehrenfeucht, 1988; Norman, 1988; MacGregor, 1989; Rhoades, Frantz, and Miller, 1990; Rhoades and Miller, 1992; Frantz and Miller, 1993; Rhoades, 1994; Ayres, Wood, Schmidt, and McCarthy, 1998). Features such as knobs, dials, textures, shapes, and structural composition (e.g., glass, plastic, wood, etc.) can suggest, invite, or prohibit certain types of behavior. Gibson (1977) and Norman (1988) referred to such features as 'affordances.' Gibson (1977) identified a number of generic categories of object affordances including 'support,' 'walk-on-able,' 'sit-on-able,' 'grasp-able,' 'climb-on-able,' and 'bump-into-able.' In a more applied context, Norman (1988) noted that affordances are those properties of objects, both perceived and actual, that suggest or imply how the object can be used. Essentially, product affordances allow people to draw upon previously learned skills, rules, and problem-solving strategies to interact with a novel product, as opposed to relying on external sources of information such as written instructions or warnings. From an accident prevention standpoint, product features that provide strong cues about the use of a product are advantageous provided they do not promote or instigate inappropriate and unsafe usage of the product. Thus, one approach to hazard identification and analysis is to study the affordances of products to determine potential unintended or unsafe uses of products.

Product design reviews may be performed by individuals involved in product development, marketing, sales, service, etc. On occasion, individuals from these different areas can identify hazards unforeseen to product designers, due to their different roles and experiences. Product design reviews can be useful also when employing a warnings or product safety consultant to evaluate the design of the product and/or its warning. Typically, part of the consultant's job is to review the product and ask questions about the evolution and present design of the product and its warnings, which may prompt designers to reconsider many issues.

Review of potentially applicable standards and regulations. Consulting voluntary consensus standards and government regulations is a typical activity in the product design process as well as the warnings design process. From the standpoint of identifying and analyzing hazards, sometimes it is useful to consider standards that do not apply directly to the product. For example, consider an amusement device that incorporates a mini-trampoline as a means for launching users onto an inflated pad. Although there may be no standard for such a unique product, there is an American Society for Testing and Materials (ASTM) standard for trampolines which might provide some assistance in identifying hazards and potential accident scenarios. Furthermore, as it happens, an older version of the ASTM standard for trampolines (ASTM, 1977) did cover mini-trampolines and provided a reference to a specialized publication on the topic. The point of this discussion is that, on occasion, the investigation of standards or regulations can expose the experiences and findings of other groups of people regarding potential product hazards or accident scenarios.

13.3.2 Techniques for Facilitating Hazard Analysis

It is important to note that the different types of analysis and approach described above can be performed in various ways. For example, formal product testing or experimentation may be appropriate. In other situations, group meetings, brainstorming sessions, focus group studies, and experienced-based or analogous reasoning can facilitate some of the previously mentioned approaches. In some situations, more empirical approaches may be particularly useful for identifying hazards as well as for developing strategies for reducing accidents and injuries. These include field studies of product use or wear, surveys related to product use or users, and analysis of accident data related to a certain product or group of products. Some examples of field studies and surveys related to reported or expected use patterns of products include Lehto and Foley (1991) and Frantz, Miller, and Lehto (1991).

Another activity that facilitates hazard identification and, more specifically, hazard analysis, is the search for and review of published documents related to the use or safety of a product or similar type of product. Unfortunately, information on product hazards is widely dispersed and not collectively indexed by any one body and the search for documents in electronic and paper form is an increasingly complex task often requiring special skills. To assist in the process of searching for and obtaining documents, we offer the following general suggestions. First, consider searching for information using two different approaches: focusing on the product or type of product, and focusing on the potential accident scenario or injury outcomes. Second, keep in mind the variety of forms safety and hazard-related information might take, including journal articles, technical reports, government reports, medical case studies and position papers, conference proceedings, accident reports, trade magazine articles, newsletters, handbooks, and textbooks. Third, consider the following types of sources:

- News publications related to product safety and liability, such as the *Product Safety and Liability Reporter* (Bureau of National Affairs) or the *Consumer Product Safety Guide* (Commerce Clearing House)

- Medical literature and databases such as 'MEDLINE'®.

- Documents and literature from government agencies such as the Consumer Product Safety Commission (CPSC), the Food and Drug Administration (FDA), the Occupational Safety and Health Administration (OSHA), or the National Institute of Occupational Safety and Health (NIOSH)
- Trade or industry association libraries, publications, recommended practices, and standards
- Safety organizations, associations, and institutes such as the National Safety Council in Itasca, IL
- Compilations and indexes of human factors/ergonomics literature such as *Ergonomics Abstracts* or conference proceedings
- Electronic databases which index journals such as *Journal of Safety Research, Ergonomics, Professional Safety, Ergonomics in Design, Human Factors, Injury Prevention,* or *International Journal for Consumer and Product Safety*
- World wide web sites for agencies and organizations that deal with various aspects of safety.

It is important to recognize that the preceding list of sources is not exhaustive—other sources are available and under development. Also, note that it can be difficult to find all the information you need on-line, primarily because this type of information is not collectively indexed by any one source, and it may be necessary to consult a number of databases.

To conclude this section, it is important to reiterate that there is no single method or approach that is best suited to identifying product hazards, and that frequently it can be very difficult to foresee hazards associated with intended uses or unintended misuses of the product. As a result, a combination of different types of analysis, be they formal or informal, is often a good way to gain a thorough understanding of product hazards and potential accident scenarios.

13.4 DEVELOPING PROTOTYPE WARNINGS

After initially analyzing the hazards associated with a product, the next major activity is to develop prototype warnings. Interestingly, even after a thorough investigation of product hazards, the process of developing warning prototypes frequently reopens the analysis of a hazard(s) and/or reveals new accident scenarios that might be associated with a previously recognized hazard. Thus, the reader should not be misled by the discrete sequential presentation of information in this chapter. In practice, there is overlap between these stages and often recurrent activities. Figure 13.2 is a basic model of the warnings design process illustrating common feedback loops.

In very practical terms, the development of warning prototypes involves determining what to say and then how to say it. Though concise, this statement is too abstract to be of great value. To be of greater assistance, the remainder of this section describes a strategy for breaking down the actual development of a warning prototype into several stages: (1) determining warning messages to convey; (2) determining the mode of communicating messages; (3) preparing text or pictorial prototypes; and (4) determining the temporal and spatial location of warnings. This basic process is not only designed to develop complete, accurate, and comprehensible warnings, but it is also designed to accommodate the practical difficulties associated with various people working intermittently on a project that can span weeks, months, and in some cases even years. Again, the discreteness and formality of these stages varies greatly depending on the situation.

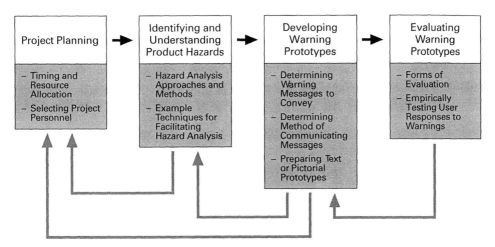

Figure 13.2 Basic model of warnings design process along with selected topics addressed in this chapter and common feedback loops.

13.4.1 Determining Warning Messages to Convey

Once hazards and accident scenarios have been identified the next issue is what, if any, messages to convey about them. To assist in determining what messages to convey in a warning, often it is useful to generate an exhaustive list of hazard-related concepts or messages that one might wish to describe to a product user, as if there were an infinite amount of time for discussion and interest on the part of the user. Generating such a list can be facilitated in a variety of ways. For example, some of the hazard-analysis activities mentioned previously may be useful in compiling a list of possible messages to convey. Another approach is to consider different characteristics of hazards, such as the nature of the hazard, the severity of the potential injury, the consequences of encountering or not avoiding a hazard, and the possible means of avoiding injury.

Still another approach is to probe persons knowledgeable of or experienced with the product about identified product hazard(s) to determine what hazard-related information they feel might be helpful to users. The following example questions illustrate this analytical approach to gathering information about hazards and accident scenarios.

- Describe the information, at a general level, you wish to convey to users/operators regarding the hazard
- Is there anything in particular that you wish to convey regarding who can be injured (e.g., installers, users, operators, assemblers, bystanders, service personnel, etc.)?
- Are there special circumstances or conditions that may increase the likelihood of injury?
- What can be done to avoid exposure to the hazard or prevent injury?
- What are the potential consequences of encountering the hazard or not taking precautionary measures?

Once there is a good general understanding of the potential messages that might be conveyed in a warning label, the next step is usually to discuss, investigate and determine which messages to actually provide. It should be noted that for virtually all products, the exercise of considering all possible messages regarding all possible hazards or risks associated with a product will lead to many messages that will not be included, either directly or indirectly. For example, an analysis of a bottle of champagne may identify hazards associated with dropping a bottle on one's foot or being cut by broken glass, yet ultimately one may decide to provide only a message related to flying corks. Aside from the impracticality of providing warnings about every possible risk associated with a product, the literature on warnings describes a host of problems associated with overusing warnings. Some of these problems include reducing attention to warnings generally, reducing attention to individual messages within warnings, reducing recall of certain warning messages, reducing the believability/credibility of warnings, and increasing the amount of time required to locate specific information in a label (see Kantowitz and Sorkin, 1983; O'Conner and Lirtzman, 1984; Driver, 1987; Purswell, Krenek, and Dorris, 1987; Dorris, 1991; McGrath and Downs, 1992; Stewart and Martin, 1994; deTurk, 1995; Edworthy and Adams, 1996; McCarthy, 1996; Krenek and Purswell, 1998).

What follows are some factors to consider in selecting messages to include in a warning, and reducing and consolidating information to a practical amount. Note that not all factors that may be important are mentioned here and some of the factors provided may not be important in all situations.

One factor to consider in determining what warning messages to convey via a product warning is the nature of the audience receiving the message. Using the product life stages analysis described previously, the types of people who might be interacting with a product at various points in its life cycle can be identified. Selecting particular messages to convey involves evaluating relevant characteristics of these potential users.

Some characteristics of potential users include their knowledge of and experience with the product or type of product, their perceptions of product hazards, and the perceived benefits of engaging in various types of behavior. In other words, what are expected users likely to know, think about, or do with the product in the absence of any written warnings? What expectations or perceptions do people have about hazards which may exist when using the product? How might these hazards evidence themselves? What consequences are associated with the hazards? These perceptions may have been formed through first- or second-hand experiences with a similar product, deduced from their knowledge of other related products and situations, learned through knowledge they have gained from reading or hearing warning messages, or gained from reading information provided with another product.

To assess consumer/operator use, misuse, knowledge, and perceptions of a product, a variety of methods may be used, including: considering the expected level of training or experience of users; gathering input from company (i.e., product manufacturer) personnel who have experience interacting with customers/users; investigating the companies' history with similar products/users; surveying users of similar products; examining used products for indications of usage patterns; unobtrusively observing the interaction between products and users; and conducting in-depth interviews of users and potential users to determine users' existing knowledge about a product (e.g., how it works and how to adjust/maintain it). Note the similarities of these strategies to those related to hazard analysis activities. In many situations, user investigations can serve the purpose of identifying hazards, potential misuses, and accident scenarios, as well as determining what messages to convey and, ultimately, how to convey them.

Investigating the characteristics and knowledge of product users is also useful when considering a second factor related to what warning messages to convey; namely, what is the likelihood that people can infer multiple messages from minimal text or graphics given the features of the product or situation and the knowledge base of users? This question may arise when there is a desire to provide a succinct warning message due to space constraints, yet there may also be a desire to provide explicit statements about the nature of the hazard, the consequences of not avoiding the hazard, and the means of avoiding the hazard. In such a situation, having investigated the user population makes it easier to decide which topics to state explicitly and which ones to imply.

Other factors to consider in determining what messages to provide and/or what messages to state explicitly include the following:

- The existence of a standard, recommended practice, or regulation recommending or requiring certain statements, given the characteristics of a product
- The ability to meaningfully group many individual statements or messages into fewer 'chunks' of information that can be more readily processed and recalled
- The ability to spatially or temporally separate messages and present individual messages at the most relevant points so as to avoid inundating users with large amounts of information
- The likelihood and severity of injury associated with a hazard, although these are seldom telling factors, without considering many other aspects of the product and users
- The extent to which users are able to identify and react to a potentially hazardous situation in time to avoid injury or product damage
- The extent to which a hazard is specific to the product, as opposed to being more generic in nature
- The availability of other sources of information (e.g., owner's manuals, training and education programs, material safety data sheets, or warnings and instructions provided with other products used in conjunction with the product in question)
- The extent to which the context of product use or the environment in which the product is used will allow people to readily infer risks associated with the product
- The potential negative consequences associated with providing messages. These may come in the form of overloading users with information, falsely alarming people and causing inappropriate or unnecessary actions, and diluting the credibility of other warning messages accompanying the product, as well as the credibility of warnings generally.

A final point regarding the process of determining which warning messages to convey is that difficulties agreeing on what warning messages to provide or which messages to state explicitly can be an indication of the need to re-evaluate or further analyze the nature of the hazard and the means of avoiding accidents and injuries.

13.4.2 Determining the Method of Communicating Messages

After determining what messages to convey in a warning, another consideration is how to convey them. In terms of written or pictorial warnings accompanying products (e.g., sticky-back labels), there can be a multitude of other issues to consider in determining how to communicate messages. Just a few issues that arise include: where to locate the messages; whether to use text or pictorials; whether to include multiple languages; whether to use color; whether to place messages in the mold of a product as opposed to applying a label; how to affix a warning to a product (i.e., whether to adhere, tag, sew, weld, or rivet it, etc.); typeface and typesize; and what type of label stock, laminate, and adhesive to use. The remainder of this section focuses on two of these issues, the use of pictorials and the location of the warnings. Readers interested in the issue of multiple languages may wish to refer to other sources such as Ross (1992), Vignali (1995), Alves-Foss (1996) and ANSI (1998b). Those interested in other formatting issues such as typeface, typesize, layout, text arrangement, and type style may refer to ANSI Z535.4 1998. (ANSI, 1998b).

Issues to consider in the development and use of symbols/pictorials

One issue that arises frequently in the consideration of warnings is the use of non-verbal, graphic warnings commonly referred to as pictorials, symbols, pictograms, or pictographs. The following is a brief list of issues to consider in the design of symbols or pictorials.

- When designing pictorials, consider the ultimate size of the images as viewed by the receiver. The size of the visual image created by the pictorial affects legibility of individual components of the pictorial and can affect interpretation of the entire pictorial
- Due to the difficulty in relying upon a pictorial alone to convey messages accurately and completely, often it is necessary to supplement a pictorial with text
- It can be particularly difficult to convey a fairly complicated sequence of events related to an accident or leading to an injury with one or more pictorials
- Symbols can be useful for conveying information quickly to a knowledgeable audience who is not reading information, but executing routine behavior (e.g., stop signs at intersections and road signs with arrows depicting to the curvature of the road ahead)
- The use of perspective in pictorials may hinder accurate interpretation of the message
- Symbols may be preferable to text for conveying spatial information (e.g., road signs with arrows pointing to the left or right)
- Care should be taken when using symbols/pictorials to convey hazards to children. Among other things, there is the possibility that children may attempt to act out their interpretation of the graphics. For example, Trommelen (1996) illustrated several symbols under consideration in The Netherlands for conveying information about children suffocating with plastic bags. One of the symbols under consideration portrayed a bag over a child's head. Such a symbol, though perhaps understood by adults, has the potential for suggesting unsafe behavior to children at risk of injury
- It is important to determine the purpose of a pictorial during its initial design, and prior to any testing that might occur. This includes identifying what messages a pictorial is intended to convey, and the function of the pictorial. For example, is the primary function of the pictorial to instruct, inform, persuade, remind, or attract attention? In some situations, a pictorial may be designed as a legible reminder of a hazard or precaution as opposed to a stand-alone stimulus that instructs viewers on the safe use of the product
- Small changes in pictorial design can significantly impact the interpretation of the pictorial
- Pictorial designers and others who know the intended meaning of a pictorial may have difficulty judging how well typical users will comprehend the pictorial and they may have difficulty identifying alternative interpretations of the pictorial. Note that this is less likely to be a major problem with text because words and syntax are more universally known and understood than the elements of a particular pictorial.

The preceding points are by no means a complete list of aspects to consider in deciding whether or not to use a pictorial or actually producing one. Additional information, both practical and theoretical, on the use of symbols, pictorials, or pictograms is provided by the American National Standards Institute (ANSI, 1994, 1998b,c); Dreyfuss (1972); FMC Corporation (1985); Frantz et al. (1991); Miller, Lehto, and Frantz (1994); Peckham (1994); Twyman (1985); and Westinghouse Electric Corporation (1981).

Determining the temporal and spatial location of warnings

For some products, the task of deciding where to place a product warning is fairly straightforward due to the product's design or its packaging. For others it is not. As Frantz and Rhoades (1993) noted, common guidelines recommend that warnings should be placed so that they are readily visible to the intended viewer and alert the viewer to the potential hazard in time to take appropriate action, but not so far in advance that the message is forgotten (Cunitz, 1981; FMC Corporation, 1985; Westinghouse Electric Corporation, 1985; ANSI, 1998a,b). While these guidelines may place temporal and spatial boundaries on the location of a warning, they may not be sufficiently explicit to yield an effective warning. Furthermore, they do not provide guidance for those situations where more than one location meets the placement criteria.

To address this issue of several possible locations for a warning, Frantz and Rhoades (1993) described a task analytical approach and evaluated its merits for determining the location of a warning label accompanying a two-drawer file cabinet. In brief, a task analysis involves decomposing a task into subtasks or elements (e.g., Table 13.1). Typically, the elements of the task are then arranged in chronological order. The level of detail for the various elements depends on the particular purpose of the analysis. With respect to the design of warnings, a task analysis helps to identify critical task elements where exposure to a hazard exists and to identify task elements preceding the exposure which can be considered as candidate warning locations. Each candidate warning location can then be evaluated to determine if it allows the user sufficient time to avoid the hazard without being too temporally distant from the hazard. Beyond these existing criteria, breaking down the tasks associated with using or interacting with a product facilitates a more careful and systematic examination of the cognitive and behavioral capabilities, expectations, and activities at various points, in relation to hazards or anticipated accident scenarios. Among other things, the cognitive facet of a task analysis involves examining the task elements and predicting the extent to which users will be seeking any external information such as instructions or warnings as opposed to applying prior knowledge to the use of the product.

A final note regarding placement of warning messages involves the location of warning statements relative to other non-safety-related, procedural information accompanying a product. More specifically, contrary to what one might expect, spatially separating all safety-related messages and precautions from messages related to correct and effective use of a product may not be the most effective means of conveying warnings. For example, studying two consumer products, Frantz (1993, 1994) found that test subjects were more likely to read and heed warning messages that appeared in the 'directions for use' section of a product's label than the 'precautions' section. Friedmann (1988) reported similar results. These findings were consistent with Lehto's (1992a,b) generic warnings design recommendation to integrate warnings into the task and hazard related context.

13.4.3 Preparing Text or Pictorial Prototypes

Many people incorrectly believe that the single major activity in the development of a warning is preparing or writing draft text and/or pictorials. Obviously, the other topics in this chapter illustrate a great many other activities associated with developing product warnings.

In everyday language, this section deals with the issue of 'how to say it' as opposed to what messages to provide, who to tell, where to tell them, and when to say it. In considering how to state, describe or illustrate messages, it is important to consider again the characteristics of the intended receivers of the message. The earlier chapters in this book and earlier sections in this chapter address many topics related to the influence of receiver characteristics on the design and effectiveness of various warning stimuli and, as a result, a lengthy discussion of user characteristics is not provided here.

On the surface it might appear that preparing text and pictorial prototypes is a distinct activity that occurs before an evaluation stage and after the previous investigatory activities, and the location of this section within this chapter may be somewhat misleading. However, this is merely a consequence of attempting to describe the warning development process in some order. In practice, the preparation of the draft warnings may occur during, or as a result of, some of the investigative activities, or may involve some form of evaluation and may ultimately bring the project to a close without further formal evaluation. It is rather difficult to establish definite boundaries or provide a step-by-step procedure. As such, what follows is a discussion of several points to consider when actually drafting text or pictorials.

The first point to consider is that individuals trained in communicating information and, in particular, communicating warning information (e.g., human factors/safety specialists, graphic artists, technical writers), are typically well suited to expedite the process of developing comprehensible warning messages. Design engineers, or similar people who are extremely knowledgeable about a product and its design, may thoroughly understand all facets of a hazard, but may have difficulty relating to an audience with very little knowledge of a hazard, and it can be difficult for attorneys, trained and experienced in communicating information in a legal fashion, to shift away from that style. This is not to suggest that design engineers or attorneys are incapable of developing satisfactory warnings. Rather, the point is to recognize the differences in training and perspective that individuals have and assign tasks accordingly to provide for a more efficient and effective development process.

A second point to consider is the potential value of soliciting input from product users as to how they might communicate certain messages. Asking potential users to suggest or generate text to convey the desired warning messages can be helpful. This can be done in individual interviews or in group settings. It can be a particularly useful strategy after initially attempting to draft the warning and identifying one or two areas of difficulty. For an example of the extensive use of product users in the development of text and pictorial signs used in an automotive service setting, see Eberhard and Green (1989).

A third point to consider is that there are guidelines and recommendations available to assist in developing comprehensible text. A few sources include Kanouse and Hayes-Roth (1980); Felker, Pickering, Charrow, Holland, and Redish (1981); FMC Corporation (1985); Kieras and Deckert (1985); Lehto (1992a,b); Backinger and Kingsley (1993); and ANSI (1998b). Some example recommendations provided by the previous sources include: (1) using consistent terminology to refer to common referents; (2) using explicit modifiers to clarify the message; (3) avoiding adverbs that are difficult to define or interpret;

(4) using active rather than passive verbs (e.g., 'Keep hands away from blade' rather than 'your hands must be kept away from blade'); and (5) using action verbs rather than nouns created from verbs (e.g., 'use' rather than 'utilization'). While these guidelines can be of value, it is important to note that many of the recommendations stem from and are applicable to typical prose as opposed to the very succinct, sometimes truncated sentences that often are required for product warnings and, as a result, may not be applicable in all situations.

In addition to general writing guidelines, there may be more specific data on the comprehensibility of various phrases and symbols or pictorials, depending on the type of product. For example, Lehto and House (1997) studied phrases commonly used in precautionary labeling for hazardous chemicals, and Collins, Lerner, and Pierman (1982) studied workers' interpretations of symbols for common hazards encountered in occupational settings. In particular, Lehto and House (1997) found that 'action' statements were rated most comprehensible and readability indexes were of questionable validity. While there are large studies which focus on measuring comprehension or interpretation of warning stimuli, there is a growing number of smaller empirical studies with varying research objectives that may also provide guidance in deciding how to convey warning messages (Laughery, Wogalter, and Young, 1994; Miller *et al.*, 1994).

13.4.4 A Note About Other Types of Warning Stimulus

Although the thrust of this chapter relates more to the development of prototypical warnings accompanying products, it is important to note that warning stimuli may come in a variety of forms, including signs and signals emitted by products (e.g., odors, sounds, vibration, etc.). These warning signs and signals often have advantages over warning labels for reminding and alerting users when they deviate from normally safe behavior patterns. The interested reader may wish to refer to Lehto (1991, 1992a,b, 1993) for a more detailed discussion of various forms of warning stimuli, their relationship to a theoretical model of human behavior, and the likely impact of various warning stimuli given different types of human performance/tasks.

13.5 EVALUATING WARNINGS

13.5.1 Forms of Evaluation

This section addresses the topic of warning evaluation. Because this chapter has focused on the development of warnings, one might assume that this section is limited to evaluating warning prototypes resulting from the previous stages. However, as this section will illustrate, strategies for evaluating warnings can apply to existing warnings as well.

At the outset, it is important to understand that the evaluation of a warning can take many different forms. For example, a prototype warning may be evaluated for its *accuracy* in describing a hazard or prescribing necessary and sufficient conditions for avoiding a hazard. This evaluation might be appropriate, for instance, for a warning describing exposure to a chemical. Another way in which a warning might be evaluated is to determine the extent to which the warning complies with standards. While compliance with standards may be straightforward for some products, it can be a complicated and difficult process for products that have more than one applicable standard, particularly those instances where compliance with one standard either precludes compliance with another standard, or where compliance with multiple standards may result in confusing warnings. Other forms of evaluating warnings include: determining the consistency of warning messages across other products that present similar hazards; comparing warnings against generic human factors research or data (e.g., to assess legibility); conducting durability tests; analyzing the practicality and feasibility of the warning message; measuring user responses to the warning; and review and analysis of warning prototypes by a warning specialist. An implication of this variety of evaluation criteria is that the activities involved in warnings evaluation vary tremendously. For example, evaluation activities might include: contacting standards organizations and regulatory bodies; conducting empirical testing; simulating compliance with the warning; informal or structured in-depth interviews with individuals or groups; reviewing warnings for similar products; or consulting with warning specialists.

For the sake of clarity, it should be noted that, in some cases, a warning may be evaluated sufficiently without conducting any empirical testing. On the other hand, empirical testing may be necessary in order to evaluate a given warning sufficiently. It has been the authors' experience that the need for formal testing and evaluation is dependent greatly on the extent of the investigations and information gathered from the previous stages. For example, if the product and its hazards have been analyzed thoroughly, the characteristics of users have been considered, and the input of potential users has been included in the design process, the need for formal empirical testing may be greatly reduced or eliminated. Again, it is important to recognize overlap between stages in the warning development process. There are times when the iterative process of developing a warning prototype is essentially the test of the warning. To help assess the need for formal empirical testing, a list of points to consider is provided later in this section.

13.5.2 Empirically Testing User Responses to Warnings

Testing peoples' responses to warnings often requires specialized knowledge regarding the design and conduct of experiments and surveys that is well beyond the scope of this chapter. The focus here will be on a few relatively common topics related to testing warnings.

To begin a discussion of empirically testing warnings, it is noteworthy to point out that there may be two basic objectives to the testing. One objective might be to test a warning in an absolute sense. For example, the objective of the test might be to answer the question: what percentage of people are likely to understand a particular warning? Another objective might be to test a warning relative to other alternatives: do people notice warning A more than warning B? A difficult issue to consider is that design decisions intended to increase effectiveness for one group of users may lower effectiveness for others. For example, by avoiding technical terms, perceived understandability may increase for naive or inexperienced workers, but at the cost of reducing the conveyed meaning of the message to experts.

A second noteworthy point is that there are different types of dependent measures that can be appropriate in different situations. Example characteristics of a warning that can be tested include: comprehensibility, memorability, conspicuity, legibility, believability, persuasiveness, perceived hazardousness, behavioral compliance with the warning, and behavioral intentions. Typically, the nature of the product and the activities associated with warning prototype development help identify the key characteristics to measure. A few factors to consider in determining the types of warning responses to measure include: the ability to obtain measurements that are reliable and valid; the extent to which a measurement might provide meaningful feedback to improve the warning; and an assessment of which aspects of the warning present the most uncertainty as to how people will respond. For example, there is often more uncertainty as to whether people will correctly interpret a pictorial than text.

A final point to note about empirical testing is that, given the various design features and attributes of a warning, it is practically impossible to test all aspects of a warning. For example, a warning can vary in terms of its signal word, text message content, size, location, use of pictorial, type of pictorial, location relative to other information, etc. For each of these characteristics there may be many alternatives from which to choose. As an example, consider the case where there are two possible signal words that are being considered, six different text messages, three different sizes, four different locations, and five different pictorials. In such a case, there are up to 720 possible combinations of these attributes. Testing each of these combinations against the other clearly is not practical.

Suggestions related to comprehensibility testing

Comprehension is frequently the measure of choice when evaluating a warning, partly because the results of such testing tend to facilitate improved communication effectiveness. Other measures, such as the extent to which people notice warnings and actually comply with warnings during use of a product, can be useful, but often they are affected by many factors other than the design of the warning. From a practical perspective, comprehensibility is also a desirable measure because, in some situations, prototype warnings can be tested without having to actually have or use prototype products that are unavailable or inoperable. To assist in making decisions regarding the need to formally measure user responses to product warnings, following is a list of factors to consider with respect to the issue of evaluating the comprehensibility of a warning. The importance of empirically testing a warning generally increases under any of the following conditions:

- A pictorial is used in the absence of text
- A pictorial is used in conjunction with essentially unrelated text or text that provides minimal assistance in interpreting the pictorial. In other words, the pictorial is the only means of conveying a general concept
- An abstract or generic pictorial is used to convey information about a specific precaution or procedure
- Text messages are very brief or truncated and it is unclear whether readers may infer sufficient details from the text
- The text or pictorials describe, illustrate, or refer to objects or hazards that may be difficult for readers to identify or understand
- A warning is expected to become an industry practice or standard (voluntary or mandatory standard).

When testing warning comprehensibility, it is important to provide test subjects with a fair understanding of the context of the warning. For example, the subjects need to have a general idea of what the product is, what it does, and the general location of the warning. Without such information, subjects may not be able to make reasonable assumptions and inferences when interpreting the warning. As an aside, it should be noted that the context in which warning information is provided can have a large effect on the interpretation of both symbols and text.

When determining test subjects' understanding of warning text or pictorials, it is important to probe subjects using multiple questions or standard probing phrases to elicit complete responses. Incomplete answers to open-ended questions can be particularly difficult to interpret and it is not uncommon for subjects to provide answers that are not directly related to the questions. Without probing, unreliable data may result, difficulties may arise in selecting among warning alternatives, and it may not be clear how a warning may be changed to increase its comprehensibility.

In addition to ascertaining subjects' interpretations of an individual prototype, it can be useful to provide subjects with more than one alternative and ask them to rank order the alternatives or state their preference for the warning that best conveys the intended message. This type of comparative assessment of warnings can be useful in identifying the elements of text or pictorials that are difficult for people to interpret. It should be noted, however, that the value of preference data is limited because it is obtained *after* subjects know what the intended message is and because people have only one warning to interpret when actually using the product. Caution should be exercised in relying on peoples' preferences for different warning alternatives due to the possibility that people may prefer a warning because of certain familiar features, even though they may not accurately interpret the element in the context of a particular warning. For example, people may express a preference for a symbol that includes the commonly used circle with a slash, regardless of their ability to actually correctly interpret the meaning of the symbol.

13.6 CLOSING REMARKS

This chapter demonstrates that the design of product warnings involves much more than simply writing a warning. Rather, ideally it should be a design process that is premised on the identification and understanding of hazards. It is also an iterative process that involves developing prototypes that are then subjected to some form of evaluation. It should be clear that these activities make the design of a warning analogous to the design of the product itself.

REFERENCES

ALVES-FOSS, J. (1996) Multiple languages in warning signs: how the number of languages and presentation of layout influence perception of hazard and memory. Master's dissertation, University of Idaho.

ANSI (1994) ANSI Z129.1, *American National Standard for Hazardous Industrial Chemicals—Precautionary Labeling*. New York: American National Standards Institute.

ANSI (1998a) ANSI Z535.2, *Environmental and Facility Safety Signs*. American National Standards Institute. Arlington, VA: National Electrical Manufacturers Association.

ANSI (1998b) ANSI Z535.3, *Criteria for Safety Symbols*. American National Standards Institute. Arlington, VA: National Electrical Manufacturers Association.

ANSI (1998c). ANSI Z535.4, *Product Safety Signs and Labels*. American National Standards Institute. Arlington, VA: National Electrical Manufacturers Association.

ASTM (1977). ASTM F381-77, *Standard Consumer Safety Specification for Components, Assembly, and Use of a Trampoline*. Philadelphia, PA: American Society for Testing and Materials.

AYRES, T.J., WOOD, C.T., SCHMIDT, R.A., and MCCARTHY, R.L. (1998) Risk perception and behavioral choice. *International Journal of Cognitive Ergonomics*, 2, 35–52.

BACKINGER, C.L. and KINGSLEY, P.A. (1993) DHHS Publication No. FDA 93-4528, *Write it Right*. Rockville, MD: Center for Devices and Radiologic Health Office of Training and Assistance.

BAGGETT, P. and EHRENFEUCHT, A. (1988) Conceptualizing in assembly tasks. *Human Factors*, 30, 269–284.

CHAPANIS, A. (1980) The error-provocative situation: a central measurement problem in human factors engineering. In TARRANTS, W. (ed.), *The Measurement of Safety Performance*. New York: Garland Press, pp. 99–128.

CLEMENS, P.L. (1982) Systems safety application. *Hazard Prevention*, March/April, 11–18.

COLLINS, B.L., LERNER, N.D., and PIERMAN, B.C. (1982) Report No. NBSIR 82-2485, *Symbols for System Safety*. Washington, DC: National Bureau of Standards.

CUNITZ, R.J. (1981) Psychologically effective warnings. *Hazard Prevention*, 17, 12–14.

DETURK, M.A. (1995) *Developing and Defending Product Warnings: A Communication Theory Perspective*. Washington, DC: Bureau of National Affairs.

DORRIS, A.L. (1991) Product warnings in theory and practice: some questions answered and some answers questioned. In *Proceedings of the Human Factors Society 35th Annual Meeting*. Santa Monica, CA: Human Factors Society, pp. 1073–1077.

DREYFUSS, H. (1972) *Symbol Sourcebook*. New York: McGraw-Hill.

DRIVER, R.W. (1987) A communication model for determining the appropriateness of on-product warnings. *IEEE Transactions on Professional Communication*, 30, 157–163.

EBERHARD, J. and GREEN, P. (1989) Report No. UMTRI-89-26, *The Development and Testing of Warnings for Automotive Lifts*. Ann Arbor, MI: University of Michigan Transportation Research Institute.

EDWORTHY, J. and ADAMS, A. (1996) *Warning Design: A Research Prospective*. London: Taylor & Francis.

FELKER, D.B., PICKERING, F., CHARROW, V.R., HOLLAND, V.M., and REDISH, J.C. (1981) *Guidelines for Document Designers*. Washington, DC: American Institutes for Research.

FIRENZE, R.J. (1978) *The Process of Hazard Control*. Dubuque, IA: Kendall-Hunt.

FMC CORPORATION (1985) *Product Safety Sign and Label System*. Santa Clara, CA: FMC Corporation.

FRANTZ, J.P. (1993) Effect of location and presentation format on attention to and compliance with product warnings and instructions. *Journal of Safety Research*, 24, 131–154.

FRANTZ, J.P. (1994) Effect of location and procedural explicitness on user processing of and compliance with product warnings. *Human Factors*, 36, 532–546.

FRANTZ, J.P. and MILLER, J.M. (1993) Communicating a safety-critical limitation of an infant carrying product: the effect of product design and warning salience. *International Journal of Industrial Ergonomics*, 11, 1–12.

FRANTZ, J.P., MILLER, J.M., and LEHTO, M.R. (1991) Must the context be considered when applying generic safety symbols: a case study in flammable contact adhesives. *Journal of Safety Research*, 22, 147–161.

FRANTZ, J.P. and RHOADES, T.P. (1993) A task analytic approach to the temporal and spatial placement of product warnings. *Human Factors*, 35, 719–730.

FRIEDMANN, K. (1988) The effect of adding symbols to written warning labels on user behavior and recall. *Human Factors*, 30, 507–515.

GIBSON, J.J. (1977) The theory of affordances. In SHAW, R. and BRANSFORD, J. (eds), *Perceiving, Acting, and Knowing: Toward an Ecological Psychology*. New York: Wiley.

HAMMER, W. (1972) *Handbook of System and Product Safety*. Englewood Cliffs, NJ: Prentice-Hall.

HAMMER, W. (1993) *Product Safety Management and Engineering*, 2nd Edn. Des Plaines, IL: American Society for Safety Engineers.

KANOUSE, D.E. and HAYES-ROTH, B. (1980) Cognitive considerations in the design of product warnings. In MORRIS, L.A., MAZIS, M.B., and BAROFSKY, I. (eds), *Product Labeling and Health Risks*. Banbury Report No. 6. Cold Spring Harbor Laboratory.

KANTOWITZ, B.H. and SORKIN, R.D. (1983) *Human Factors: Understanding People–System Relationships*. New York: Wiley.

KIERAS, D.E. and DECKERT, C. (1985) Technical Report No. 21 [TR-85/ONR-21], *Rules for Comprehensible Technical Prose: A Survey of Psycholinguistic Literature*. Ann Arbor, MI: University of Michigan.

KRENEK, R.F. and PURSWELL, J.L. (1998) Warning label content strategies: an approach to the integration of user knowledge and designer judgment. *International Journal of Cognitive Ergonomics*, 2, 53–60.

LAUGHERY, K.R., WOGALTER, M.S., and YOUNG, S.L. (1994) *Human Factors Perspectives on Warnings*. Santa Monica, CA: Human Factors and Ergonomics Society.

LEHTO, M.R. (1991) A proposed conceptual model of human behavior and its implications for the design of warnings. *Perceptual and Motor Skills*, 73, 595–611.

LEHTO, M.R. (1992a) Designing warning signs and warning labels: Part I—Guidelines for the practitioner. *International Journal of Industrial Ergonomics*, 10, 105–113.

LEHTO, M.R. (1992b) Designing warning signs and warning labels: Part II—Scientific basis for initial guidelines. *International Journal of Industrial Ergonomics*, 10, 115–138.

LEHTO, M.R. and FOLEY, J.P. (1991) Risk-taking, warning labels, training, and regulation: are they associated with the use of helmets by all-terrain vehicle riders? *Journal of Safety Research*, 22, 191–200.

LEHTO, M.R. and HOUSE, T.E. (1997) Evaluation of the comprehension of hazard communication phrases by chemical workers. In *Proceedings of the 13th Triannual Conference of the International Ergonomics Association*. Tampere, Finland.

LEHTO, M.R. and SALVENDY, G. (1991) Models of accident causation and their application: review and reappraisal. *Journal of Engineering and Technology Management*, 8, 173–205.

MCCARTHY, R.J. (1996) Hazard analysis and risk-based design methodologies. In *Failure to Warn: Product Warnings, Instructions and User Information*. Chicago: American Bar Association.

MCGRATH, J.M. and DOWNS, C.W. (1992) The effectiveness of on-product warning labels: a communication perspective. *For the Defense*, October, 19–24.

MACGREGOR, D.G. (1989) Inferences about product risks: a mental modeling approach to evaluating warnings. *Journal of Products Liability*, 12, 75–91.

MILLER, J.M., LEHTO, M.R., and FRANTZ, J.P. (1994) *Warnings and Safety Instructions*. Ann Arbor, MI: Fuller Technical Publications.

NORMAN, D.A. (1988) *The Psychology of Everyday Things*. New York: Basic Books.

O'CONNER, C.J. and LIRTZMAN, S.I. (1984) *Handbook of Chemical Industry Labeling*. Park Ridge, NJ: Noyes Publications.

PECKHAM, G. (1994) *The Hazard Pictorial Library* [Reference manual and software]. Milford, PA: Hazard Communication Systems, Inc.

PURSWELL, J.L., KRENEK, R.F., and DORRIS, A. (1987) Warning effectiveness: what do we know? In *Proceedings of the Human Factors Society 31st Annual Meeting*. Santa Monica, CA: Human Factors Society, pp. 1116–1120.

RHOADES, T.P., (1994) The use of affordance and categorization approaches to evaluate selected occupational movement behaviors at auto hauling terminals. Doctoral dissertation, University of Michigan, Ann Arbor, MI.

RHOADES, T.P., FRANTZ, J.P., and MILLER, J.M. (1990) Emerging methodologies for the assessment of safety related product communications. *Proceedings of the Human Factors Society 34th Annual Meeting*, Vol. 2. Santa Monica, CA: Human Factors Society, pp. 998–1002.

RHOADES, T.P. and MILLER, J.M. (1992) Methods used on atypical climbing systems. In KUMAR, S. (ed.), *Advances in Industrial Ergonomics and Safety*, Vol. IV. London: Taylor & Francis, pp. 1021–1028.

ROLAND, H.E. and MORIARTY, B. (1983) *System Safety Engineering and Management*. New York: Wiley.

ROSS, K. (1992) When are multi-lingual warning labels appropriate? *Product Liability International*, 105.

SEIDEN, R.M. (1984) Product safety assurance. In SEIDEN, R.M. (ed.), *Product Safety Engineering for Managers: A Practical Handbook and Guide*. Englewood Cliffs, NJ: Prentice-Hall, pp. 1–36.

STEWART, D.W. and MARTIN, I.M. (1994) Intended and unintended consequences of warning messages: a review and synthesis of empirical research. *Journal of Public Policy and Marketing*, 13, 1–19.

TROMMELEN, M. (1996) Harmonization of safety-related information on child-care. In ROGMANS, W.H.J., VENEMA, A., and RUEF, B. (eds), *Safety Labelling: The Role of Product Information as a Tool for Increasing Consumer Safety*. Amsterdam: European Consumer Safety Association, pp. 83–92.

TWYMAN, M. (1985) Using pictorial language: a discussion of the dimensions of the problem. In DUFFY, T.M. and WALLER, R. (eds), *Designing Usable Texts*. Orlando, FL: Academic Press, pp. 245–312.

USDL/OSHA (1994) Publication No. OSHA 3071 [reprinted], *Job hazard analysis*, US Department of Labor, Occupational Safety and Health Administration. Washington, DC: Government Printing Office.

VIGNALI, R.M. (1995) Foreign language warnings and the duty to warn. *Risk Management*, 83–91.

WESTINGHOUSE ELECTRIC CORPORATION (1981) *Product Safety Label Handbook*. Trafford, PA: Westinghouse Printing Division.

PART FIVE

Forensics

The last section describes the legal aspects of warnings in the USA. Implications of statutory and case law for warnings design, including potential consequences of failure to warn, are discussed. The chapter on the human factors expert witness provides insight on how litigation is aided by research and analysis.

CHAPTER FOURTEEN

The Law Relating to Warnings

M. STUART MADDEN[1]

Pace University School of Law

Product manufacturers have a duty to provide instructions and warnings sufficient to permit a product to be used safely or to enable a user to make an informed choice not to use the product. The nature of this duty has been shaped by decisions in state and Federal courts in the US over several decades. Although a product may be properly designed and manufactured, the seller may be subject to liability if during its foreseeable use the product has an unreasonable potential for injury that is not readily apparent to the user and carries no warnings of the risk or instructions as to safe use. Responsibility for providing adequate warnings may be found under principles of strict liability, negligence and warranty. Warnings must be communicated by means of positioning, lettering, coloring and language that will convey to the typical user of ordinary intelligence the information necessary to permit him/her to avoid the risk and, as appropriate, to use the product safely.

14.1 INTRODUCTION

A product manufacturer has an obligation to provide warnings and instructions sufficient to permit product users to use a product safely, or to make an informed choice not to use the product. The contours of this informational duty have been shaped by thousands of decisions brought in state and Federal courts over the US over the last several decades.

The court decisions fashioning this informational obligation have been developed hand in hand with the American Law Institute's (ALI) Second and Third Restatements of Torts. The ALI, a private body comprising attorneys, judges and law professors, has committed itself to preparing Restatements of numerous areas of law, intended to consolidate and rationalize rules and doctrines followed by courts in the majority of United States jurisdictions and, where courts have adopted conflicting doctrines, select the 'better' rule of law. The discussion in this chapter shows that the Restatement (Second) of Torts[2] and the Restatement (Third) of Torts: Products Liability[3] both guide and have been guided by the growing body of court decisions on a manufacturer's warning obligations.[4]

---
[1] Charles A. Frueauff, Research Professor and Distinguished Professor of Law, Pace University School of Law.
[2] Restatement (Second) of Torts §§ 388, 395, 402A (ALI 1965).
[3] Restatement (Third) of Torts: Products Liability (ALI 1998).
[4] See Victor E. Schwartz, The Restatement (Third) of Torts: Products Liability—A Guide to Its Highlights 9 & n.14 (National Legal Center for the Public Interest 1998).

Even when a product is unerringly designed and manufactured, injury or damage occasioned by its reasonably foreseeable use may subject the seller to liability. Such liability may be found if the product has a potential for injury that is not readily apparent to the user and carries no warnings of the risk or, where appropriate, instructions as to the use of the product without harm. A claim that a manufacturer failed to provide adequate warnings or instructions is probably the most prevalent element in modern products liability litigation. For the purposes of the discussion to follow, warnings and instructions are distinguished along these lines: 'warnings' call attention to a danger, while 'instructions' are intended to describe procedures for effective and reasonably safe product use. Thus, a product's warning may be adequate, while its instructions are deficient and actionable, or the reverse may be true.

Although the warning responsibilities of manufacturers, retailers and other product marketers vary, this chapter gives the greater weight to discussing the obligations of manufacturers. Ordinarily the manufacturer is responsible for the design of the product, has the greatest presumptive expertise as to its risks, and almost always is the author of its warnings and instructions. Thus the decisions are uniform in imposing the greatest informational obligations upon the manufacturer.

Since the mid 1960s, the predominant approach to the legal analysis of products warnings and instructions has been to gauge their sufficiency in terms of three doctrinal categories: negligence, warranty and strict liability in tort. In a significant modern development, the Restatement (Third) of Torts: Products Liability § 2(c) puts such doctrinal categories aside in favor of a functional definition of a warning or instruction defect. Section 2(c) states: '(c) a product is defective because of inadequate instructions or warnings when the foreseeable risks of harm posed by the product would have been reduced or avoided by the provision of reasonable instructions or warnings by the seller[,] . . . and the omission of the instructions or warnings renders the product not reasonably safe.'[5]

To be adequate under any theory of liability—doctrinal or functional—a necessary warning, by its size, location and intensity of language or symbol, must be calculated to impress upon a reasonably prudent user of the product the nature and extent of the hazard involved. As stated by one Federal trial court in the products liability context, the issue of warning adequacy poses the question of whether a warning legend 'is communicated by means of positioning, lettering, coloring and language that will convey to the typical user of average intelligence the information necessary to avoid the risk[.]'[6]

Together the judicial decisions indicate that a seller will have a duty to provide warnings as to unreasonable risks or personal injury or property damage arising from the use or consumption of a product. This duty arises where the risk of such injury or damage is substantial and the seller knows or should know that the user is less informed concerning

---

[5] The Restatement (Third) treatment of a seller's informational and marketing obligations tracks the decisional law, and endeavors to rationalize, rather than to depart from, the established risk/utility analysis of product 'defect' adopted in the majority of jurisdictions. See Restatement (Third) of Torts: Products Liability § 2(c) cmt. a: 'Most courts agree that, for the liability system to be fair and efficient, the balancing of risks and benefits in judging product . . . marketing must be done in light of the knowledge of risks and risk-avoidance techniques reasonably attainable at the time of distribution. To hold the manufacturer liable for risk that was not foreseeable when the product was marketed might foster increased manufacturer investment in safety. But such investment by definition would be a matter of guesswork . . .'

[6] Stanley Industries, Inc. v. W.M. Barr & Co., Inc., 784 F. Supp. 1570, 1575 (S.D. Fla. 1992), citing M. Stuart Madden, The Duty to Warn in Products Liability: Contours and Criticism, 89 W. Va. L. Rev. 221, 234 (1987). The court here uses the phrase 'average intelligence' not in the statistical sense of the term 'average,' but rather as 'ordinary' intelligence. Throughout this chapter the use by courts and by the author of the term 'average' will likewise mean 'ordinary.'

that risk than the seller. The warning itself must be communicated to permit risk avoidance and, as appropriate, safe product use.

14.2 WHERE DUTY TO WARN ARISES

14.2.1 Strict Liability and Negligence

Negligence

A manufacturer has a duty to give adequate warnings of any substantial risk when it 'knows of or has reason to know' that in the absence of such warnings the product is likely to be dangerous for the use supplied.[7] This duty to warn is triggered where the potential for harm from the use of the product without warnings or instructions is 'significant' or unreasonable.[8]

Determining whether the risk is unreasonable requires a balancing of the seriousness of harm, and the probability that the harm will occur if appropriate steps are not taken, against the cost or burden of taking precautions.[9] The manufacturer's duty to warn under negligence principles attaches when it knows or should know of a product's hazards. The 'should know' aspect of this warning obligation arises from the accepted rule that in evaluating a failure to warn claim, 'the manufacturer is held to that degree of skill and of knowledge of developments in the art of the industry then existing when the product was manufactured.'[10] Thus the manufacturer 'is held to the knowledge and skill of an expert' in understanding and anticipating product risks.[11] Accordingly, the manufacturer, charged with the 'should know' standard, carries the burden 'of discovering the product's dangers to the foreseeable user and providing the warning concerning those dangers.'[12]

The duty to warn under conventional negligence principles turns upon the reasonable foreseeability of harm by use of or exposure to the product in the absence of warnings. As one court observed: 'If there is some probability of harm sufficiently serious that ordinary men would take precautions to avoid it, then failure to do so is negligence.'[13] It is the *harm* that must be foreseeable, rather than the precise *means* by which that harm may eventually occur.[14]

Strict Liability in Tort

Comment *i* to Restatement (Second) of Torts § 402A provides that a product will be considered to be defective and unreasonably dangerous when it is 'dangerous to the extent beyond that which would be contemplated by the ordinary consumer who

---

[7] Restatement (Second) of Torts § 388.
[8] See Suchomajcz v. Hummel Chem. Co., 524 F.2d 19 (3d Cir. 1975) (two minors killed and four injured from experimentation with firecracker 'kits' ordered by mail from advertisement in 'Popular Mechanics').
[9] Restatement (Second) of Torts § 291 states: 'Where an act is one which a reasonable man would recognize as involving a risk of harm to another, the risk is unreasonable and the act is negligent if the risk is of such magnitude as to outweigh what the law regards as the utility of the act or the particular manner in which it is done.'
[10] Smith v. FMC Corp., 754 F.2d 873, 877 (10th Cir. 1985).
[11] Borel v. Fibreboard Paper Products Corp., 493 F.2d 1076 (5th Cir.), cert. denied 419 U.S. 869 (1974).
[12] Foremost-McKesson Co. v. Allied Chem. Co., 680 P.2d 818 (Ariz. Ct. App. 1983).
[13] Bean v. Ross Mfg. Co., 344 S.W.2d 18, 25 (Mo. 1961).
[14] See, e.g., Spruill v. Boyle-Midway Inc., 308 F.2d 79 (4th Cir. 1962) (insufficient warning given to mother in a household where 14 month old infant perished from chemical pneumonia following ingestion of furniture polish).

purchases it,'[15] the so-called 'consumer expectation' test of strict liability. Most courts have held that even under principles of strict liability, a product should be considered unreasonably dangerous only through reference to some form of risk–utility analysis. A typical risk–utility evaluation, which will vary in its particulars from state to state, involves consideration of the seriousness of the risk, the number of persons who will be exposed, the ability of the user to avoid the hazard, and the feasibility of safety measures, be they design or informational, that the manufacturer might take.

How does a claim in strict products liability differ from a claim brought on a theory of negligence? Most commentators and courts have stated that the strict liability inquiry pertains to the condition, or dangerousness, of the product, while the negligence evaluation focuses on the reasonableness or unreasonableness of the seller in marketing the product in its final condition.[16] Even if one accepts the distinction between a 'product based' (dangerousness) inquiry and an inquiry that is 'conduct based' (reasonableness of the seller), it is more helpful to identify the evaluation that is common to both theories. Analysis identical under both strict liability and negligence theories is the evaluation of the type of harm that may be caused in the absence of adequate risk communication, measured in terms of the degree of risk, the severity of injury, and the number of persons likely to be affected, that represents an unreasonable danger. A generally accepted standard is that a dangerously defective article is one 'which a reasonable man would not put into the stream of commerce if he had knowledge of its harmful character.'[17] From the above it becomes clear that in the context of failure to warn liability, the functional characteristics of strict liability and negligence theories are almost indistinguishable.[18]

The ALI's comments accompanying Restatement (Second) of Torts § 402A provide guidance both as to the distinctions and similarities between negligence and strict liability in tort. Comment *j* states the seller's obligation to inform the consumer or user of hazards of which the seller either knew or should have known at the time of initial sale. Even in strict liability, therefore, 'a seller is under a duty to warn of only those dangers that are reasonably foreseeable,' a standard which by its grounding in foreseeability 'coincides with the standard of due care in negligence cases.'[19]

A strong informed consent rationale pervades warnings analysis, with a representative expression of when the duty to warn arises under principles of strict liability in tort is stated by the courts as 'whenever a reasonable man would want to be informed of the risk in order to decide whether to expose himself to it.'[20]

One helpful analysis places the claims of unreasonable danger and inadequate warning in strict liability in the context of two tests as to whether a product is reasonably safe: (1) whether the product's utility outweighs the risk to its user, and (2) if the utility outweighs the risk, whether the risk has been reduced insofar as possible without a material diminution of the product's utility. When a product succeeds in passing the first standard, it must pass the second test, for a product will not be considered reasonably safe if the same product could have been either made or marketed more safely, with no substantial lessening of utility. Warnings cases in strict liability reflect most clearly the second

---

[15] Restatement (Second) of Torts § 402A cmt. *i*.
[16] See, e.g., In re Air Crash Disaster at Washington, D.C., 559 F. Supp. 333 (D.D.C. 1983) (in strict liability, the merchant selling an unreasonably dangerous product is liable for injuries proximately caused therefore, regardless of fault).
[17] Phillips v. Kimwood Mach. Co., 269 Or. 485, 525 P.2d 1033 (1974).
[18] See, e.g., Opera v. Hyva, Inc., 86 A.D.2d 373, 450 N.Y.S.2d 615, 618 (1982): 'Where the theory of liability is failure to warn or adequately instruct, negligence and strict products liability are equivalent causes of action.'
[19] Borel v. Fibreboard Paper Products Corp., 493 F.2d 1076, 1088 (3d Cir. 1973), cert. denied 419 U.S. 869 (1974).
[20] Moran v. Johns-Manville Sales Corp., 691 F.2d 811, 814 (6th Cir. 1982).

standard, for they advance the proposition that 'regardless of the overall cost–benefit calculation the product is unsafe because a warning could have made it safer at virtually no added cost and without limiting its utility.'[21]

14.2.2 Warranty

The absence of adequate warnings or instructions on a product may in some circumstances support a finding that the product marketed in this condition is not merchantable, and is in breach of § 2-314 of the Uniform Commercial Code. The court so concluded in the celebrated decision of Borel v. Fibreboard Paper Products Corp.[22] in which the plaintiff, an industrial insulation worker who contracted the diseases of mesothelioma and asbestosis as a result of 33 years of exposure to respirable asbestos, brought an action against certain manufacturers of insulation products containing asbestos. The warranty count of the complaint alleged that the defendant's products were unreasonably dangerous and unmerchantable, because of defendant's 'failure to provide adequate warnings of the foreseeable danger associated with them.'[23]

In another action illustrative of how a seller's failure to provide adequate warnings may be found to render the product unmerchantable, plaintiff brought suit alleging breach of an implied warranty of merchantability against the seller of aerosol deodorant that, after application, ignited on his skin as he lit a cigarette. The court agreed with plaintiff's contention that an implied warranty of merchantability applied to the contents of the deodorant can as well as to the can itself. The court further agreed that a failure to warn of dangerous propensities of either could render the product unmerchantable and that plaintiff's contention that the warnings on the can were inadequate therefore posed factual questions for the jury.[24]

Even when a product is, strictly speaking, fit to perform its intended function, the manufacturer's failure to warn the buyer of adverse side effects may constitute a breach of the implied warranty of merchantability. In one representative action, a purchaser of potato sprout suppressant alleged a failure to warn in both tort and warranty against the seller of that product, used for dusting seed potatoes before storage to retard sprouting. The bags in which the product was contained cautioned only that there might, after planting the following season, be 'a slight delay in emergence.'[25] While the product apparently succeeded in retarding emergence, it also, evidence showed, caused erratic emergence, multiple sprouting, and small potatoes. The court affirmed that goods are not fit for their ordinary purpose within the meaning of UCC § 2-314 if the manufacturer fails to warn of adverse 'side-effects which [result] from its use.'

14.3 THE EFFECT OF OBVIOUSNESS OF THE DANGER

The majority rule is that there exists no duty to warn of obviously hazardous conditions.[26] Authority consistent with the conclusion that a manufacturer need not warn of hazards

---

[21] Beshada v. Johns-Manville Prods. Corp., 447 A.2d 539, 545 (N.J. 1982).
[22] Supra note 19.
[23] Supra note 19 at 1086.
[24] Reid v. Eckerds Drugs, Inc., 253 S.E.2d 344 (N.C. Ct. App. 1979), cert. denied 257 S.E.2d 219 (1979).
[25] Streich v. Hilton Davis, Div. of Sterling Drug, 692 P.2d 440, 442-443 (Mont. 1984).
[26] See, e.g., Fanning v. LeMay, 230 N.E.2d 182 (Ill. 1967) (slipperiness of shoes when wet); Ward v. Hobart Mfg. Co., 450 F.2d 1176 (5th Cir. 1971) (placing hand in operating meat grinder).

that are of common knowledge has involved slingshots, BB guns, darts, chairs on casters for invalids, kerosene used by industrial workers, and the activity of diving from a roof into a four-foot-deep swimming pool.

The law has been often, but inadequately, summarized that there should be no recovery for failure to warn where the hazard posed by the product was obvious to the ordinary user or consumer. The position taken in the decisions comprising this body of law is stated by one court in this language: 'A manufacturer cannot manufacture a knife that will not cut or a hammer that will not mash a thumb or a stove that will not burn a finger. The law does not require him to warn of such common dangers.'[27] A consistent position is suggested by the comments to Restatement (Second) of Torts § 388 which provide that the supplier's duty to warn others of hazardous propensities of a product applies 'if, and only if, he has no reason to expect that those for whose use the chattel is supplied will discover its condition and realize the danger involved.'[28] Decisions consistent with this approach include a denial of recovery to the plaintiff upon a finding of the obviousness of the hazard of using a power saw without the guard in place, using kerosene near a source of ignition, and putting one's hand in a meat grinder during operation.

The doctrine denying recovery for injuries caused by product hazards that are obvious or known to the user or consumer has been applied even where the injured parties are children, embracing a logic that prompted one court in an action caused by slingshot to state: 'Ever since David slew Goliath young and old alike have known that slingshots can be dangerous and deadly.'[29] Comparable results have been reached in actions involving minors' use of BB guns, pointed darts, and denatured alcohol.

Some have argued that automatic preclusion of liability based only upon alleged obviousness of the danger ill serves the cost internalization objective underlying strict liability in tort, i.e., the concept that the price of a product should reflect its total cost to the manufacturer, including the cost of liability judgments levied against it, when its products cause injuries.[30] Pursuant to such cost internalization, a manufacturer will 'spread' the risk by procuring liability insurance, enabling it to pay such liability judgments as are won against it by injured consumers, while spreading the cost of such insurance through the upward adjustment of product costs. Those suggesting that tort law should sustain incentives to manufacture safer products rather than create manufacturer incentives to produce products the dangers of which are obvious to all[31] therefore propose that the obviousness of a product's danger should not be an absolute defense, but rather should constitute but one of the factors in determining whether the product is defective.[32] As put by one court in a suit brought by the injured operator of a conveyor belt against the manufacturer of a dressing applied to the belt before the accident: 'There is no valid

---

[27] Jamieson v. Woodward & Lothrop, 247 F.2d 23, 26 (D.C. Cir. 1957). Jamieson involved the plaintiff's purchase of an elastic exerciser that was essentially 'an ordinary rubber rope, about the thickness of a large lead pencil, about forty inches long, with loops on the ends.' Plaintiff was injured when the extended exerciser slipped and struck her in the eye.

[28] Restatement (Second) of Torts § 388 cmt. k. See also Bartkewich v. Billinger, 247 A.2d 603 (Pa. 1968): '[W]e hardly believe it is anymore necessary to tell an experienced factory worker that he should not put his hand into a machine that is at that moment breaking glass than it would be necessary to tell a zookeeper to keep his head out of a hippopotamus' mouth.'

[29] Borjorquez v. House of Toys, Inc., 133 Cal. Rptr. 483, 484 (Cal. Ct. App. 1976).

[30] The authors of Restatement (Second) of Torts § 402A stated plainly that the 'purpose of [strict liability in tort for defective products] is to ensure that the costs of injuries resulting from products are borne by the manufacturers that put such products on the market, rather than by the injured persons who are powerless to protect themselves.'

[31] Consider, e.g., the old fashioned exposed roller/wringer first generation washing machines.

[32] Dorsey v. Yoder Co., 331 F. Supp. 753, 759 (E.D. Pa. 1971).

reason for automatic preclusion of liability based solely upon obviousness of danger in an action founded upon the risk-spreading concept of strict liability in tort which is intended to burden the manufacturers of defectively dangerous products with special responsibilities and potential financial liabilities for accidental injuries.'[33]

14.4 CAUSATION AND DISREGARD OF WARNINGS

14.4.1 Generally

In the failure to warn claim, as in other products liability causes of action, the plaintiff's proof must establish causation. In its most elementary form, such proof will show that, had the seller supplied an adequate warning, the injured claimant would have altered his or her behavior so as to avoid injury.[34]

As expressed by one court, 'the evidence must be such as to support a reasonable inference, rather than a guess, that the existence of an adequate warning may have prevented the accident before the issue of causation may be submitted to the jury.'[35] In that action an automobile manufacturer avoided liability for an asserted breach of duty to warn of the risk that its automobile with a standard transmission may lurch forward or backward if started without engaging the clutch. The court concluded that plaintiff's argument that a warning would have prevented the injury was 'mere speculation.' Congruent authority is found in an action arising from plaintiff's injuries following an effort to prime an automobile carburetor by application of gasoline poured from a quart jar. Affirming the trial court's judgment for the automobile manufacturer, the appellate court approved the introduction at trial of evidence tending to show that the plaintiff was careless and would have disregarded any warning in the vehicle's manual against such do-it-yourself initiatives.[36]

Two presumptions, both bearing on causation, have gained widespread approval in duty to warn litigation. The first, applicable where some warning is, in fact, given, is stated in comment *j* to Restatement (Second) of Torts § 402A. It provides that '[w]here a warning is given, the seller may reasonably assume that it will be read and heeded....' The reciprocal presumption adopted by many courts is that when no warning has been given, a plaintiff may benefit from the presumption that had a warning been given, it would have been read and heeded.[37] Such a presumption has the meritorious effect of obviating the need for speculative testimony concerning whether plaintiff would have heeded a warning.

The defendant may overcome plaintiff's claim by showing either: (1) that even with an adequate warning the plaintiff would have acted in an identical way, and thus would have suffered the injury; or (2) that the independent acts, negligent or otherwise, of a third party were the proximate cause of the plaintiff's injury. In an illustrative example involving a claim arising from injuries sustained in mounting a multi-piece tire rim, reversing the trial court's denial of summary disposition for the tire manufacturer, an appellate court found that the plaintiff's claim was precluded by his own testimony 'that

---

[33] Olson v. A.W. Chesterton Co., 256 N.W.2d 530, 537-38 (N.D. 1977).
[34] Van Buskirk v. Carey Canadian Mines, Ltd., 760 F.2d 481 (3d Cir. 1985).
[35] Conti v. Ford Motor Co., 743 F.2d 195, 198 (3d Cir. 1984).
[36] Warner v. General Motors Corp., 357 N.W.2d 689, 693–95 (Mich. Ct. App. 1984).
[37] Nissen Trampoline Co. v. Terre Haute First Nat'l Bank, 332 N.E.2d 820, 826 (Ind. Ct. App. 1975), rev'd on other grounds 358 N.E.2d 974 (1976).

if had he read a warning with respect to the danger he would still have followed precisely the same repair procedures.'[38]

The manufacturer may also attempt to overcome the plaintiff's proof of causation by showing that the misconduct act of a third party, often the employer, or in the case of pharmaceuticals, the physician, operated as the 'efficient intervening cause' of the injury.[39] Accordingly, in one suit brought by an employee injured when instructed by his employer to clean out a tank that still smelled of gasoline purchased from the defendant, the court held that authorization by the employer to use these cleaning procedures when gasoline vapors were still present in the tank constituted a break in causation sufficient to relieve the seller of liability.[40]

14.4.2 Unintended or Unforeseeable Use of Product

Generally stated, a manufacturer is required to produce a product that is reasonably safe for its intended use. In addition, where foreseeable misuse of the product may create an unreasonable risk of injury or damage, the manufacturer must provide warnings adequate to permit the user to avert the hazard.

Not very long ago, the law permitted the manufacturer to presume that its product would be devoted to its normal use, and if it was safe when so used, it would 'not [be] liable in damages for injury resulting from an abnormal or unusual use not reasonably anticipated.'[41] Under this view it was held that it was not a foreseeable use of an automobile that it might be involved in collisions, that a hood designed for use as harness equipment might be used to support a man pruning trees, or that a consumer might splash cleaning fluid into her eye.

More recent decisions have created a products liability remedy for injuries occasioned by product misuse where no remedy existed before. A galvanizing influence in this development has been comment k to Restatement (Second) of Torts § 395, which states that '[t]he manufacturer may ... reasonably anticipate other uses than the one for which the chattel is primarily intended.' Under this approach the key determination is whether the use to which the product has been put is one that the seller ought reasonably to have foreseen, and as to which he should be liable for any injury caused, is 'whether the plaintiff was acting within a commonly known area of conduct.'[42] By such a common conduct standard, therefore, a kitchen chair used by a consumer to reach a high shelf was found to be in foreseeable use when the backrest failed to support her weight, causing injury. The common conduct standard likewise would support the conclusion that the common, while unfortunate, use of an automobile is its involvement in collisions.[43]

A line of authority requires a manufacturer to anticipate the environment in which a product will be used, and the risks of misuse, however unorthodox, that may inhere in such an environment. For example, in Spruill v. Boyle-Midway, Inc.,[44] the court held that the manufacturer is 'expected to anticipate the environment which is normal for the use of his product, and where ... that environment is the home, he must anticipate the

---

[38] Spencer v. Ford Motor Co., 367 N.W.2d 393 (Mich. Ct. App. 1985).
[39] Bennison v. Stillpass Transit Co., 214 N.E.2d 213 (Ohio 1966).
[40] Bennison, id.
[41] McCready v. United Iron & Steel Co., 272 F.2d 700, 703 (10th Cir. 1959).
[42] Note, Foreseeability in Product Design and Duty to Warn Cases—Distinctions and Misconceptions, 1968 Wisc. L. Rev. 228, 233.
[43] Volkswagen of America, Inc. v. Young, 321 A.2d 737, 745 (Md. 1974).
[44] 308 F.2d 79 (4th Cir. 1962).

reasonable foreseeable risks of the use of his product in such an environment... [even] though such risks may be incidental to the use for which the product was intended.' Thus, to use one widely appreciated example, the manufacturer of clothes must foresee that the wearer may, unwittingly, bring the garment into contact with cigarettes, stove burners, or other sources of ignition. The manufacturer will be liable for any injury occasioned by the garment's unreasonable flammability in such a setting, notwithstanding the fact that bringing the fabric into contact with an ignition source is surely not an intended use of the product. It is, nevertheless, a foreseeable misuse.

When, in contrast, the misuse of the product is of such a nature as to have been not reasonably foreseeable, the logic of precluding a plaintiff's recovery for an injury that occurred by nonforeseeable misuse of the product is that the manufacturer is not required to produce a product that is wholly incapable of injuring the user. Accordingly, as suggested by one court hypothesizing an accident in which an automobile leaves the road, coming to rest in a river: 'It could scarcely be argued,' the court states, 'that the manufacturer should have produced an automobile which would float.'[45]

14.5 PERSONS TO BE WARNED

14.5.1 Generally

The general rule is that the seller of a product that may pose a risk of injury if not accompanied by adequate warnings as to the risk and, as appropriate, instructions for safe use, has a duty to warn the purchaser of the hazards. Moreover, where the seller can reasonably foresee that the warning conveyed to the immediate vendee will not be adequate to reduce the risk of harm to the likely users of the product, the duty to warn has been interpreted to extend beyond the purchaser to persons who foreseeably will be endangered by use of or exposure to the product. Included are members of the public who might be injured as a result of lack of adequate warning.

The leading expression of the weighing process that should accompany a manufacturer's determination of whether additional warnings should be given beyond those available to the immediate purchaser is stated by the court in Dougherty v. Hooker Chemical Co.[46] That court called for the balancing of the following considerations: 'the dangerous nature of the product, the form in which the product is used, the intensity and form of the warnings given, the burdens to be imposed by requiring warnings, and the likelihood that the particular warning will be adequately communicated to those who will foreseeably use the product...'

The question concerning to whom it should be given requires evaluation of the harm likely to occur in the product's use without warnings, the reliability of any intermediary to whom the warning is given, the nature of the product involved and the burden on the manufacturer in disseminating the warning.[47] In some circumstances the class to which the duty is owed and that to which the warning should go are not coextensive. The professional user and the medical-pharmacological learned intermediary doctrines represent two such types of situation. The bystander doctrine represents another. The court in Sills v. Massey-Ferguson, Inc.[48] framed the bystander issue well in its disposition of an

---

[45] Dyson v. General Motors Corp., 298 F. Supp. 1064, 1073 (E.D. Pa. 1969).
[46] 540 F.2d 174 (3d Cir. 1976).
[47] See, e.g., Frederick v. Niagara Mach. & Tool Works, 107 A.D.2d 1063, 486 N.Y.S.2d 564, 565 (1985).
[48] 296 F. Supp. 776 (N.D. Ind. 1969).

action brought against the manufacturer of a lawnmower for an injury suffered by a bystander struck in the jaw by a bolt picked up and thrown by the lawnmower. Identifying the duty of a manufacturer of a product that creates a hazard to give effective warnings to those who may foreseeably be affected by it, the court recognized that such a warning need not necessarily go to the person injured. It also recognized that, on the facts before it, 'it would be admittedly difficult for a manufacturer to warn the general public' of the lawnmower projectile phenomenon. The appropriate warning in such a setting, the court concluded, would be one 'adequate and sufficient ... [to] apprise the reasonable person of the dangers at hand.' This would probably be one as to safety precautions 'given to the user of the mower ...'

For prescription pharmaceuticals and medical products, including biological products available only through a prescribing physician, the general rule is that a manufacturer satisfies its warning obligation by conveying adequate cautionary information to the medical community, ordinarily the prescribing physician.[49] In some circumstances, however, even the sale of a prescription pharmaceutical may trigger a duty to provide warning information directly to the user. One insightful treatment holding that a manufacturer's warnings to even a highly skilled medical intermediary may be inadequate to prevent liability arose from an action brought by a woman who alleged that she suffered a stroke caused by her use of oral contraceptives.[50] In that action the court recognized that in most circumstances involving prescription drugs the manufacturer satisfies its duty to warn by conveying the necessary and appropriate information to the treating physician. It proceeded, however, to examine the patient–physician relationship in the administration of oral contraceptives in light of the Restatement (Second) of Torts § 388 comment n, which requires that the manufacturer's reliance on an intermediary must be reasonable. The court discovered that unlike the ordinary circumstances of patient–physician consultation common to the authorization of most prescriptions, with oral contraceptives there existed (1) 'heightened participation of patients relating to use,' (i.e., the patient often identified the type of prescription product she desired); (2) 'substantial risks,' (i.e., stroke); (3) the ease and practicability of direct warnings from the manufacturer to the user; and (4) the women's limited prescribing ('annual') and oral ('insufficient or ... scanty ...') contact with the physician to justify reliance on manufacturer communication to the medical community alone. On these grounds, the court concluded that the manufacturer of oral contraceptives had a duty to provide adequate warnings and instructions directly to recipients of oral contraceptive prescriptions.

The manufacturer's dilemma of what, if any, information must be stated to persons other than the immediate purchaser arises as well where the manufacturer sells in bulk. This question has been treated in actions on claims involving products such as chemicals and natural gas, with a resolution that can be stated generally as providing that, for products sold in bulk, the wholesaler discharges its duty to warn by conveying adequate warning to the immediate purchaser. If, on the other hand, the products sold by the bulk seller are already packaged, 'ordinary prudence may require the manufacturer to put his warning on the package where it is available to all who handle it.'[51]

---

[49] See generally M. Stuart Madden, 1 Products Liability (2d) § 23.12 (1988 & 1995 Supp.)
[50] MacDonald v. Ortho Pharmaceutical Corp., 475 N.E.2d 65 (Mass. 1985), cert. denied 106 S. Ct. 250 (1985).
[51] Jones v. Hittle Service, Inc., 549 P.2d 1383, 1393–94 (Kan. 1976). See Hubbard-Hall Chem. Co. v. Silverman, 340 F.2d 402 (1st Cir. 1965) (manufacturer of insecticide required to place adequate warnings on the bags in which it was sold, including, arguably, international symbols of toxicity, such as the skull and crossbones, where the evidence showed that English warnings might not be understood by semi-literate farm laborers).

14.5.2 Allergic or Idiosyncratic Users

Where a manufacturer's product is safe for use by most persons likely to come into contact with it, but is likely to create an allergic or highly unusual reaction in only a small proportion of the population, special issues arise as to the manufacturer's duty to warn the allergic or idiosyncratic individual.

Several decisions have adverted to the consumer expectation standard of Restatement (Second) of Torts § 402A comment *i*. The comment *i* language has prompted conclusions by some courts that an allergic or hypersensitive reaction to a product is not that of an ordinary or normal consumer. Reference is also made to comment *j*, which provides that the manufacturer should provide a warning where 'the product contains an ingredient to which a *substantial number* of the population are allergic, and the ingredient is one whose danger is not generally known . . .'[52]

Illustrative decisions adopting a 'substantial' number or an 'appreciable' number test include Kaempfe v. Lehn and Fink Products Corp.[53] in which the court stated that there exists no manufacturer duty to warn unless the consumer is one of a 'substantial number or of an identifiable class of persons who were allergic to the defendant's products.' The court in the latter case found that under this standard there should be no recovery where the evidence showed only four complaints out of approximately 600 000 units of spray deodorant sold, a determination required, the court emphasized, not only by 'the weight of authority but also by common sense application of the negligence doctrine.'

Particularly when only a very small proportion of the population is put at risk, the severity of the illness or injury to which the warning would be directed is properly a factor in determining whether the manufacturer has a duty to warn. Such was the logic of the court in Davis v. Wyeth Laboratories, Inc.,[54] where an action involving the risk to participants in a mass polio immunization program of contracting the disease was argued to be only one in a million. The court likened a consumer's right to be advised of product risks to a patient's personal autonomy interest in only submitting to such procedures as to which he or she has given 'informed consent.' Phrasing its approach in terms of whether the potential risks and the potential benefits of a medical pursuit would present an ordinary patient with a 'true choice judgment,' the Davis court adopted this approach: 'When, in a particular case, the risk qualitatively (e.g., of death or major disability) as well as quantitatively, on balance with the end sought to be achieved, is such as to call for a true choice judgment, medical or personal, the warning must be given.'

14.5.3 Professional Users

The rule is stated generally that there is no duty to give a warning to members of a trade or profession against dangers generally known to that group.[55] Adherence to this approach is demonstrated by decisions holding that there is no duty to warn about the dangers of high exposure to benzene when the individual exposed to the benzene is a

---

[52] Id., cmt. *j*: '. . . or if known is one which the consumer would reasonably expect not to find in the product, the seller is required to give a warning against it, if he has knowledge, or by the application of reasonable, developed human skill and foresight should have knowledge, of the presence of the ingredient and the danger.' Id.
[53] 21 A.D.2d 197, 249 N.Y.S.2d 840 (1964), affirmed 20 N.Y.2d 818 (1967).
[54] 399 F.2d 121 (9th Cir. 1968).
[55] Lockett v. General Elec. Co., 376 F. Supp. 1201, 1209 (E.D. Pa. 1974), affirmed 511 F.2d 1394 (3d Cir. 1975).

professional tank stripper whose job required contact with comparably hazardous cargo, and that there is no duty to warn an experienced stuntman about the hazards of jumping from a height of 323 feet into an air cushion rated for 200 feet.

One obvious rationale for distinguishing the so-called professional user doctrine from the doctrines discussed above, concerning the duty to warn about known or obvious dangers, is that the product sold to or coming into contact with the professional frequently may be sold only to members of that trade. Plausibly one can maintain that the producer of bulk quantities of rodenticide, sold only to seed and feed stores in bags not smaller than 100 pounds, can expect that the users would be acquainted with the safe use of rodenticide, and that a manufacturer should not have to warn a farmer about the dangers of drinking concentrated herbicide.[56]

One early articulation of this approach was offered in Helene Curtis Industries, Inc. v. Pruitt,[57] in which the plaintiff was injured while using defendant's hair preparation administered by a friend. The manufacturer defended successfully that its product was plainly marked 'For Professional Use Only' and that its warnings and other cautionary information were sufficient for the safe administration of the product by beauticians.

The professional user exception has been invoked successfully when the injured individual or individuals have had first-hand knowledge of the characteristics of a product, even absent direct professional experience. One such holding was an affirmed lower court finding for a defendant concrete manufacturer in an action brought by two men who purchased the defendant's concrete for use as a foundation for an addition to their home and suffered chemical burns from contact with the product. Noting that '[b]oth plaintiffs had experience in working with concrete' and had clothed themselves to provide protection from the risk of, among other things, chemical burns, the court held that a manufacturer had no duty to warn of risks thus known to the user.[58]

Courts have not, however, reflexively denied recovery to experienced workers on the basis of a presumptive familiarity with any hazards associated with their trade or craft. Leading authority has proposed a rule providing that where the pertinent product safety information has not reached the individual who will use the product or be exposed to the peril, it is of no moment that the worker's supervisor or employer may have superior knowledge or information of the hazard. Thus, in Jackson v. Coast Paint & Lacquer Co.,[59] an action was brought by a painter severely burned when the epoxy paint with which he was painting the inside of a railway tank car ignited. The evidence showed that while the plaintiff's employer may have been familiar with the risk that vapors accumulating in a confined area could create the risk of explosion when coming into contact with a spark, the plaintiff himself was not. Reversing a verdict for defendant, the court stated that '[t]he adequacy of warnings must be measured according to whatever knowledge and understanding may be common to painters who will actually open the containers and use the paints; the possibly superior knowledge and understanding of painting contractors is irrelevant.'

As in Jackson, the record in Borel v. Fibreboard Paper Products Corp.[60] required the conclusion that, however familiar others might have been with the hazards of inhalation of respirable asbestos, the dangers of such exposure were not sufficiently apparent to insulation workers to relieve the manufacturers of the duty to warn.

---

[56] Ziglar v. E. I. DuPont de Nemours & Co., 280 S.E.2d 510, 515 (N.C. Ct. App. 1981).
[57] Helene Curtis Indus., Inc. v. Pruitt, 385 F.2d 841 (5th Cir. 1967), cert. denied 391 US 913 (1968).
[58] Gary v. Dyson Lumber & Supply Co., 465 So. 2d 172 (La. Ct. App. 1985).
[59] 499 F.2d 809 (9th Cir. 1974).
[60] Supra note 19.

14.6 ADEQUACY OF WARNINGS

Once there is determined to be a duty to warn, the task of the finder of fact is often to evaluate whether the warnings or instructions as were provided were adequate. To be adequate in the legal sense, the warning or instruction must be adequate, if followed, to render the product reasonably safe for its intended and foreseeable uses. Evaluation of the adequacy of a warning requires consideration of at least (1) the dangerousness of the product, (2) the intensity and form of the warnings given, (3) the burden upon the manufacturer of providing warnings, and (4) the likelihood that the particular warning will be adequately communicated to those who will foreseeably use the product.[61]

Thus, measuring the adequacy of a warning requires consideration of both its form and its content. The *form* of an adequate warning, be it rendered in a separate tag, integrated into the printed material on the product's container, or otherwise communicated, must first be such that it could reasonably be expected to command the attention of the reasonably prudent user, and to alert that user of (1) the risks involved in using the product, and (2) the means of avoiding or lowering those risks. The *content* of an adequate warning, in turn, must be of such a nature as to be 'comprehensible to the average user and to convey a fair indication of the nature and extent of the danger to the mind of a reasonably prudent person.'[62]

Consistent with the above, a warning may be *inadequate* if (1) its physical characteristics, including its size and placement, are so small or obscure that the reasonable consumer would not read it, or (2) if it fails to inform the reasonable consumer of the pertinent hazard and the means for its avoidance.[63] For example, concerning the dual prongs of the latter requirement, if the hazard to be avoided is venomous snakes in the grass, a sign saying simply 'Keep off the Grass' would be inadequate for its failure to describe with sufficient impact the nature of the risk as well as for its failure to inform the visitor of any means of safe passage. Concerning impact alone, such an understated warning would surely fail, in the expression of one court, to convey an 'intensity sufficient to illuminate the mind of a reasonable [person].'[64] On the other hand, if the sign said 'Use Foot Bridge,' it might be adequate in terms of advising the reader of the means of avoidance of the risk, yet it would also fail our hypothetical duty to warn again for its failure to impress the reader with the fact that 'a minor departure from instructions might cause serious danger. . . .'[65] Finally, were the sign to state, in an idiom popular in parking regulation, 'Don't Even Think of Stopping Here!,' the message would arguably convey the prohibitory message to the reader with sufficient emphasis. Again, it would fail as a warning for its want of information as to the nature and extent of the risk.

The warning's conspicuousness, prominence, and size of print, in comparison to the print size employed for other parts of the manufacturer's message, must be 'adequate to alert the reasonably prudent person.'[66] For example, a manufacturer's notice, printed on the label of bottles of its furniture polish in print of size and color identical to that used

---

[61] Dougherty v. Hooker Chem. Corp., 540 F.2d 174, 179 (3d Cir. 1976).
[62] Harless v. Boyle-Midway, 594 F.2d 1051, 1054 (5th Cir. 1979) (action brought in products liability following death of 14 year old boy who attempted to use pressurized propellant recreationally).
[63] See Brown v. Gulf Oil Co., Prod.Liab.Rep.(CCH) ¶ 10,474 (Tenn. Ct. App. 1985).
[64] D'Arienzo v. Clairol, Inc., 310 A.2d 106 (N.J. Super. Ct. 1973).
[65] Phillips v. Kimwood Mach. Co., 525 P.2d 1033, 1041 n.17 (Or. 1974).
[66] First Nat'l Bank in Albuquerque v. Nor-Am Agric. Prods., Inc., 537 P.2d 682, 692 (N.M. Ct. App. 1975) (action against manufacturer of disinfectant used to treat seed, later ingested by a hog, caused injuries to central nervous systems of children eating the meat of the animal).

for the balance of the manufacturer's message, was held insufficient to avoid liability for the death of an infant who died of chemical pneumonia after ingesting only a small quantity of the product.[67] Additional authority confirms that the evaluation of the impact of a warning and its consequent effect on the user or consumer involves '[q]uestions of display, syntax, and emphasis.'[68]

A leading decision finding a manufacturer liability for failure to warn of the 'extent' and 'gravity' of the risks posed by exposure to the manufacturer's product is Borel v. Fibreboard Paper Products Co.[69] This was an action by an insulation worker against manufacturers of insulation materials containing asbestos to recover for alleged breach of duty to warn adequately of the risks of asbestos-related disease. Reviewing warnings that cautioned, in part, that protracted inhalation of respirable asbestos 'may' be harmful, and advised workers to 'avoid breathing [asbestos] dust.' The court responded sharply: '[N]one of these so-called 'cautions' intimated the gravity of the risk: the danger of fatal illness caused by asbestosis and mesothelioma or other cancers. The mild suggestion that inhalation of asbestos ... 'may be harmful' conveys no idea of the extent of the danger.'

Warning language that is ambiguous, obtuse, or a hedge of the manufacturer's acknowledgement of the hazards associated with the product will be found to be inadequate to communicate the extent and the seriousness of the harm. In one action implicating a prescription drug in a patient's loss of vision with the potential for permanent blindness due to optic neuritis, the warning under review stated only that administration of the drug 'may produce decreases in visual acuity which appear to be due to optic neuritis.' That statement, in light of information available to the manufacturer indicating a 'permanent loss of vision [to patients] in a significant number of instances,' impressed the appellate court as being 'highly ambiguous.'[70]

Modern tort standards of personal injury, premises, and governmental liability are uniform in recognizing that the effectiveness of hazard warnings can be vitiated by habituation or by overwarning. The latter has sometimes been referred to as the 'cry wolf' phenomenon.[71] Even if the supplier warns of a risk in the most gripping language, actions taken by the manufacturer or by persons working on its behalf can erode the efficacy of an otherwise adequate warning. For example, in Incollingo v. Ewing,[72] the court held that the plaintiff should be able to introduce evidence that 'detail men' working on behalf of a pharmaceutical manufacturer 'overpromoted' the attributes of the drug in their presentations to the medical community at large and to such an extent as to lessen the impact of the manufacturer's cautionary written material. Such authority may be harmonized readily with the conclusion of another court that it is the duty of the pharmaceutical manufacturer to instruct its detail men 'at least, to warn the physicians on whom

---

[67] Spruill v. Boyle-Midway, Inc., 308 F.2d 79 (4th Cir. 1962). The warning had nothing to attract special attention to it except the words 'safety note' and the language advising that the product 'may be harmful, especially if swallowed by children.'
[68] D'Arienzo v. Clairol, Inc., 310 A.2d 106 (N.J. Super. Ct. 1973).
[69] Supra note 19.
[70] Ross v. Jacobs, 684 P.2d 1211 (Okl. Ct. App. 1984).
[71] See general Berquist v. United States National Weather Service, 849 F. Supp. 1221, 1228 (N.D. Ill. 1994) ('[T]he defendants have identified a NWS policy to avoid the 'cry wolf' syndrome, that is, a policy to strive for the highest rate of severe weather detection while maintaining 'the lowest possible false alarm rate in the issuance of warnings.' The social considerations underpinning this NWS policy are obvious, for it requires the NWS to balance greater safety with greater effectiveness.'); Ingredient Communication Council, Inc. v. Daniel Lungren, 4 Cal. Rptr. 2d 216, 222 (Cal. App. 1992)('William Viscusi, an economics professor, also extolled the superiority of the ICC warning system on the grounds that it avoids overwarning and the dilution of impact phenomenon caused by standardized warnings on too many products.').
[72] 282 A.2d 206 (Pa. 1971).

they regularly call of the dangers of which [the manufacturer] has learned, or in the exercise of reasonable care should have known.'[73]

Should a manufacturer consider bi-lingual warnings or universal symbols? An increasing number of courts have found that the creators or users of signs or warnings should take into account use demographics in determining whether or not to warn exclusively in English, or to consider use of universal symbols. The concern that solely-English signage may fail to provide sufficient information to persons for whom English is not their principal language has arisen in litigation involving persons of Latino origin. An early and influential decision addressing this question was Hubbard-Hall Chemical Co. v. Silverman,[74] in which plaintiffs, natives of Puerto Rico, were injured while administering defendant's pesticide Parathion. Finding for plaintiffs' survivors, the court stated: 'We are of the opinion that the jury could reasonably have believed the defendant should have foreseen that its admittedly dangerous product would be used by . . . persons like plaintiffs' intestates, who were . . . of limited education and reading ability, and that a warning [even if it complied with Federal statutory requirements] would not, because of its lack of a skull and bones or other comparable symbols or hieroglyphic, be "adequate" . . .'[75]

A further example, Stanley Industries, Inc. v. W.M. Barr & Co.[76] arose from a fire started when two employees, both natives of Nicaragua, increased the risk of spontaneous combustion by failing to store properly rags soaked in defendant's Kleanstrip Boiled Linseed Oil being used to oil wooden tables. The Federal trial court stated that the question of causation in such a claim is '[h]ad the seller provided an adequate warning, would the injured plaintiff have changed his behavior so as to avoid injury?'[77] The court found that plaintiff deserved jury consideration of its claim that the manufacturer should have accompanied its product with warnings in Spanish or universal symbols with this statement: 'Given the advertising of defendant's product in the Hispanic media and the pervasive presence of foreign-tongued individuals in the Miami workforce, it is for the jury to decide whether a warning should at least contain universally accepted cautionary symbols.'[78]

14.7 CONCLUSION

The preceding shows clearly that in the US private litigation pursuant to state common law, or state products liability statutes codifying that common law, have played a dominant role in molding a manufacturer's duty to provide adequate warnings and instructions. All too frequently, however, products liability trials involving warnings issues have proceeded in blithe disregard of any scientific examination of how warnings have worked, or have failed to work, in a particular setting. Many manufacturers craft warnings without subjecting them to prior testing or trial for effectiveness, and indeed until very recently most warnings tracked industry custom rather than any fresh and scientifically based knowledge of risk communication.

---

[73] Sterling Drug, Inc. v. Yarrow, 408 F.2d 978, 992 (8th Cir. 1969) (action involving blindness allegedly caused by administration of drug Aralen for treatment of rheumatoid arthritis).
[74] 340 F.2d 402 (1st Cir. 1965).
[75] Id., 340 F.2d at 401. See also Campos v. Firestone Tire & Rubber Co., 485 A.2d 305, 310 (1984) (seller may be obliged to introduce pictorial symbols to warnings intended for labor groups including persons of unskilled or semi-skilled categories, of whom many may not speak English).
[76] 784 F. Supp. 1570 (S.D. Fla. 1992).
[77] Id., 784 F. Supp. at 1574, citing M. Stuart Madden, supra note 49, at 270.
[78] Id., 784 F. Supp. at 1576.

From the perspective of tort law, the imperative is that the warning design with the most efficient integration of risk avoidance and informed choice elements should prevail. Whether the risk communication is in the form of posters on the wall of the employee lunchroom, package tags, labels or symbols, the guiding inquiry is whether the warning design has a proven effectiveness in its intended application and with the targeted audience.

To an ever increasing degree, courts hearing personal injury cases involving claimed inadequacy of cautionary information consider expert evidence of human factors experts in evaluation of the warning obligation and the adequacy of warnings given.[79] In the modern litigation environment, it would be foolhardy for a manufacturer to evaluate the effectiveness of its hazard warnings without a scientifically supportable evidentiary basis.

---

[79] Often the issues of risk and causation are of sufficient complexity that litigants from both sides will present expert testimony as to risk, significance of existing visual clues, and appropriateness of warnings. For example, in Watt v. United States, 444 F. Supp. 1191 (D.D.C. 1978), plaintiff brought a successful suit under the Federal Tort Claims Act, 128 U.S.C. § 2674, for injuries she sustained in a fall on steps in a fountain area at a Smithsonian Institution museum. In support of her claim that 'the pedestrian walkway around the fountain was unreasonably dangerous because of the difficulty presented in visually detecting the presence of the platform steps,' 444 F. Supp. at 1193, plaintiff called a psychologist and human factors specialist who testified that 'the variegated granite presented a visual dilemma because of its texture and because the sharp edges of the rectangular platform blocks with grouting between them [did] not provide a contrast to the edge of the step . . . until [a pedestrian was] only one stride away.' Id.

See also Ramm v. W. & D. Machinery, 1994 US Dist. LEXIS 11336 (N.D. Ill. 1994). This case involved plaintiff's claims, among others, that an envelope making machine contained inadequate warnings of the risk of catching one's hand in a pinch point between two rotating horizontal cylinders. In resisting defendant's motion for summary judgment, which was ultimately granted, plaintiff relied upon the deposition opinion of Harold Wakely, 'a human factors engineer, who state[d] that the "machine's" warning stickers were placed too low, were located too far from the cylinders in question, and did not provide an adequate description of the injury that could occur.' Id. 1994 US Dist. LEXIS 11336, *8.

CHAPTER FIFTEEN

The Expert Witness

KENNETH R. LAUGHERY

Rice University

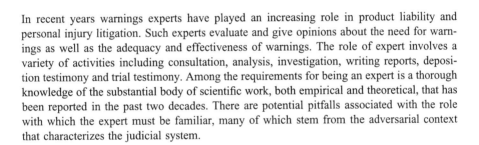

In recent years warnings experts have played an increasing role in product liability and personal injury litigation. Such experts evaluate and give opinions about the need for warnings as well as the adequacy and effectiveness of warnings. The role of expert involves a variety of activities including consultation, analysis, investigation, writing reports, deposition testimony and trial testimony. Among the requirements for being an expert is a thorough knowledge of the substantial body of scientific work, both empirical and theoretical, that has been reported in the past two decades. There are potential pitfalls associated with the role with which the expert must be familiar, many of which stem from the adversarial context that characterizes the judicial system.

15.1 INTRODUCTION

Recent years have witnessed a growing involvement of human factors specialists, psychologists, communications specialists and others in the role of expert witness in personal injury and product liability litigation in the USA. One of the topics on which a good deal of this activity has focused is warnings. Some of the questions typically addressed by the expert working in this arena are:

(1) Is there a need for a warning?
(2) Is an existing warning or warning system adequate?
(3) What would an adequate warning system be?
(4) Would an adequate warning system make a difference?

In this chapter several aspects of this role will be presented and discussed along with some of the issues associated with it.

15.2 THE WARNING EXPERT'S ROLE

What is the role of the warnings expert in litigation? While the emphasis in this chapter will concern 'being an expert,' not the legal or judicial underpinnings of the role, it is appropriate to state some formal definitions or rules regarding the answer to this question. There are also some informal aspects of the role to be noted.

15.2.1 The Formal Role

There are three Federal Rules of Evidence to be noted (Slater, 1993).

Rule 702. If scientific, technical, or other specialized knowledge will assist the trier of fact to understand the evidence or to determine a fact in issue, a witness qualified as an expert by knowledge, skill, experience, training, or education, may testify thereto in the form of an opinion or otherwise.

Rule 703. The facts or data in the particular case upon which an expert bases an opinion or inference may be those perceived by or made known to the expert at or before the hearing. If of a type reasonably relied upon by experts in the particular field in forming opinions or inferences upon the subject, the facts or data need not be admissible in evidence.

Rule 403 (the 'balancing' rule). Although relevant, evidence may be excluded if its probative value is substantially outweighed by the danger of unfair prejudice, confusion of the issues, or misleading the jury, or by considerations of undue delay, waste of time, or needless presentation of cumulative evidence.

From the rules, the role of the expert is to educate the judge and/or jury (the trier of fact) with regard to information that is beyond their 'common sense' or personal experience. The expert must be impartial and must not demonstrate an interest in the case outcome. Thus, while the expert is employed by one side (plaintiff or defense), and although the expert may form attitudes or opinions about the distribution of fault or blame, the role calls for neutrality in this regard.

15.2.2 The Informal Role

While the formal role of the expert is to advise or educate the judge and/or jury, generally he/she does not simply examine the facts of a case and then formulate and state opinions. Experts engage in a number of other types of activity in the course of working on a case. These activities include serving as a consultant, analyst, investigator, researcher and report writer.

It is not uncommon for an attorney to contact a potential expert without a clear notion of what the warnings issues are or what such an expert has to offer. This circumstance may arise as a result of the attorney having been told he/she 'needs' a warnings expert, as a result of the attorney reading about warnings expertise and its place in litigation, or a variety of other reasons. In such instances the expert must function as a consultant in advising and educating the attorney as to the nature of such expertise and what he/she can and cannot do. For example, it may be possible to examine a set of circumstances and to determine that a warning was needed when none was provided or that a given warning was inadequate; however, it may not be possible to develop an exemplar of an adequate warning given the available information. The inhalation hazard associated with a chemical solvent may not be addressed adequately in a warning, but the development of a complete warning system for such a product would require information about other hazards such as ingestion and skin contact.

Similarly, often the warnings expert is an advisor, analyst, investigator and even researcher (data collector). Frequently the warnings issues in a case are not so simple as whether or not a warning statement on a product label or a sign is adequate. Rather, the variety of media through which such information was, could have been, or should have been communicated must be determined and assessed. Further, what knowledge or information the target audience already had is relevant. The latter information is an example of where an expert may on occasion function as a researcher in collecting such data.

Another point regarding the expert role concerns report writing. Frequently, experts are asked or required to submit a written report in which they indicate their opinions and the basis of those opinions. Given the adversarial context of litigation, one can expect that such reports will receive extensive examination and critique, including scrutiny by others with similar expertise.

15.2.3 The Expert's Influence

The expert witness is in a potentially powerful position with regard to influence in litigation. The influence is based primarily on two aspects of the role. First, the expert is generally interacting with people who know much less about the area of expertise. This greater knowledge, of course, is the reason the expert is involved in the first place. Thus, except for similar experts who may be employed on the opposite side of a case, there is no one qualified to challenge or evaluate at a scientific/technical level the opinions of the expert.

The second source of influence stems from the fact that the expert can give opinions. This aspect of the role differs from the fact witness who provides information but is not permitted to render an opinion. For example, the fact witness cannot begin a sentence (when testifying) 'It is my opinion that . . . ,' while the expert can. Further, as noted in Rule 703 cited above, the expert does not necessarily have to cite or provide the basis for opinions, although recently the judicial system in the US has been moving in the direction of requiring the expert to provide more information as to the scientific basis of opinions.

With this potential influence come temptations and responsibilities. These aspects of the role will be addressed later in the chapter.

15.3 DEFINING THE WARNING EXPERT

What are the qualifications to be a warning expert? Many people give expert testimony on the need for, adequacy of and effectiveness of warnings. Not all are qualified.

15.3.1 Who/What Is a Warning Expert?

Rule 702 states that one may be qualified as an expert by knowledge, skill, experience, training or education. There is no requirement of an advanced academic degree or specific type of experience. An experienced auto mechanic with limited formal education could be an expert witness on some subject of auto repair. On the other hand, one need not have designed or worked on a type of environment or equipment in order to serve as an expert. A university professor in mechanical engineering may have never designed an internal combustion engine (or any other engine), but he/she may be qualified as an expert on the basis of education and knowledge. Ultimately, the issue of whether a person is qualified to testify as an expert is decided by the court (the judge), not the jury.

Recently there has been growing concern as to what constitutes expertise in this context. A broad analysis of this issue is not within the scope of this chapter. For a general discussion of the matter see Slater (1993), and for an interesting and challenging perspective see Huber (1990). Of concern here is the issue of warning expertise.

Qualifications

What education, knowledge, experience, etc. should a warning expert have? Clearly a degree in warnings is not a criterion since no such thing exists. A perspective on this issue may be gleaned from noting what a warning is. It is a *communication* that has the purpose of communicating information such as hazards, consequences and instructions. Also, it has the intent to *influence* people's behavior appropriately. Warnings may be written, they may be spoken, or they may be coded (non-verbal) sounds; hence, they are also displays. From this perspective, developing and/or evaluating a warning requires information about hazards, consequences and appropriate forms of behavior as well as knowledge about how such information should be communicated/displayed in order to influence behavior.

This set of requirements may seem like a tall order to find in a given individual, but a closer examination reveals a more meaningful definition of the warning expert. First, information about hazards, consequences and appropriate behavior are generally not the bailiwick of the warning person; rather, these are matters about which such experts make assumptions based on information from engineers, toxicologists, safety professionals, etc. The warning person does not have to have technical expertise in the biological reactions to breathing the vapors of a chemical solvent or the center of gravity and handling dynamics of an off-road vehicle. These are inputs gleaned from other experts and/or technical literature. The relevant expertise for the warning person is communications/displays and human behavior. Thus, psychologists (especially those with training in human cognition), human factors specialists and ergonomists, and people in the field of communications represent the most likely potential pool of people with such expertise.

Whatever the general category from which a warning expert may be drawn (psychologist, ergonomist, communications theorist, etc.) there are some specifics the expert should have in his/her bag of knowledge and tools. First is a thorough familiarity with the substantial body of research literature that has emerged over the past two decades. This literature, which has been presented and discussed in other chapters in this book, provides a good basis for understanding the issues associated with warning design and effectiveness. The bag of tools should contain also expertise in some of the relevant methodologies such as hazard, fault-tree and failure-modes analyses, task analysis, display design, and data collection and analysis techniques. Perhaps one of the most important requirements for the warning expert is a level of knowledge about human cognition; that is, how people process information. The work of Lehto and Miller (1986) has been important, in part, for its emphasis on the cognitive perspective.

15.3.2 Who/What Is Not a Warning Expert

The above outline of criteria for warning expertise can be supplemented with a few comments about who/what is not a warning expert. Examples would be as follows.

> A mechanical engineer with knowledge of vehicle design and dynamics and the hazards associated with certain uses of vehicles, while he/she may be an outstanding engineering expert, is not an expert on warning about these hazards.

A toxicologist or physician with knowledge about illnesses or diseases associated with exposure to chemicals may be an excellent expert on these topics, but he/she is not qualified on the basis of that knowledge to be a warning expert.

A person who has written a few (or perhaps many) warnings in his/her career that by most criteria are not very good is not necessarily a warning expert.

These examples, of course, are not intended to be critical of mechanical engineers, toxicologists or physicians. Rather, their purpose is to add perspective to what a warning expert is, or should be. Experience indicates that people with a wide variety of credentials and experiences have been permitted to give expert testimony about warnings in the US courts. The qualifications of many could be questioned seriously.

15.3.3 The Court's Judgment

How can the court distinguish between the legitimate warning expert and someone who does not have such expertise? As noted earlier, such distinctions are ultimately made by judges. These decisions are not simple, especially when one considers the fact that they are being made by a person who himself/herself is not expert in the vast array of subject matter, including warnings, that are addressed by expert witnesses in courtrooms. It is well beyond the scope of this chapter to address this decision making issue. However, it is important for the potential warning expert to realize that such decisions are being made and that information will be sought to serve as a basis for the decision. Following are some questions, the answers to which could/would serve as a basis for such decisions.

What is the content area of the person's education and/or knowledge? Is it in an area such as psychology, human factors, ergonomics or communications?

What is the person's level of knowledge about or familiarity with the technical literature on warnings?

Has the person done research and/or published on the topic of warnings? Has the research been funded? By whom? Have the publications appeared in peer-reviewed journals, proceedings and books?

Has the person had experience in designing warnings?

Has the person had a role in relevant national organizations, advising government agencies, consulting with and or working for industry on relevant projects, serving as an editor or reviewer of scientific literature, etc.

As noted earlier, all of the various criteria implied by these questions do not have to be met to qualify as a warnings expert. However, in order to be accepted and to function as an expert in this field, at least some of these credentials will have to be met.

15.4 GETTING INVOLVED AS A WARNINGS EXPERT

A question frequently asked is 'how do I get involved as a warnings expert in litigation?' Given that one has the qualifications, there are several things that might be done or efforts that might be made to get going. First, however, one should be sure about jumping in.

There are both opportunities and pitfalls associated with the expert role; some of these will be discussed in the next section of this chapter. Probably the best way of learning more about the expert witness role is to talk to people who have done it. Also, there are more formal opportunities to pick up such information. For example, workshops, panel sessions and symposia on the topic have been part of the program at recent meetings of the Human Factors and Ergonomics Society. Indeed, workshops and training programs on the subject of the expert role are sponsored by various organizations and can provide valuable information.

Four ways of getting involved that can prove useful are as follows. (1) Let people currently serving as experts know you are interested. They can pass your name along to attorneys looking for an expert. (2) Advertise your services. There are publications accessed by the legal community where you can pay for such listings. (3) List your name and credentials with organizations who market experts. There are numerous organizations who do such marketing for a share of your fee. (4) Contact attorneys or law firms and let them know you are interested and what your credentials are. This approach generally is not as effective as one might think.

It should be noted that there are pros and cons with the above approaches. The second, advertising, poses some potential hassles in the context of adversarial litigation. It is not uncommon for an attorney representing the adversary of the person who hired you to try to make you appear as a 'hit man/woman' on the basis of such 'selling yourself'. The third approach involves another entity taking a cut of the fee, which may mean you will have to work for a lower fee for yourself. Nevertheless, these are some of the ways one can get into the role of expert.

A relevant point here concerns the apparent communications network among attorneys. If you do get involved in a case and do a good job, chances are you will get more opportunities through this network.

15.5 ISSUES, PROBLEMS, TEMPTATIONS

There are numerous potential issues, problems and temptations associated with being an expert witness. In this section a few such matters will be touched on. There are others. Anyone thinking about getting involved as an expert witness should make the effort to learn more about them, and anyone already working as an expert is well advised to keep them in mind in carrying out such work. The adversarial context in which the expert functions is quick to capitalize on errors, usually at some cost to the expert.

15.5.1 Ethics

Most codes of ethics set out principles that are applicable to the role of the warning expert in litigation. The codes promulgated by the American Psychological Association and the Human Factors and Ergonomics Society are examples. The expert should be familiar with these and/or other applicable codes and follow them carefully.

15.5.2 Boundaries

One of the success rules of the expert game is knowing what you know and knowing what you don't know. It is important, indeed critical, for the expert to stay within the boundaries of his/her expertise. If you are a warning expert, limit your analyses and opinions to warning issues; do not address matters associated with engineering design, child development, or a host of other potentially 'nearby topics.' This rule may seem straightforward, but it is not—for various reasons. First, often boundaries are fuzzy. For example, the psychologist serving as a warning expert may (hopefully) know a great deal about human cognition. When testimony from different fact witnesses is contradictory as to the circumstances of some accident event, it may be tempting to offer opinions about that testimony based on knowledge of human memory. A good rule of thumb is do not do so unless you really are expert on memory and you have specifically been asked to evaluate and form opinions about those issues. A second reason why boundaries are sometimes violated is that the expert gets asked questions in a deposition or in trial about peripheral issues such as a design feature of some piece of equipment. While the warning expert may feel confident that he/she can provide a correct answer, this is a temptation to be avoided. Most often such a question will not have been posed innocently; rather, it is likely to be an effort to maneuver the expert out onto a limb that can then be sawed off from behind. A third reason the boundary rule can be difficult to follow is related to the fact that experts represent a cost in litigation, and the attorney who hired you may understandably want to keep costs under control. If you provide opinions on the other issues, the attorney may not need to hire other experts, thus cutting expenses. Be careful about providing such 'favors.'

15.5.3 Consistency

In addition to defining the boundaries of one's expertise, it is also important to be consistent within those boundaries. An opinion today about some warning issue that differs from an opinion tomorrow about some similar warning issue is not likely to go unnoticed—at least not for long. This also may seem like an easy rule to follow. For example, if one has expert knowledge about the criteria for warning design and the factors that influence when warnings will and will not be effective, and if one applies this knowledge uniformly in different cases, then it would seem relatively easy to be consistent in formulating opinions. Not so. First, situations or circumstances are not usually the same or different; rather, they vary as shades of gray. The nature of the accidents and injuries/illnesses as well as the products and people involved may vary in numerous dimensions which, in turn, makes the analyses and formulation of opinions complex. Indeed, on a subject such as warnings, this complexity is one of the reasons an expert is needed. A second reason why being consistent can be difficult is that the opposing attorney may through artful questioning attempt to get the expert to be contradictory. Here again is the adversarial nature of the litigation context to which the expert must be sensitive. Third, consistency can be a challenge because the warning expert may have opportunities to work on the defense side of some cases and the plaintiff side of other cases. Generally, in cases involving warning issues, the defense attorney will be happy if the expert's opinion is that warnings were adequate and/or the warnings were not a causal factor in the injury/illness. The plaintiff's attorney, on the other hand, is looking for opinions that the warnings were inadequate and this inadequacy was a causal factor. Warning experts who work only for defendants or plaintiffs will have less difficulty being consistent, but they will face other challenges regarding integrity and impartiality. It is important to keep in mind that one is a warning expert, not a defense of plaintiff expert. The real solution to the consistency challenge is to be true to the empirical and theoretical science of warnings and not to bend to the demands of the litigation context.

15.5.4 Being Current

Related to the consistency issue is the challenge of staying current with the empirical and theoretical science in the area of warnings. This is not a minor challenge given the increase in research activity during the 1980s and 1990s. Also, this time period has experienced significant efforts in the development of standards and guidelines for warnings. It is imperative that the would-be warning expert as well as the established expert be knowledgeable about the current state of the science. The growth and development of our knowledge about the design and effectiveness of warnings is probably the one legitimate basis for changes in our opinions about how warnings should be designed and how they function.

15.5.5 Fees

How much is a warnings expert worth? Many factors determine the answer to this question, including the qualifications of the expert and the needs of the attorney (supply and demand, as our free-market economist colleagues would say). Regarding the attorney's needs, it should be noted that often, but not always, warnings are secondary to design as an issue in a case. As a general rule, attorneys may be willing to pay more for engineering experts than for warning experts, and they usually do pay more for medical experts simply because physicians demand more and their testimony is necessary for the case. Surveys in the 1996–97 time period have shown a range of fees for human factors and ergonomic (including warnings) experts, with lows of $75–100/hour, and highs in excess of $300/hour (Lovvoll, 1997). Some experts have different fees for different activities— if they do, it is generally charging higher rates for deposition and trial testimony.

15.5.6 The Adversarial Setting

There are many things that might be said about the adversarial nature of the setting in which the warning expert functions. Several such points have already been noted. A few general 'rules of the game' should be kept in mind. (1) The expert's role is to advise or educate the jury. While the expert is employed by one side of the case, he/she must be impartial and unbiased. Experts may or may not do a good job, but they are not supposed to take the role of trying to win a case. The expert does not win or lose. (2) The plaintiff attorney's role is to *win* the case for the plaintiff. He/she is biased and does win or lose. (3) The defense attorney's role is to *win* the case for the defendant. He/she is biased and does win or lose. (4) The attorneys in a case will do whatever is necessary within the boundaries of the law and acceptable practice to win. (5) The attorney for the other side will make every effort to discredit you and your testimony. This effort will include getting you to contradict yourself, including researching your work in past cases to identify inconsistent opinions. It may also include employing other warning experts whose opinions differ from yours. (6) The attorney for the other side will make an effort to discredit the entire domain of warning expertise (e.g., Hardie, 1994). This effort may include the argument that the issues of warning design and effectiveness are within the province of the jury; that is, these are issues that jurors (lay people) are capable of evaluating without the help of experts. The effort may also include the argument that there is no 'hard science' associated with warnings, and it simply represents people offering opinions for money.

The above examples of 'rules' relevant to the adversarial litigation setting may at first seem excessively critical of attorneys and the adversarial system. That is not the intent. Rather, these are characteristics of many or most of the circumstances associated with the context in which the warning expert functions, and there are potentially serious pitfalls if one is not aware of them. Again, the expert's best defense against the various challenges inherent in the role is to be true to the empirical and theoretical science of warnings and to avoid being caught up in the adversarial nature of the proceedings.

15.5.7 More Likely Than Not

There are many aspects of the law, or laws in various jurisdictions, and of the legal proceedings that influence the role of the warning expert. One of these aspects that occurs in many jurisdictions is the more-likely-than-not rule. This rule concerns the probability that an adequate warning would have changed the outcome. Obviously, in the domain of warnings the plaintiff wants to argue that an adequate warning would have made a difference, while the defendant wants to argue that an adequate warning would not have changed the outcome. More specifically, the more-likely-than-not rule requires that the probability be greater than 0.5 that the warning would have changed the outcome. Thus, in such circumstances warning experts, in order to be helpful, must form opinions that the probability of the warning having the desired effect is greater or less than 0.5, depending on which side he/she is working for. Quantifying warning effectiveness in this fashion requires at a minimum that the expert be familiar with the empirical findings in the scientific literature and be able to generalize those findings to the circumstances of a specific case. For example, if the injured person looked for safety information and/or read the warning and thought he/she was behaving safely, the expert may be in a stronger position to opine that an adequate warning would have prevented the injury.

15.5.8 Nothing Is Secret

The rules of discovery are complex. The point to be noted here is that the expert's credentials, past experiences in other cases, and specific work on a case must all be revealed on request in working on a case. There are variations of this rule in different states, but in general the expert should assume that everything he/she does in a case must be revealed to the other side. This information would include anything provided or revealed by the attorney who hired him/her. It would include all notes, conversations, activities, publications reviewed, etc. related to the case. It can also include work in other past and present cases. Thus, the warning expert, like all experts, needs to be aware that 'nothing is secret.'

15.6 THE WARNING EXPERT'S FUNCTIONS

As noted earlier, the role of the warning expert may include a variety of activities. This section presents a summary of the various activities that may be involved. Every case is different to some extent, and the expert's role will in part be defined by the circumstances of the case. Figure 15.1 presents an overview of the activities.

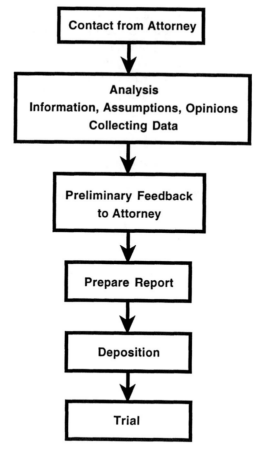

Figure 15.1 Expert activities.

15.6.1 Initial Contact and Decision

Involvement in a case usually begins when the expert is contacted by an attorney—a phone call or a letter (most likely a fax since attorneys usually seem to be in a hurry). This contact will include a brief description of the product involved (if it is a product case), the nature of the accident or illness, and the issues about which the attorney is seeking expert help. If the issues and expertise match and if the attorney and expert agree on procedures, fees, etc., the expert's activities in the case will begin.

One type of information the expert should solicit in the initial contact is the time frame. What does the attorney expect/need by when? Requirements can and do span the full gamut in this regard. Occasionally, a plaintiff's attorney will not yet have filed the lawsuit, and is looking for advice about whether or not there is a warning issue to be pursued. In such circumstances there may be ample time for the expert to carry out his/her work. Alternatively, a defense attorney may have just deposed a plaintiff's warning expert and decided he/she also needs to employ such a person. Here the time frame may be quite short and demand a great deal of work in a brief period. It is not uncommon also for attorneys, either plaintiff or defense, to be in a rush situation simply because they have delayed their own work on the case and/or the court has limited the available time for discovery. The point is that the expert needs to take into account the time requirements of a case in deciding whether or not to get involved.

Additional points to be noted about the initial contact and decision concern formal agreements and retainers. Some experts require an agreement in the form of a contract; others do not. Generally, such contracts are 2–3 page documents that spell out the scope of the work and billing and payment procedures. It is signed by both the attorney and expert. Some experts require a retainer up front; others do not. Typically, retainers are in the range of $800 to $2500. In some cases confidentiality agreements are signed with the purpose of limiting the use and distribution of information obtained while working on the case.

15.6.2 Analysis

As shown in Figure 15.1, the warning expert's analysis may include a variety of activities. In one sense the figure misrepresents the analysis activities in that such efforts continue throughout involvement in a case. Analysis may continue right up to the time of testimony in court or until the case settles as new information becomes available that is relevant to the warnings issues. Nevertheless, early work on a case focuses on analysis of information and the formulation of opinions.

Gathering information

Typically, the information examined by the warning expert comes from two sources: the attorney and the expert himself/herself. By the time the attorney employs the warning expert, often a great deal of information is available. This information may include:

- the complaint, which will give some of the alleged facts of the case and can be useful background;
- accident reports;
- information about the product (if a product was involved) including the actual product, an exemplar, photographs, manuals, package inserts, etc. as well as information about the product's history and development, including warnings;
- information about the job and work environment if the injury/illness was job related;
- statements and/or depositions of fact witnesses, including the plaintiff (if available);
- reports and/or depositions of other experts; and
- standards and guidelines such as those put out by government agencies, the American National Standards Institute (ANSI), the American Society for Testing Materials (ASTM) or Underwriters Laboratories (UL).

There may be occasions when the warning expert is involved very early that he/she has an opportunity to recommend to the attorney some of the kinds of information that will be useful in evaluating the warnings issues. Such information could include:

- results of hazard analyses (failure mode, fault tree, etc.) that have been done;
- procedures and criteria involved in developing the existing warning system;
- relevant past safety behaviors of the plaintiff; and
- safety history of the product or environment (US Consumer Product Safety Commission data, accidents that have occurred at this particular highway intersection, etc.).

Information relevant to the warning expert may be gathered by the expert also. One kind of information, of course, is the relevant scientific literature that addresses the warning issues involved. In addition, the expert may carry out tests, surveys, and other procedures (e.g., focus groups) to gather relevant information. Relevant information for the warning expert would include:

- are the hazards, consequences and appropriate modes of behavior 'open and obvious?'— that is, does the appearance and/or function of a product or environment provide the warning information?
- what do people already know about the relevant hazards, consequences and appropriate modes of behavior?
- how do people use a product? and
- do people notice, understand and respond to a warning system?

There are a few potential problems that can arise during the information gathering phase to which the expert should be sensitive. One is that relevant information may be available that is not provided by the attorney. This omission could occur because the attorney did not realize the information was relevant or because the attorney withheld it thinking it was harmful to his/her case. The latter circumstance should not be tolerated, and can be cause for the expert to withdraw from the case. An example would be a highway accident in which the driver (plaintiff) was drunk and the blood alcohol content analysis was withheld from the expert. On the defense side an example might be a letter or memorandum from the product manufacturer's file indicating that the warning system was deliberately downplayed so as not to negatively impact sales. While such withholding of information is rare, it does occur, and it can be embarrassing to the warning expert. It can also change the expert's opinions.

Another information problem is the absence of relevant information. In a death case there may be little or no information about whether the plaintiff noticed or understood the warnings or what the plaintiff knew or understood about the hazards, consequences and correct modes of behavior. Still another problem is contradictory information. Multiple witnesses to an accident may report different and contradictory accounts. These problems and others must be taken into account by the expert.

Assumptions versus opinions

The distinction between assumptions and opinions is important for the warnings expert. In developing opinions about the need for, adequacy of, and effectiveness of a warning system, the expert will typically review and evaluate a great deal of material such as noted above. Enroute to formulating opinions, assumptions will be made about a number of things. Examples are:

- what were the hazards and consequences associated with the product being used?
- what were the hazards and consequences associated with the activity being carried out, such as performing a task?
- what warnings were provided and how were they provided?
- what did the relevant people know about the hazard and consequences?
- where, when and how did the accident, injury and/or illness occur?

Note that the above questions refer to matters that are essentially factual; that is, it is not a matter of the warning expert's opinion as to what warnings were provided or what someone knew. The answers to the questions, the assumptions, may be made with varying degrees of confidence depending on the quality of available information, but they are still assumptions. It should also be noted that assumptions made by the warning expert may actually be based on opinions of other experts. The hazards and consequences associated with some chemical solvent may be defined on the basis of the expert opinions of a toxicologist. To the warning expert, they are assumptions.

One of the reasons the assumptions–opinions distinction is important is that it helps the warning expert define the boundaries of expertise and the basis of opinions. Obviously different assumptions can lead to different opinions, and it is common for the expert to be asked for opinions based on different assumptions about the facts. Indeed, questioning along this line can be a strategy of an opposing attorney to 'push' the expert to find out under what circumstances his/her opinions would vary. Clearly one can be asked to make assumptions, however invalid, that would lead to different opinions. If, for example, one is asked to assume than an injured person already had been informed about the hazards and consequences of using a product, one might reach different conclusions about the potential effectiveness of a warning.

Formulating assumptions

There are several types of information or sets of facts that the warning expert typically needs to be able to formulate opinions. Some of the more important information categories are represented in the above questions regarding assumptions. As mentioned earlier, two of the major difficulties in formulating assumptions are missing information and contradictory information. An understanding of what the brain damaged or deceased person knew about hazards is important to the warning expert, but it is not available from that person. Assumptions regarding this knowledge may have to be based on testimony of coworkers and family members as well as records of training, experience and past performance of the person. The nature of a product warning system may not be available information after an accident that destroyed the product. Was the label still legible? Was the manual still available and/or was it passed along when the product was sold in the second hand market? Answers to these and other similar questions are not always readily available. Contradictory information presents different but sometimes equally difficult problems for the warning expert. The different perspectives and memories of fact witnesses constitute a common problem that was noted earlier. It would not be surprising to find two employees who were witnesses to a workplace accident where one described a warning sign on the wall of the work environment and the other denied such a sign was ever present.

There are no simple answers or solutions to these problems for the warning expert who is trying to figure out what are the appropriate assumptions to make regarding the important facts and circumstances that underlie his/her opinions. Rather, what follows are a few suggestions to keep in mind when carrying out the analysis.

(1) Try to be clear about what information is needed or what are the issues about which one needs to make assumptions in order to formulate opinions about warnings.

(2) Recognize that the analysis carried out by an expert is in part a sleuthing activity in which one is not only examining information that is provided but also searching for information that is needed.

(3) Do not hesitate to ask questions or request information.
(4) Be prepared to recommend that steps be taken to obtain information that is not currently available. This may involve interviews, tests, surveys, etc.
(5) Be prepared to express opinions based on two or more different sets of assumptions.

The fifth point warrants an additional comment. Typically, the plaintiff's attorney will be happy with one set of assumptions while the defense attorney will prefer a different set of assumptions. It is likely that during deposition or trial testimony the expert will be asked his/her opinions given the different assumptions. Ultimately, the expert is responsible for the opinions, not the assumptions, and the opinions should be true to the empirical and theoretical science of warnings. If different assumptions warrant a different opinion, give it.

Formulating opinions

There are several categories of issues about which the warning expert is typically asked to express opinions. These categories are:

- was a warning needed?
- was the warning adequate?
- would the warning have made a difference?

In addition, the expert may be asked the basis of the opinions. In the discussion that follows it should be recognized that when the term warning is used, it is really referring to a warning system, potentially consisting of several warning components.

Is a warning system needed?

There are a number of considerations that feed into the answer to this question. The first, of course, is whether or not a hazard exists. No one would suggest warning about a nonexistent hazard. A second consideration is whether or not the hazard is open and obvious. In the law, generally it is not necessary to warn about a hazard that is open and obvious. Most would probably agree that a sprocket driven chain probably is an obvious pinch point hazard and that the inhalation hazards associated with a solvent are not open and obvious. But many hazards are not so easily classified on this dimension. A commonly accepted practice is that if it is not obvious that it is obvious, warn. A third consideration is whether or not people already know about the hazard through previous training or experience. If so, a warning may not be needed. Fourth, is a warning needed as a reminder? There may be circumstances where hazards are known, but due to other circumstances, such as high task loading, a reminder is needed. These are factors that the warnings expert considers in formulating opinions about the need for warnings.

Is a warning system adequate?

This question, of course, is central to the work of the warning expert. Most of the chapters in this book have addressed issues related to answering this question. Factors to be taken into account and criteria to be applied have been discussed at length. One aspect of the question that has not been addressed but is important to the warning expert, however, concerns the definition of adequacy. In the context of litigation, a warning is either adequate or inadequate. It is a one-of-two-states opinion that the expert must provide. In reality of course, the goodness or badness of a warning system is a continuum: they may be good or very good; they may be bad or very bad. So the expert's task includes deciding where to draw the adequacy cutoff. It is difficult at best to define how such decisions should be made. Some rules of thumb that may be helpful, but not always applicable, follow.

(1) The warning system must be considered as a whole. A poor warning in a manual that may or may not be seen may not render the warning system inadequate if a good warning is on the product. On the other hand, for many products an adequate on-product component is necessary for the system to be adequate. The point is that the different components of the system do not necessarily get equal weight in judging adequacy.

(2) Minor violations of criteria or guidelines may not be a basis for inadequacy. For example, using the signal word 'caution' instead of 'warning' when the latter is appropriate to the hazard, is probably not a valid reason alone for declaring a warning inadequate. Indeed, the research on this issue seems to indicate it does not matter.

(3) The opinion that the warning system probably would not have changed the outcome is not a basis for declaring the system adequate. The issue of whether or not the warning is adequate from a design perspective is not the same question as would it have made a difference. Obviously, these two issues are related, but they are also, to some extent, independent.

Would the warning system have made a difference? Like the question of adequacy, this issue is central to the work of the warning expert. It goes to the issue of cause. If a warning system was inadequate but it had no bearing on the accident, injury or illness in question, then its inadequacy is irrelevant. A frequent defense in warning cases is that warnings are not effective in influencing people's behavior, and, therefore, their adequacy or inadequacy is irrelevant. Like the factors that influence how good a warning is, the factors that influence how effective a warning is have been discussed at length in the earlier chapters. Frequently the effectiveness issue is a difficult one for the warning expert. No one would argue that bad warnings will be effective, and most people would not argue that good warnings are not more likely to be effective than bad warnings. Further, few if any would suggest that even the best warning system will be effective 100% of the time. The problem comes about in formulating opinions about how effective the warning system would be. It is generally difficult to quantify effectiveness. Categorical judgments may be made, such as 'more likely than not,' but they must be made with careful consideration of the best information available. This information includes the design of the warning system, characteristics of the target audience, circumstances of the task or activities of the people involved, and the empirical scientific literature relevant to the issue.

It has been increasingly important for experts in all disciplines to be able to provide the basis of opinions. Recent court decisions have placed a greater requirement that there be a scientific basis for expert opinions. The warnings expert is no exception. The substantial empirical and theoretical literature on warnings provides a major basis for opinions, and the expert needs to be familiar with it and to cite it.

Oftentimes in working on a case questions arise for which the existing technical literature contains no directly relevant information or data. The question may concern a specific risk perception issue (e.g., what do people know about this hazard or its consequences?), a specific warnings issue (e.g., is the existing warning understandable to this target audience?), or a variety of other specific issues. In such circumstances, it may be desirable to collect some relevant data through an experiment or a survey. An excellent example of such an effort was reported by Senders (1994) in which a survey was carried out to determine how people would connect a gas heater.

Such efforts are the exception rather than the rule. Experts are more likely to base opinions on more broadly based empirical data and/or theory. There are numerous reasons why such data collection activities are not carried out.

(1) Time. Often warnings experts get involved late in a case and the time fuse is short. There may not be time to collect data.

(2) Cost. Cost will depend on the extent of the effort, but getting a significant or meaningful amount of data can be expensive.

(3) Doing it right. Doing the 'quick-and-dirty' study often has utility in solving applied problems. In the adversarial litigation context, however, one may well expect to come under attack for methodologies that are less than perfect, even though they are appropriate to the circumstances. Be prepared to defend such efforts.

(4) 'Wrong' results. Perhaps one of the most disconcerting reasons why data specific to the issues of a case often are not collected is that lawyers are reluctant to support such efforts, because not knowing the results in advance they fear the outcome may provide ammunition for the opposing side.

Despite the above constraints, and despite the fact that often it is not necessary, more scientific expertise could be brought to the litigation process if such specific data collection efforts were incorporated on a more regular basis.

15.6.3 Preliminary Feedback to Attorney

At some point after some of the analysis has been completed, the expert will provide feedback to the attorney regarding preliminary opinions. Actually, this process may be more interactive than the sequence of steps reflected in Figure 15.1 might suggest. In any case, the attorney is going to want to know what you think so that he/she can decide whether to continue to employ you in the case. If your opinions are not supportive of his/her case, your work will be finished. This, of course, is a potential pitfall for the expert in that a job and a fee is lost as a result of 'not telling him/her what he/she wants to hear.' The alternative is filled with even greater pitfalls.

It is important that the feedback be as frank and as complete as possible. For example, suppose you are working for a plaintiff and you decide the warning system for a product is inadequate overall, but that one of the components (the manual) is good. It is critical that the attorney understand that if asked questions about the manual during testimony you will have to acknowledge it is good. Similarly, if working for a defendant and the warning system is poor but you conclude that even a good warning would not have been effective, you will still have to acknowledge that the warnings were not good.

It is also important that the attorney understand what assumptions you are making as a basis for your opinions. Where there are questions about the validity of the assumptions or alternatives to be considered, the attorney needs to know how your opinions would be influenced by such alternatives.

15.6.4 Preparing Reports

Reports are a part of the discovery process. Essentially they are intended to provide the opposing attorney with information about who the expert is, what he/she has done, and what his/her opinions are. Reports are not required in all cases. Generally, cases in the Federal courts require a report as well as a current curriculum vitae and a list of cases in which the expert has given testimony during the past four years. Report requirements vary, and often no report is required.

Another dimension on which report requirements may vary is specificity. In some instances a report may be brief and very general, stating opinions only in the broadest terms. For example, such a report might state only that the warning was inadequate and that had a good warning been provided the accident would have been prevented. In other instances, reports must be specific and complete. If a warning is judged to be inadequate, the specifics of its inadequacies must be spelled out and the basis for such an opinion provided. Further, in some circumstances if an opinion is not provided in the report, the expert may not be permitted to express that opinion in trial. Thus, it is imperative that the warning expert understand report requirements early in his/her work on a case, since clearly they have implications for the level of analysis carried out before the report is prepared.

Typically, the opinions in the report of a warning expert will address the issues posed earlier: was a warning needed, was the warning system adequate, and would the warning system have made a difference? As already noted, these issues may be addressed in varying levels of detail and, depending on the case, with varying degrees of emphasis. For example, if the opinion is that a warning was not needed, the other two issues are irrelevant.

A final point on reports. They should be prepared with great care. Odds are they will be scrutinized extensively and every point subjected to questioning and perhaps challenged.

15.6.5 Deposition Testimony

The next step is likely to be a deposition. A deposition is part of the discovery process in which each side has an opportunity before trial to examine the other side's evidence—including the expert's opinions. The procedure usually involves the attorney for the opposing side examining the expert in a question–answer format. There may be more than one questioning attorney, such as when you are working for the plaintiff and there are multiple defendants. The expert is under oath, the procedure is recorded, and the testimony is considered part of the formal record of the case; that is, it can be used later during trial. Thus, it is important for the expert to be well prepared and consistent. Contradictions between deposition testimony and trial testimony are not likely to go unnoticed and can discredit the expert. The deposition like the trial is adversarial, and the opposing attorney will be attempting to establish such things as:

- questions or shortcomings regarding the expert's credentials;
- flaws in the analyses carried out;
- contradictions in or concerns about the opinions; and/or
- the basis for the opinions, scientific data, theory, experience.

Generally, the tone of depositions is quite professional. Attorneys come prepared and they get on with the business at hand. There are exceptions, and at times bad manners and hostile behavior emerge. It is critical that the expert not get caught up in the argumentative or emotional aspects of the situation. Indeed, it should be realized that more often than not such situations are probably part of the opposing attorney's strategy to confuse the expert or to get him/her to take positions or express opinions that are contradictory or that cannot be supported. The opposite of bad manners and hostility can also occur—overfriendliness. Watch out if the attorney starts a question with 'Dr., you will agree with me won't you sir that . . .'

As already noted, credentials include past experience as an expert, and such experience is a legitimate subject of discovery. If you have worked as an expert in the past and provided deposition and trial testimony, it is not uncommon for the opposing attorney to have researched your previous work. Extensive questioning may focus on opinions in earlier cases. Such situations is one of the reasons why consistency is so important.

15.6.6 Trial Testimony

The last step in the overall process of being an expert is to testify in court. The expert is questioned by the attorneys representing both sides. Credentials are established and opinions are expressed, including the basis for the opinions. There are three aspects of the warnings expert's role in the courtroom to be noted here. First, he/she is in the position of communicating the nature and results of an analysis including methodology and data, as well as opinions that may be technical in nature, to an audience of lay people—the jury. Communicating technical information and opinions to a jury can be challenging. Metaphors, visual aids and demonstrations can be exceptionally helpful, and should be developed as part of the warnings expert's bag of communication tools. Charts that list the criteria for warnings and examples of good and poor warnings can help the jury to understand the expert's opinions.

The second aspect of the testimony in the courtroom is directed more to the warnings expert than to other kinds of expert. It concerns attitudes or information the jury may have that has to be overcome or changed. One point concerns information. By the time the warnings expert begins testimony, the jury will usually have heard descriptions of the accident or illness and presentations about the hazards associated with the product. Often a role of the warnings person is to provide an analysis in terms of how a product was used and what the injured party knew or did not know about product hazards *at the time of the accident*. In short, the warning expert must help the jury analyze the issues in the proper context, not in terms of what everyone in the courtroom knows now. Another point concerns attitudes. People often have a predisposition to believe that if someone gets hurt, it is because he/she made a mistake. The warning expert needs to help the jury take a more systems oriented view of systems involving people.

The third aspect of courtroom testimony facing the warning expert is that juries do have experience with and knowledge about warnings. However correct or incorrect, complete or incomplete this knowledge is, it exists. It is appropriate to assume that most juries will have a limited understanding of the display design and communication principles that are relevant to warning design, and they will not appreciate where warnings fit into the overall safety scheme. Thus, another role of the warning expert is to expand the jury's understanding of warnings and their role in safety so that the expert's opinions can be better appreciated and accepted.

15.7 CONCLUSIONS

As the body of scientific literature in the field of warning design and effectiveness has grown and developed, so apparently has the role of warning issues in product liability and personal injury litigation. As these issues continue to be addressed in litigation there will be a continuing need for capable experts in the subject matter. The role of the warning expert can be challenging, but also it is important. And it is important that it be done well.

In this chapter some of the techniques, procedures and challenges associated with the role of warning expert have been presented and explored. It should be noted, however, that most of the topics presented could be addressed in much greater depth than the scope of this chapter permitted. Indeed, anyone considering taking on the role of warning expert would be well advised to consider and study further the various requirements and responsibilities that come with the role.

Finally, it is increasingly common for each side of a case to have a warning expert and for these experts to disagree. Such circumstances are to be expected and are not a reason to abandon the role of warning experts in litigation. It should be noted that engineers, physicians, toxicologists and economists also disagree and work on both sides of cases. The important point is that the people who serve as warning experts be qualified and that they do their best to provide high quality expertise to our judicial system.

REFERENCES

HARDIE, W.H. (1994) Critical analysis of on-product warning theory. *Product Safety and Liability Reporter*, February 2, 145–163.

HUBER, P.W. (1990) Pathological science in court. *Daedalus (Journal of the American Academy of Arts and Sciences), Risk*, 119, 97–118.

LEHTO, M.R. and MILLER, J.M. (1986) *Warnings, Vol. I, Fundamentals, Design, and Evaluation Methodologies*. Ann Arbor, MI: Fuller.

LOVVOLL, D.R. (1997) 1997 HFES salary survey. *Human Factors and Ergonomics Society Bulletin*, 40, 1–3.

SENDERS, J.W. (1994) Warning assessment from the scientist's view. *Ergonomics in Design*, 6–7.

SLATER, A.D. (1993) Federal standards for admissibility of expert testimony and the applicability of privileges to communications with experts and materials generated by experts. In *The Role of Expert Witnesses in the 1990's and Beyond*. Falmount, MA: Seak, pp. 16–51.

Subject Index

abstractness 158, 160
accident data 6, 297
accuracy 305
adequacy 291–2, 303, 316, 325, 327–9, 344–5
adherence *see* compliance
admiration *see* likability
adversarial 331, 333, 338
advertising 9, 29, 87, 103, 105–7, 235, 255
affordances 296
age differences 207–9
agreement 95
alcohol 28, 31, 75, 93, 107, 113, 235, 274
alcoholic beverage warnings 197, 204
allergic users 325
ambiguity 35
American Institute of Graphic Artists (AIGA) 285
American Law Institute (ALI) 315, 318
American National Standards Institute (ANSI) 265, 302–4, *see also* ANSI Z535
American Society for Testing and Materials (ASTM) 266
annoyance 138
ANSI Z535 72, 127, 133, 157–8, 161, 166, 171–3, 180, 267, 275–6, 279, 280–1, 283–4, 286–8
archival 75
argument strength 91
asbestos 319, 326, 328
attention 18, 28–33, 130
attention capacity 18
attention capture 124–31
attention maintenance 124, 131–8
attitude change 90
attitudes 18, 40–3, 189–213, 245, 257
auditory 104, 129–30, 137, 164, 178–80, 291
availability 197–9
avoidance 231, 299, *see also* denial
awareness 7, 150, 204, 234

behavior 18, 53–5, 210, 245–51, 253–8, 296, *see also* compliance
behavioral intentions 53–79, 190, 221, 227, 236, 245–57, 300

beliefs 18, 40–3, 189–213
believability 43, 306
benchmarking 294
benefits 228
benign experience 205
biases 197
bi-lingual *see* language
boomerang effect 222, 231, 238
border 30, 58, 127, 173, 175
bottleneck 18–20, 23
brevity 35, 111, 157
brightness contrast 125–6
bypassing stages *see* skipping stages
bystanders 323–4

capacity 18, 123
catastrophic event deductive analysis 294
causation 321
cautionary intent *see* behavioral intentions
caveat emptor 5
CD-ROM 115
ceiling effects 64–5
central cues 90
channel 17, 94, 99–117
character compression 132
character spacing 133
chemical hazard 67
chemistry laboratory paradigm 32, 59, 67–8, 246, 248, 253
children's interpretation 302
C-HIP model 15–23, 140–1
cigarettes *see* tobacco
clarity 328
clutter 110, 124, 161, 247, 256
cognitive modeling methods 296
coherence 34
collateral information 110–12
color 28, 30, 110, 126–7, 134, 172, 175, 178–9, 229, 248–9, 256–7, 268, 275–7, 280–1
color contrast 126
combustible 153
commercials *see* advertising

common dangers 320
communication model 3, 9, 16, 255
communication-persuasion model 86–7, 223, *see also* persuasion
competitor product analysis 294
completeness 111
complexity 112
compliance 44–51, 53–77, 232, 245–58, 300, 306, *see also* behavior
compliance cost *see* cost of compliance
composite model *see* C-HIP model
comprehension 18, 20, 31–40, 149–68, 281–3, 302–7
computer simulation 252, *see also* virtual reality
computers 144
confederate 232
confidence 233
configuration *see* shape
conflicting prior behavior 306
conformity 89
consciousness 7, *see also* awareness
consensus *see* social influence
consequences 154, 196, 225, 227–30, 299
consistency 109
consolidated model *see* C-HIP model
conspicuity *see* salience
Consumer Product Safety Commission (CPSC) 3, 6, 266, 277, 287, 298
content 293, 327–8
context 126, 167, 301, 307
control 54, 58
controlled substance 271
corrective messages 115–16
cost of compliance 72, 74, 200–1, 228–9, 232, 253–4, 257, 297
cost of non-compliance 201, 229
costs and benefits 191, 200–1
counter-advertising 115
court 335
credentials 333, 335, 339
credibility 87, 89, 90, 92, 301
criteria 10–11, 300
critical confusion 128, 158
critical incident technique 294
cry wolf *see* false alarms
cued recall 38
cues 7, 90–2, 113, 130, 149–50, 170–2, 224, 295, 344
cultural differences 176

debiasing 199
deception 59, 116
decision making 237–8, 253, 301
defense against hazards 4–5
demand characteristics 31, 60
demographics 89, 206, 250, 329
denial 222, 224, 231, 238
Department of Transportation (DOT) 266–7
designing warning labels *see* developing warning labels
detection time 29–30
developing instructions 296, 304
developing warning labels 291–3, 298, *see also* labels

direct experience 235–6
directions for use 291–2, 299–306, *see also* instructions
direct-to-consumer (DTC) 116
disclosures 105–6, 115
discomfort 229
discounting *see* denial
discovery 339, 347
disposal 102
disregard 321
distance 131
distinguishability *see* legibility
distraction 130–1
doctor–patient communications 114
driving skills/ability 198
Drug Enforcement Administration (DEA) 271
drug labeling 298
drugs 268, 270
dual code theory 158
dual process 106, 224, 231
durability 134, 306
duty to warn 291, 315–16, 325

education level 210
effectiveness 54–8, 227–8, 256, 300–6, 329, 345
efficiency 58
elaboration 113
elaboration likelihood model (ELM) 90–2
electrical hazard 70–1
embedding 230, 247–8
emotion 222–4
empirical testing *see* testing
encoding 112
energy analysis 295
environmental conditions 125, 134, 301
Environmental Protection Agency (EPA) 3, 136
epidemiological 62, 75–6
ethics 54
evaluation *see* testing
ex post facto 61–2
exit 284
expectations 190–1, 200, 211, 221, 238, 317–18
expected utility 201
experience 7, 202–3, 251, 299–300, 306, *see also* familiarity
experienced users 325–6
expert witness 329–49
expertise 87, 91, 92, 232, 258, 336
explicitness 35, 94, 154, 157, 201, 227, 238, 249, 250
eye movement (tracking) 21, 28–9, 35, 107, 111

face-to-face communication 104, 113–15, 255
failure modes and effects analysis 294
failure to warn 291
false alarms 43, 138, 301, 328
familiarity 109, 137, 141, 180, 202–6, 221, 226–7, 233–4, 238, 248, 250–1, 257, 299–300
Fatal Accident Reporting System (FARS) 6
fault tree analysis 294
fear 222–49, 256
Federal Trade Commission (FTC) 105, 116, 272–3, 287
feedback 21–2, 140–1, 190
field studies 31, 61–2, 71–2, 106, 246, 253, 296–7

figure-ground 28–9, 125–7, 134, 302
film *see* video
fire safety 284
first-time users 101–2
flammable 153
flash rate 125
floor effects 64–5
fluorescent color 126
FMC 279
focus groups 35–6, 166–7, 297
follow-up assessment 144
font 133–4
font size *see* print size
food 268–70
Food and Drug Administration (FDA) 3, 266–70, 287, 298
footnotes *see* supers
foreign language 301, *see also* language
formal 111
format 101, 135
frequency 109, 130, 233, 300, *see also* experience
funding 275

gender *see* sex
geography 302
glare 125
grammar 302
graphic 286
guarding 5
guidelines 12, 143, 157, 301, 304
gustatory sense 124

habituation 22, 140–1, 165, 180–1, 205, 233
hazard 3–7, 42, 282, 293
hazard analysis 6, 293, 295, 297, 300
hazard control hierarchy 4, 5, 13, 16, 224, 231
hazard impression 172–8, *see also* hazardousness
hazardousness 4, 92, 109, 172–3, 180, 192–9, 206, 226–7, 229, 230, 251–2, 257, 300, *see also* hazard impression, perceived risk
health warnings 272–3
heuristics 91–2, 197, 231–3, 238
Heuristic-Systematic Model (HSM) 91–2
hidden hazard 8
highlighting 126
history 12
human error 5
hybrid model *see* C-HIP model

icons *see* symbols
idiosyncratic users 325
illumination 125, 134
illustrations *see* symbols
immediate vendor 323
implied warranty of merchantability 319
incidental exposure 30–2, 58–9, 230
individual differences 233, 236, 250, 257
inferences 155
information 285
information acquisition *see* knowledge acquisition
information influence 88
information processing model 3, 9, 16, *see also* C-HIP model
information seeking 141, 205, 303
informed choice *see* informed consent

informed consent 315, 318
informed decisions 8
injury control 5
injury experience 235–6
injury likelihood *see* likelihood
injury severity *see* severity
input–output matrix 87
inserts 111
Institutional Review Board (IRB) 59
instruction manual *see* product manual
instructions 136, 228, 231, 235, 247, 304, 316
integrating 108, 110, 116, 136–7, 247
intelligibility 138
intended carefulness *see* behavioral intentions
intentional exposure 59
interactions 58
interactive warnings 30, 230, 239, 247
interference 130
intermediaries 323–4
intermediate measures 27–50
intermediate stages 27–50, 99–240
international 286
International Organization for Standards (ISO) 158–61, 286–8
Internet 115–16
interviews 295–6
involvement 90
irradiation 134
iterative design 168

job hazard analysis 294
job safety analysis 294

kinesthetic *see* touch
knowledge acquisition 33, 149–51, 168
knowledge gap 150–1, 154

label design 304
label size 136
labeling regulations *see* regulations
labels 108, 110–11, 268, 270–4, 277, 301–3
laboratory studies 61, 67
language 151–2, 156, 163, 166, 210, 249, 288, 302, 316, 329
latency *see* reaction time
lateral masking 127
law 12, 315–30
layout 301–2, *see also* format
leading 132
leaflets 104
learned intermediaries *see* intermediaries
learning *see* training
legibility 131, 133, 135–6, 156, 161–2, 302
length 301, 304
letter height *see* print size
levels of performance 213, 299, 300
likability 88, 232
likelihood 4, 191–3, 195, 225–7, 236, 250–2, 301
likelihood of complying *see* behavioral intentions
Likert-type rating scales *see* ratings
limited capacity 7
limited space 10, 136
limited time 10
linear model 19, 21

literacy 157
literature search 297–8
litigation 12, 316, 331
location 40, 102, 104, 107, 112, 128–9, 135, 137, 139–40, 247–8, 256, 303, *see also* placement
locus of control 42, 253
long-term memory 7
looking behavior 21
loudness 130–1
lower case 132

Manual on Uniform Traffic Control Devices (MUTCD) 267
masking 130
mass media 107, 254–7, *see also* multiple modalities
matching 39
Material Data Safety Sheet (MSDS) 10, 152, 301
mechanical hazard 69–70
media 17, 93, 95, 100, 115, 255
mediating response 223
medical devices 268
memory 7, 18, 31–2, 38–40, 112, 123, 230, 256
mental model 112–13, *see also* schema
message framing 199
message processing 19
meta-analysis 245
method-oriented taxonomy 61
Mine Safety and Health Administration (MSHA) 266
misuse 300, 322
modalities 17–18, 100, 305
modeling 170, 232, 254, 258
more likely than not 339, 345
motivation 18, 221–39
Mr Yuk 250
multiple features 127–8
multiple languages 288, 302, *see also* language
multiple locations 129
multiple measures 65, 76
multiple methods 55
multiple modalities 95, 106, 139, 249, *see also* modalities
multiple sources 94–5
multiple voice warnings 139

National Electronic Injury Surveillance System (NEISS) 6, 198
National Fire Protection Association (NFPA) 94, 266, 284, 287
National Institute of Occupational Safety and Health (NIOSH) 298
National Institute of Standards and Technology (NIST) 267, 278–9
need for cognition 90
negation *see* prohibition
negative transfer 168–9
negligence 317–18
no-warning control 58, 195
noise 130, 138
non-linear processing 21–2
normative influence 88
noticeability 18, 44, 108, 235, 248, 327
novice users *see* first-time users

object resemblance 160
observation 62, 63, 66, 71, 73
obviousness of danger 319–20, *see also* open and obvious
Occupational Safety and Health Administration (OSHA) 3, 5, 9, 152, 266, 274–6, 287, 298
odor *see* olfactory
older adults 135, 142
olfactory 124, 305
on-product warnings 103, 345
open and obvious 8, 10, 344, *see also* obviousness of danger
open-ended test 38, 167, *see also* testing
operating instructions 136
operator manual *see* product manual
opinion 3, 332
optimism *see* over-confidence
order 114
organization 106, 156
organizing information 296, 304
OTC advertising 116
outcome-related expectations 191, 200
over-confidence 197–8, 235
overloading 139–40
over-promotion 328
over-the-counter (OTC) drugs 111, 268
over-warning 140, 301, 328
owners manuals *see* product manuals

parallel process 224, 231
parsing 112
participants 60, 179, *see also* target population
patient package insert (PPI) 234
perceived effectiveness 201
perceived hazard *see* hazardousness
perceived risk 251, 299–301, *see also* hazardousness
peripheral cues 90, *see also* cues
personal relevance *see* relevance
personal risk 235
personality 209
personalization 228, 235
personnel resources 304
persons to be warned *see* target population
persuasion 90, 113, 223, 230–2, 238, 300, 306
pesticides 10
phone survey 35
physical characteristics 247
physical trace 62, 73–4
pictograms *see* symbols
pictorials *see* symbols
picture superiority 158
placards *see* signs
placement *see* location
pop-out effect 28
positive transfer 168–9
posters *see* signs
post-task questionnaire 21
post-use 102–4
power 88–9
practicality 10
practical considerations 291–308
precautionary intent *see* behavioral intentions
precautions 299–300
prediction 55–7, 62, 65–6, 77, 246

pre-existing knowledge 7–8, 21–2, 32, 94, 296, 300
preferences 308
pre-purchase 100–1
prescription products 268, 324
presence of warnings 246
presumptions 321
pre-testing 36
print *see* font
print advertising 107
print size 106, 108, 131–2
prior injuries 252
prior knowledge *see* pre-existing knowledge
prioritizing 10, 140
proactive symbols 250
probability 299
probing 308
procedural knowledge 299–300
procedure 228
processing shortcuts 22–3
product damage 71
product design review 296
product labels *see* labels
product liability 315
product life stages analysis 296
product manuals 107, 129, 248
product/system centered analysis 294
products liability 316, 329
prohibition 135, 160
prohibition symbols 308
prominence *see* salience
prospective memory 181
protection motivation model 224, 252
prototypes 149, 296, 298, 306–7
punishment 225
purchase decisions 93
purchase intentions 178, 208
purchasing 35, 75, 101, 104, 154

qualification 347
questionnaires 56, 77, 304

radiation 275
railroad markings 277–8
ranking 308
ratings 41, 43, 56
reaction time 29–30, 66, 127, 301
reactive symbols 250
readability 34, 131, 155–6, 305–6
recall *see* memory
receivers 11, 109, 191, 202, 210
recognition *see* memory
recommendations *see* guidelines
redundant cues 124, 179, *see also* cues
regulations 12, 266–8, 278, 287–8, 297–301
relevance 90, 109–10, 141, 222, 234–5, 238
reliability 246
reminders 7, 109, 113, 221, 344, *see also* cues
remote users 323
residual risk 18
resource allocation 292
responsibility 8
Restatement (Second) of Torts 315–17, 320, 322, 324–5

Restatement (Third) of Torts 315–16
retrieval 113, 170
retroreflective 125
right-to-know 8
risk 4, *see also* hazardousness
risk analysis 293–4, *see also* hazard analysis
risk communication 190
risk perception 6, 41–2, 191, 197, 252, *see also* hazardousness
risk taking 209–10, 221, 236–8
risk/utility evaluation 317–18
role model *see* modeling
rotating warnings 181
rules of evidence 332

saccharin 75
safety 265, 275–9, 283, 287–8
safety equipment 253
salience 20, 28, 30, 33, 103, 106–8, 123–4, 128, 137, 139, 143, 235, 256, 306
schema 112
scientific support 329–30
scoring 38–9, 64, *see also* testing
script theory 205, 233–4
seat belts 5, 59, 76, 231, 256
second stimulus 129
selecting messages 300–1
self reports 30, 62–3, 111, 300
self-efficacy 200–2, 231, 258
sensation seeking 236, *see also* risk taking
sensitivity 65–6, 138
sensory limitations 210
sensory modality *see* modalities
separation 136–7, 247
sequential model 19
serif 133
severity 191–3, 195–6, 199, 226–7, 236, 249–52, 299–301, 328
sex 206–9, 235, 252–3, 300
shape 40, 107, 110, 127, 173, 178, 229–30, 280, *see also* symbols
short-term memory 7, *see also* memory
side-effect information 114
signal word 30, 36, 110, 127, 176–9, 195–6, 229, 275, 307
signals 104
signs 104, 109–10, 274, 280
similarity 89
simulation 170, 296
size 110, 127, 131, 136, 302
skipping stages 22–3
skull and crossbones symbol 250, 324, 329
smell *see* olfactory
social influence 88–9, 231–2, 258
social-cognitive theory 190
sound *see* auditory
source 17, 85–96, 233
specific *see* explicitness
speech *see* voice
speed *see* reaction time
stage model 16–19
standardization 101, 141, 233
standards 265–88, 297, 301, 305
static v. dynamic models 213
statistical power 66

statistical tests 64
statistics 232
stress 142, 253
strict liability 317–18
subjective measures 246
subjective-expected utility (SEU) 224
superior knowledge 326
supers 105–7
suppression 197, 199
Surgeon General 233
surround *see* color contrast
surveys *see* questionnaires
switch stage 123
symbols 12, 30, 36–40, 45, 108–10, 127–8, 133, 157, 160–3, 173–6, 211, 246, 250, 257, 275, 279–86, 301–8, 316, 329
system 5, 9, 10, 117
systematic processing 91, 231–3

tactile *see* touch
tactual *see* touch
tag 248
target audience *see* target population
target population 11, 34, 37, 135, 142, 165–6, 323–4
task analysis 303
task loading 344
taste *see* gustatory sense
technology 144
television 105–7, 255
terminology 153
testimony 347–8
testing 11, 20, 34–5, 38–40, 142, 281, 283, 293, 297, 305–8, 329
text 305–7
theoretical framework *see* C-HIP model
threat 222–31, 238, 252
threat-related expectations 191
time 104–6
time of contact 233
time stress 253
timing 292
tobacco 29, 75, 93–4, 107, 115, 196–7, 204, 233, 255, 271–3
touch 124, 139, 291, 305
toxic shock syndrome (TSS) 234
trade-off 155–7, 165
training 94, 168, 170–1, 300
transportation 267, 285
trustworthiness 87–8

understanding *see* comprehension
Underwriters Laboratory (UL) 266
unforeseeable use 322
uniform – code 319
uniform – commercial 319
unintended use 322
upper case 132
urgency 172, 179–80
usability testing 294–6
user knowledge 297, 299–302, *see also* pre-existing knowledge
user/operator centered analysis 294
utility theory 224

vagueness 154
valence 249
validity 246
value added 58
value-expectancy 190, 224, 227
variability 58, 64
verbal warnings *see* voice
video 18, 62–3, 69, 94, 115, 254
viewing 128
viewing angle 135
viewing time 303
virtual environment 77
virtual reality 252
visibility 284
visual acuity 131
visual angle 131
visual warnings 125, *see also* warning attributes
visualizability 160
vividness 230, 236
voice 104, 129, 138–9, 164–5, 249
voicing style 180

warning attributes 195–7, *see also* color, explicitness, length, location, shape, symbols
warning effectiveness *see* effectiveness 54, 306
warning labels *see* labels
warning overuse *see* over-warning
warning system 9
warning tags 301
warning vs. no warning *see* no-warning control
warnings – expectations linkage 212
warnings expert *see* expert witness
warranty 319
willingness to comply *see* behavioral intentions
willingness to read warnings 111, 226
wording *see* text

Author Index

Italics indicate citation is in reference section.

Abdel-Halim, M.H. 29, 39, *50*
Abelson, R. 205, *217*, 234, *241*
Acton, W.I. 201, *213*, 229, *239*
Adams, A.S. 58, 76 *78*, 124, 138, 139, *145*, 178, *183*, 300, *309*
Adams, J.R. *80*
Adams, S.K. 29, *46*
Adler, R. 256, *258*
Ager, J.W. 75, *79*
Ajzen, I. 55, *78*
Albarracin, D. *260*
Albaum, G.S. 42, *50*
Alexander, J. 255, *260*
Allender, L. 41, *48*, 109, *119*, 141, *145*, 180, *184*, 206, *215*, 226, *240*, 251, *259*
Allison, S.T. 67, *81*, 95, *97*, 170, *187*, 200, *218*, 229, *242*, 253, *261*
Altman, C.J. *117*, *118*
Alves-Foss, J. 302, *308*
American Institute of Graphic Artists (AIGA) 283, 285, *288*
American National Standards Institute 127, 128, 133, 137, 141, *144*, 157, 158, 160, 166, 167, 171–3, 176, *182*, 196, *213*, 267, 275, 276, 280, 283, 286, *288*, 302, 303, 304, *308*
American Society for Testing and Materials (ASTM) 297, *308*
Amir, M. 114, *117*
Anderton, P.J. 133, *144*
Andrassy, J.M. 163, *185*
Andrews, J.C. 43, *46*, 113, *117*
Arkes, H.R. 199, *213*
Armstrong, G.M. 106, *117*
Arthur, P. 160, 162, 168, *183*
Asch, S.E. *96*
Ashcraft, M.H. 170, *182*
Asper, O. 29, *46*
Averback, E. 127, *144*
Averill, J.R. *239*, 222
Axelrod, S. 113, *120*

Ayres, T.J. 209, 296, *309*
Azar, N. 43, *52*, 72, *81*

Baber, C. 37, 41, *46*
Backinger, C.L. 111, *117*, 304, *309*
Backlund, F. 109, *119*
Baddeley, A.D. 123, *144*
Baggett, P. 296, *309*
Bailey, L.A. *48*
Ballard, J.L. 107, 173, 196, *120*, *186*, *217*, 230, *241*
Bandura, A. 190, 200, 201, *214*, 231, *239*
Baneth, R.C. 136, *148*
Banks, W.W. 164, *182*
Barber, J.J. 255, *258*
Barbera, C. 30, *48*, 70, *79*, 139, *145*, 248, *259*
Barfield, D.A. 173, *187*
Barlow, A.N. 207, *217*
Barlow, T. 32, 40, *46*, *51*, 57, 63, *78*, *81*, 106, 107, *117*, *121*, 128, 136, *144*, *147*, 195, 196, 201, 207, *218*, 227, 228, *242*, 249, 251, *258*, *261*
Barofsky, I. 92, *97*
Baron, R. 237, *240*
Bartek, P.A. 112, *117*
Barzegar, R.S. 180, *182*
Bauer, R.A. 88, *96*
Baum, C. *117*
Bauman, K.E. 255, *258*
Beales, H. 100, *117*
Bean, H. *215*
Becker, M.H. 201, *218*, 225, *239*
Begley, P.B. 71, *81*, 250, *261*
Bell, M. 133, *148*
Beltramini, R.F. 43, *46*, 93, *96*, *214*
Berkowitz, M.J. 285, *289*
Berman, E. 29, *47*, 107, *119*
Bernhardt, K. 105, *120*
Berry, S. 113, 255, *259*
Berscheid, E. 88, *96*

Berstein, A. *260*
Best, A. 106, *118*
Bettman, J.R. 101, 103, 105, *118*
Beyer, R.R. *213*
Bhalla, G. 107, *118*
Blood, D.J. *218*
Boersema, T. 110, *118*, *183*
Boff, K.R. 127, *147*
Bohannon, N.K. 195, 198, *214*, 235, *239*
Boles, D.B. 30, 33, 37, 38, 44, *48*, 56, 63, 67, *79*, 107, *119*, 173, *184*, 211, *215*, 250, *259*
Booher, H.R. 37, 39, *46*
Boone, M.P. 164, *182*
Booth, M. 255, *260*
Boster, F.J. 256, *258*
Bostrom, A. 112, *117*, *118*
Bouhatab, A. 246, *258*
Boyle, J. 115, *118*
Brannon, L. 55, 63, 74, *78*
Brantley, K. 160, *182*
Braum, C. 107, *118*
Braun, C.C. 18, 54, 58, 67, *78*, 172, *182*, 208, *214*, 249, 252, 257, *258*, *261*
Brelsford, J.W. 11, *13*, 35, 37, 41, 42, *48*, *49*, *51*, *52*, 57, 71, 79, *81*, 108, 109, *119*, *121*, 127, 141, 142, *146*, *147*, 155, 166, 171, 180, 181, *185*–7, 192, 196, 202, 210, *214*, *216*, *219*, 226, 227, *240*, *242*, *243*, 250, 251, 253, *260*–2
Brems, D.J. 192, 198, 199, *214*, *219*, 226, *242*, 252, *261*
Breshahan, T.F. 42, *46*, 110, *118*, 172, 177, *182*, 196, *214*, 229, *239*
Brewster, B. 127, *146*, *148*, 166, *187*
Brinberg, D. *120*
Britton, B.K. 155, *182*
Brnich, M.J. 43, *50*, 87, *97*, 205, *216*
Brock, R.C. 107, *118*
Brock, T.C. 89, *96*
Broder, J.M. 273, *288*
Brown, J.D. 255, *258*
Brown, T.J. 126, *144*
Bruce, M. 132, *145*
Brucks, M. 100, *118*
Brugger, C. 162, 168, *183*
Bruyas, M.P. 160, *183*
Bryk, J. 42, *46*, 110, 172, 177, *182*, 196, *214*, 229, *239*
Burkett, J.R. 34, *50*
Burnkrant, R.E. 89, *96*
Bzostek, J.S. 127, *144*

Cacioppo, J.T. 90, 91, *96*, *97*, 113, *120*
Cahill, M.C. 36, 39, *46*
Caird, J.K. 159, *186*
Cairney, P.T. 39, *46*, 163, *183*
Calitz, C.J. 158, 159, 164, *183*
Canadian Inter-Mark 32, *46*
Carter, A.W. *148*
Casali, J.G. 59, *79*, 180, *184*
Casey, S. 176, *183*
Celuch, K. 202, *214*
Chaiken, S. 55, *78*, 88, 91, 92, 92, *96*, *97*, 231, *239*, *240*
Chan, L.M. 246, *259*

Chapanis, A. 127, *144*, 156, 172, 177, *183*, 294, *309*
Charrow, V.R. 111, *119*, 304, *309*
Chase, C. 256, *261*
Childers, T.L. 36, *46*
Christ, R.E. 29, *46*
Chy-Dejoras, E.A. 42, *46*, 69, *78*, 94, *96*, 252, 254, *258*
Cialdini, R. 231, *239*
Clark, E.M. 107, *118*
Clark, K.L. 126, *145*
Clark, L. 256, *261*
Clarke, S.W. 30, *48*, *51*
Cleary, P.D. 190, *214*
Clemens P.L. 293–5, *309*
Cleveland, R.J. 201, *214*, 229, *239*
Clift-Matthews, W. 180, *183*
Coate, D. 255, *260*
Cochran, D.J. 107, 110, *118*, *120*, 173, *186*, 196, 217, 230, *241*
Cochran, W.G. 64, *78*
Cohen, A. 43, *52*, 72, *81*
Cohen, H.H. 166, *184*, 198, 206, 207, *216*, *218*, 227, 233, 235, *240*, *242*, 251, 253, *261*
Cole, B.L. 129, 133, *144*
Coleman, M. 34, *46*, 210, *214*
Collins, B.L. 12, 36, 40, *46*, *49*, 108, 110, *118*, *120*, 125, 134, *146*, 159, 172, *183*, 196, 209, *214*, 230, *239*, 281, 283, 284, *288*, *289*, 290, 305, *309*
Combs, B. 41, *46*, 47, *49*, 195, *216*
Consumer Product Safety Commision (CPSC) 277, *289*
Conzola, V.C. 71, *78*, 177, *183*
Cooper, G.E. 164, *183*
Coren, S. 130, *144*, 161, *183*
Coriell, A.S. 127, *144*
Cousineu, A. 89, *96*
Cox III, E.P. 16, 17, 58, 61, 77, *78*, 258, *258*
Craig, C.S. 87, *96*
Craik, F.I.M. 59, *78*, 170, *183*
Cramer, J.A. 74, *78*
Creel, E. 35, *49*, 153, *185*
Creighton, P. 102, *121*, *148*, 181, *187*
Crowther, M. 180, *183*
Cullingford, R. 108, *118*
Cummings, J.B. 166, *185*
Cunfer, A.R. 181, *184*
Cunitz, A.R. 256, *259*, 303, *309*

Da Cruz, L. 108, *118*
Dahlstedt, S. 125, *145*
Davis, R. 108, *118*
DeTurck, M.A. 30, 31, 38, *48*, 141, *145*, 205, 206, 208, *215*, 249, 258, 300, *309*
Deckert, C. 304, *310*
Deese, J. 170, *183*
DeJoy, D.M. 9, 15, 18, 21, *23*, 41, 43, 45, *46*, 56, *78*, 191, 197–9, 213, *214*, 231, 234, 235, *239*
Dennis, I. 164, *184*
Denscombe, M. 237, *239*
Dershewitz, R.A. 62, *78*
Desaulniers, D.R. 41, *47*, 51, 56, *81*, 109, *121*, 135, 141, *145*, *147*, 180, *183*, *187*, 192, 195,

200, 206, 209, *214*, *219*, 226, *242*, *243*, 251, *261*
Deutsch, M. 88, *96*
DeVellis, B.McE. 201, *218*
Dewar, R.E. 29, 36, *49*, *51*, 127, 134, *145*, 157, 160, 161, 162, 168, 173, *183*
Dholkia, R. 90, *97*
Dietrich, D.A. 126, 136, *147*, *148*
Digby, S.E. 153, *185*
Dingus, T.A. 18, 41, *47*, 62, 69, 72–4, *78*, *79*, 139, *145*, *146*, 200, 205, *214*, *215*, 229, *240*, 248, 253, *258*, *259*
Dion, K. 88, *96*, 237, *240*
Donner, K.A. 42, *47*, *214*, 253, *258*
Donovan, R.S. 127, 147
Dorris, A.L. 41, 44, *47*, 70, *78*, 102, *118*, 300, *309*, *310*
DOT (Department of Transportation) 267
Douglass, E.I. 110, *118*
Downs, C.W. 251, *260*, 300, *310*
Drake, K.L. 177, *183*
Dreyfuss, H. 36, *47*, 303, *309*
Driver, R.W. 92, *96*, 300, *309*
Drory, A. 39, 45, *50*, 109, *120*
Drug Enforcement Administration (DEA) 271
Duffy, R.R. 30, 32, 38, *47*, 70, *78*, 139, *145*, 156, *183*, 201, 205, *214*, 230, *240*, 247, *259*
Duffy, T.M. 34, *47*
Dunlap, G.L. 177, *183*
Dunn, J.G. 41, *47*
Durvasula, S. 43, *46*, 113, *117*, *213*
Dutt, N. 126, *145*

Eagly, A.H. 55, *78*, 91, *96*, 230, 231, *239*, *240*
Easterby, R.S. 39, 40, *47*, 110, *118*, 209, *214*
Eastman Kodak Company 130, *145*
Eberhard, J. 35, 36, 37, *47*, 304, *309*
Eckes, T. 55, *78*
Edwards, M.L. 75, *78*
Edworthy, J. 58, *78*, 124, 138, 139, *145*, 164, 178, 180, *183*, *184*, 300, *309*
Ehrenfeucht, A. 296, *309*
Einstein, G.O. 181, *184*
Eisenberg, J.M. 89, *96*
Eisworth, R. *260*
Eldredge, D.H. 130, *146*
Ellis, N.C. 75, *78*
Ells, J.G. 127, *145*
Endlund, H. 114, *118*
Environmental Protection Agency (EPA) 136, *145*
European Economic Community *289*
Eustace, M.A. *260*

Farid, M.I. 42, *47*, 57, *79*
Farquhar, J.W. 255, *260*
Federal Highway Administration (FHWA) 267, *289*
Federal Railroad Administration (FRA) 277, 278, *289*
Federal Trade Commission (FTC) 272, 273, *289*
Feingold, P.C. 209, *214*
Feist, J. 55, 63, 74, *78*
Felker, D.B. 111, 112, *119*, 304, *309*
Fenaughty, A.M. *216*
Ferrari, J.R. 246, *259*

Feshbach, S. 222, *240*
Fhaner, G. 201, 229, *240*
Fidell, S. 29, *47*
Finnegan, J.P. 44, *49*
Firenze, R.J. 294, *309*
Firestone, I.J. 75, *79*
Fischer, P.M. 29, 32, 39, 40, *47*, 107, 108, 111, *119*
Fischhoff, B. 6, *12*, *13*, 41, *47*, *49*, *50*, 112, *117*, *118*, 192, 195, 197, 198, *215–17*, 226, 227, *242*, 251, *260*
Fishbein, M. 55, *78*, *260*
Fjelde, K. 29, 35, *47*
Flay, B.R. 255, *259*
Flesch, R.F. 34, *47*, 156, *184*
Fletcher, J.E. 107, 108, *119*
Flora, J.A. 255, *259*
FMC Corporation 127, *145*, 157, 160, 172, 177, *184*, 196, *215*, 228, *240*, 279, *289*, 303, 304, *309*
Foley, J.P. 72, *79*, 201, *216*, 229, 236, *241*, 251, 252, 254, 257, *260*, 297, *310*
Fontenelle, G.A. 56, *81*, *121*, *147*, *187*, 200, *219*, *243*, *261*
Food and Drug Administration (FDA) 268, 269, 270, *289*
Forbes, R.M. 57, *81*, 136, *147*
Ford, G.T. *119*
Foster, J.J. 132, *145*
Fowler, R.C. 63, *80*
Fox, R.J. 107, 108, *119*
Foxman, E.R. 107, *119*
Francis, C. *260*
Frantz, J.P. 6, 36, *47*, *50*, 69, *79*, 124, 128, 136, 139, *145*, 153, 165, *184*, *185*, 197, 201, 205, 211, 212, *215*, 226, 228, 230, 235, *240*, 247, 250, 256, *259*, 296, 297, 303, *309*, *310*
Frederick, L.J. 127, 136, *148*, 152, 166, *187*, 196, 199, 207, 211, 212
Freedman, J.L. 199, *215*
Freedman, M. 285, *289*
French, J.R.P. 88, *96*
Frey, K.P. 230, *240*
Friedmann, K. 21, *23*, 33, 37, 44, 46, *47*, 69, *79*, *145*, 198, *215*, 226, 235, *240*, 250, 257, *259*, 303, *309*
Funkhouser, G.R. 34, 36, *47*, 108, *119*

Gallup Canada Inc. 32, *47*
Galluscio, E.H. 29, 35, *47*
Gammella, D.S. 207, *217*
Gantt, M. *215*
Gardner-Bonneau, D.J. *215*
Garvey, P.M. 132, *145*
Gates, W. 115, *119*
Geiselman, R.E. 29, 38, *49*
Geller, E.S. 59, *79*, *80*, 246, *261*
Gerard, H.B. 88, *96*
Gibson, J.J. 296, *309*
Gill, R.T. 30, 32, 38, 40, *48*, 70, *79*, 139, *145*, 248, *259*
Glanzer, M. 256, *259*
Glorig, A. 130, *147*
Glover, B.L. 72, 77, *79*, *81*, 134, *146*, 160, *186*, 248, 252, *259*, *261*

Glynn, S.M. 155, *182*
Godfrey, S.S. 30, 32, 38, 41, *48*, 56, *81*, 108, 109, *119*, *121*, 141, *145*, *147*, 180, *184*, *187*, 200, 206, *215*, *219*, 226, 227, 233, 234, *240*, *243*, 251, 253, 254, *259*, *261*
Goldhaber, G.M. 30, 31, 38, *48*, 141, *145*, 205, 206, 208, *215*, 249, *258*
Gomer, F.E. 72, *79*
Gondek, K. 114, 115
Goodman, A.C. 75, *79*
Goodman, L.S. 138, *146*
Gordon, E. 206, 211, *217*, 234, *241*
Granda, R.E. 177, *183*
Gravelle, M.D. 36, *50*
Graves, K.L. 75, *79*, 109, *119*, 215
Gray, W.B. 34, *48*, 210, *215*
Greco, P.J. 89, *96*
Green, P. 35, 36, 37, 39, *47*, *48*, 304, *309*
Greenfield, T.K. 30, 32, *48*, 61, 75, *79*, 204, *216*
Grichting, W.L. 256, *258*
Griffith, L.J. 172, 177, *184*
Gross, M.M. 213
Grossman, M. 255, *260*
Gur-Arie, O. 106, *119*
Gurol, M.N. 106, *117*
Guynn, M.J. 181, *184*

Haas, E.C. 164, 180, *184*
Hadden, S. 5, *12*
Haddon Jr., W. *260*
Hakiel, S.R. 110, *118*, 209, *214*
Hammer, W. 294, 295, *309*
Hammond, A. 3, 16
Hane, M. 201, 229, *240*
Hankin, J.R. 75, *79*
Hansen, W.B. 255, *259*
Hardie, W.H. 338, *349*
Hards, R. 164, *184*
Harkness, A.R. 199, *213*
Hartley, J. 157, *184*
Hartshorn, K. 215
Hatem, A.T. 44, *48*, 69, *79*, 251, *259*
Hathaway, J.A. 41, 69, *47*, 78, *79*, 139, *145*, 200, *214*, 229, *240*, 253, *258*
Haugvedt, C.P. 90, *96*
Hayes-Roth, B. 42, *48*, 113, *119*, 304, *309*
Heckler, S.E. 36, *46*
Heimstra, N.W. 41, *50*
Heinrich, H.W. 5, *12*
Hellier, E. 124, *145*, 164, 180, *184*
Helquist, M. *260*
Hemphill, D. 155, *182*
Henderson, D.P. 75, *80*
Hermann, W. 88, *97*
Herrera, O.L. 152, *187*
Hershey, J.H. *48*
Higbee, K.L. 222, *240*
Hill, G.W. 42, *49*, 141, *146*, 177, 180, *185*, 207, *216*
Hilton, M.E. 212, *215*, 221, *240*
Hodgkinson, R. 36, *48*
Hoffmann, E.R. 36, *49*
Holland, V.M. 111, *119*, 304, *309*
Hoonhout, H.C.M. *183*
Horn, E. *48*, *119*

Horst, D.P. *213*
House, T.E. 305, *310*
Houston, M.J. 36, *46*, *105*, *106*, 119
Hovland, C.I.I. 87, *96*, *97*, 222, *240*
Howard, R.A. 112, *119*
Howard-Pitney, B. 32, 39, 43, *49*, 94, *97*
Hoy, M.G. 105, 106, *119*
Huber, J. 191, 218, 228, 242, 334, *349*
Hughes, J. 36, *48*
Hughes, P.K. 129, *144*
Hulse, S.H. 170, *183*
Hummer, J.R. 126, *145*
Hunn, B.P. 62, 73, 139, *146*, 205, *215*, 248, *258*, *259*
Hunnicutt, G.G. 204, *217*
Hunter, J.E. 55, *79*, 245, 246, *259*
Hwang, M. *215*

International Standards Organization (ISO) 158, 160, *184*, 286, *289*

Jack, D.D. 159, *184*
Jamrozik, K. 108, *118*
Janis, L. 87, *97*, 222, *240*
Janz, N.K. 225, *239*
Jarrard, S.W. 195, *219*, 228, *243*, 246, *261*
Jaynes, L.S. 30, 33, 38, 44, *48*, 56, 63, 67, *79*, 107, *119*, 173, *184*, 211, *215*, 250, *259*
Jerdee, T.H. 89, *97*
Job, R.F.S. 224–6, *240*
Johansson, G. 109, *119*
Johnson, D. 203, *215*, 251, 255, *259*
Johnson, R.P. 59, *79*
Jones, S. 196, *215*, 228, 230, *240*, *241*
Jungermann, H. 112, 119

Kabance, P. 34, *47*
Kabbara, F. *215*
Kahneman, D. 197, 199, *215*, *218*, 237, *240*
Kalsher, M.J. 30, 32, *47*, *48*, *51*, 56, 63, 70, 72, *78*, *79*, *81*, 108, 110, *121*, 124, 127, 128, 136, 139, 140, *145*, *146*, 148, 165, 166, *187*, 197, 201, 205, 211, *214*, *219*, 230, 235, *240*, *243*, 247, 248, *259*, *261*
Kane, R. 255, *259*
Kanouse, D.E. 42, *48*, 113, *119*, 154, *185*, 196, 201, 211, *217*, 227, 229, 304, *309*, *310*
Kantowitz, B.H. 300
Karnes, E.W. 35, 41, 42, *48*, *49*, 136, 141, *146*, 153, 168, 177, *185*, 195, 207, *215*, *216*, 227, *241*
Kaskutas, L.A. 30, 32, *48*, 61, 75, *79*, 204, *216*
Kasper, R.G. 41, *48*
Kejriwal, S.K. 77, *80*, 209, *217*, 236, *241*
Keller, A.D. 36, 39, *48*, *260*
Kelley, H.H. 87, *97*, 222, *240*
Kendrick, J.S. 108, *118*
Kessler, L.G. *260*
Kieras, D.E. 304, *310*
Kim, M.S. 55, *79*, 245, 246, *259*
King, L.E. 29, *50*, 267, 285, *290*
Kingsley, P.A. 111, *117*, 304, *309*
Kinnear, T.C. 106, *119*
Kintsch, W. 155, *184*
Klare, G.R. 34, *48*, 156, *184*

Klauer, K.M. 36, *50*
Klein, K.W. 67, *81*, 142, *148*, 248, *261*
Klimberg, R. 35, *50*
Knapp, M.L. 209, *214*
Koch, G.G. 255, *258*
Kogan, N. 237, *240*
Kohake, J. 162, *185*
Kolbe, R.H. 105, *119*
Kozminsky, E. 156, *184*
Krampen, M. 36, *48*
Kraus, S.J. 55, *79*
Kreifeldt, J.G. 35, *48*, 154, *184*
Krenek, R.F. 300, *310*
Krugman, D.M. 29, *47*, 107, 108, *119*
Krumm-Scott, S. 44, *49*
Kryter, K.D. 130, *146*
Kustas, M.S. 177, *183*

La Prelle, J. 255, *258*
La Rue, C. 166, *184*, 206, *216*, 227, 233, *240*
Labor, S.L. 112, *119*
Lage, E. *97*
Lambert, J.V. 34, *50*
Lamson, N. 131, *147*
Landee, B.M. 29, *50*
Langlois, J.A. 35, *48*
Lastovicka, J.L. 107, *118*, *120*
Lau, R.R. 255, *259*
Laughery, K.A. *48*, 247, *259*
Laughery, K.R. 15, 16, 28, 29, 32, 35, 37, 38, 41, 45, *48*, *49*, *52*, 56, 57, 79, 92, *97*, 108–10, *119*, 121, 127, 128, 133, 136, 141, 142, *145–8*, 153, 155, 166, 173, 180, *184*, *185*, *187*, 192, 196, 200, 201, 206, 210, *215*, *216*, 219, 226–9, 234, *240*, *242*, *243*, 247, 249, 251, 253, *259*, *260*, 305, *310*
Laux, L.F. 36, 39, 42, *49*, 92, *97*, 158, 159, *185*, 196, 202, 207, *216*, 227, *240*, 249, 253, *260*
Layman, M. 41, *49*, 195, *216*
LeClerc, F. 101, *120*
Lefton, L.A. 245, *260*
Lehto, M.R. 5, 6, *13*, 34, 36, 43, *46*, *47*, *49*, 58, 69, 72, *79*, *80*, 102, 110, *120*, 128, 139, 141, 199, 201, 212, 213, *216*, 229, 235, 236, *241*, 251, 252, 254, 257, *260*, 294, 297, 303–5, *309*, *310*, 334, *349*
Leirer, V.O. 163, *185*
Leonard, D.C. 34, *50*, 57, *80*, 155, *186*, 197, *217*, 226, *242*
Leonard, S.D. 17, 18, 34, 35, 41, 42, *48*, *49*, 136, 141, *146*, 153, 158, 159, 166, 168, 177, 180, *185*, 195–7, 207, 211, *215*, *216*, 227, *241*
Lerner, N.D. *46*, *49*, 108, 110, *118*, *120*, 125, 134, *146*, 159, *183*, 209, *214*, 281, 283, 284, *289*, 290, 305, *309*
Leventhal, H. 222, 224, 228, *241*
Levy, M. 256, *261*
Lewitt, E.M. 255, *260*
Ley, P. 112, 114, *120*
Liau, T.L. 34, *46*, 210, *214*
Liberman, A. 231, *239*
Lichtenstein, S. *13*, 41, *47*, *49*, *50*, 192, 195, 198, *215*, *216*, *217*, 226, 227, *242*, 251, *260*
Liddell, H. 135, *148*, 209, *219*
Light, L. *260*

Lipstein, B. 87, *97*
Lirtzman, S.I. 42, *47*, 57, *79*, 92, 93, *97*, 196, *216*, 300, *310*
Locander, W.B. 88, *97*
Lockhart, R.S. 59, *78*, 170, *183*
Loken, B. 32, 39, 43, *49*, 94, *97*
Loo, R. 29, *49*
Loomis, J.P. 43, *49*
Loring, B.A. 196, 201, 211, *216*, 227, 229, *241*
Lourens, P.F. 256, *260*
Lovvoll, D.R. 27, *52*, 54, 136, *148*, 166, 247, *259*, 338, *349*
Lowrance, W.W. 191, *216*, 226, 227, *241*
Loxley, S. 164, *184*
Lust, J. 202, *214*

MacBeth, S.A. 159, *185*
Maccoby, N. 255, *259*, *260*
MacGregor, D.G. 36, *49*, 296, *310*
Mackett-Stout, J. 36, *49*
MacKinnon, D.P. 38–40, *49*, 61, 75, *80*, 216
Madden, M.S. 12, 315, 316
Magat, W.A. 109, *120*, 191, *218*, 228, *242*
Magnus, P. 255, *260*
Magurno, A.B. *52*, 67, 72, *81*, 126, 127, 134, 136, 142, *146*, *148*, 152, 160, 162, 166, 168, *185–7*, 248, *261*
Maheswaran, D. 92, *96*
Maibach, E.W. 255, *259*
Main, B.W. *50*, 153, 165, *184*, *185*
Mallett, L. 43, *50*, 87, *97*, 205, *216*
Manrai, L.A. 106, *120*
Marin, G. 204, 210, *217*
Marras, W.S. *187*
Martier, S.S. 75, *79*
Martin, E.G. 41, *50*, *217*, 219, 226, 235, 236, *241*, *242*, *243*, 252, *261*
Martin, I.M. 109, *120*, 192, 198, 204, 205, *218*, 258, *261*, 300, *311*
Matthews, D. 42, *49*, 177, *185*, 195, 227, *241*
Mattson, R.H. 74, *78*
Mausner, B. 90, *97*
Mausner, J. 90, *97*
Mayer, D.L. 36, *49*
Mayer, R.N. 61, *80*, 109, *120*, 158, *185*, 204, 207, *216*, *217*
Mazis, M.B. 16, 18, 92, *97*, 100, 105, 106, 108, 109, 113, *118–21*, 204, 206, 211, *217*, 234, *241*
McBride, D.K. 138, *146*
McCann, J.M. 87, *96*
McCarthy, R.J. 300, *310*
McCarthy, G.E. 44, *49*
McCarthy, J.V. 36, *49*
McCarthy, R.L. 44, *49*, 296, *309*
McCormick, E.J. 4, *13*, 124, 125, 126, 130, 132, 134, 138, *147*, 161, *186*
McDaniel, M.A. 181, *184*
McDowell, I. 201, *216*, 229, *241*
McGrath, J.M. 251, *260*, 300, *310*
McGuire, W.J. 85, 87, 90, 92, 93, *97*, 202, *216*, 223, *241*
Mckenna, N.A. 67, *81*, 95, *97*, 170, *187*, 200, *218*, 229, *242*, 253, *261*
McNeill, B. 105, *120*
McNeill, D.L. 106, *121*

Meeker, D. 133, *145*
Meichenbaum, D. 74, *80*
Menon, G. 114, *120*
Middlestadt, S.E. 255, *260*
Midthune, D.N. *260*
Miller, J.M 36, *47*, 58, *80*, 102, 110, *120*, 124, 130, 141, *146*, 165, *184*, 212, *216*, 237, *240*, 296, 297, 303, 305, *309*, *310*, 334, *349*
Milloy, D.G. 127, *145*
Mills, B. 137, *148*
Milroy, R. 164, *186*
Minarch, J.J. 29, *50*
Mongeau, P. 256, *258*
Moore, P.A. 107, *119*
Moray, N. 141, *146*
Morgan, M.G. 112, *117*, *118*
Mori, M. 29, 39, *50*
Moriarity, S. 133, *146*
Moriarty, B. 294, *310*
Moris, L.A. *50*
Moroney, W.F. 159, *185*
Morrell, R.W. 163, *185*
Morris, L.A. 16, 18, 35, 92, *97*, 106, 108, 109, 113–15, *120*, *121*, 154, *185*, 196, 201, 204, 206, 211, *217*, 227, 229, 234, *241*
Morrow, D.G. 163, *185*
Moscovici, S. *97*
Mowen, J.C. 87, *97*
Muehling, D.D. 105, 107, *119*
Mulligan, B.E. 138, *146*
Murff, E.J.T. 58, *78*, 258, *258*
Murphy, J. 105, *120*
Murphy, S. 63, *81*, *107*, *121*, 128, *147*, 201, *218*, 227, *242*, 248, *261*
Murray, K.B. 106, *120*, 134, *146*
Murray, L.A. 160, *186*
Murray, N.M. 106, *120*
Murry, J.P. 107, *120*
Myers, D.G. 254, *260*

Naffrechoux, M. 89, *97*
National Fire Protection Association (NFPA) 284, *290*
National Research Council 190, *217*
Nelson, D.L. 158, *186*
Netemeyer, R.G. 43, *46*, 113, 117, 213
Nisbett, R.E. 230, *241*
Nohre, L. 38, *49*, 75, *80*
Nonprescription Drug Manufacturers Association 111, *120*
Norman, D.A. 296, *310*

O'Conner, C.J. 109, *121*, 300, *310*
Office of Management and Budget (OMB) 267, 278, *290*
O'Neill, B. 43, *50*, *260*
Occupational Safety and Health Administration (OSHA) 274, 276, *290*
Oldenburg, B. 255, *260*
Olsen, J.C. 112, *120*
Orwin, R.G. 75, *80*
Otani, H. 18, 149
Otsubo, S.M. 21, *23*, 30, 33, 37, 42, 44, *50*, 56, 70, *80*, 205, 206, 211, 212, *217*, 226, 227, 233, *241*, 251, 257, *260*

Oude Egberink, H.J.H. 256, *260*
Ouellette, V.L. 74, *78*
Owen, N. 255, *260*

Padgett, C.A. 255, *258*
Paine, C. 137, *148*
Paivio, A. 158, *186*
Papastavrou, J.D. 34, 43, *49*, 199, 213, *216*, 235, *241*
Park, D.C. 163, *185*
Patak, D.S. 112, *119*
Patterson, B.H. 256, *260*
Patterson, L.T. 204, *217*
Patterson, R.D. 164, *186*
Payne, J.W. 101, *118*
Peckham, G. 303, *310*
Peeler, M.O. 48
Penney, C.G. 165, *186*
Pentz, M.A. 75, *80*, 216
Perry, R.W. *50*
Peter, J.P. 112, *120*
Peters, G.A. 228, *241*
Petty, R.E. 90, 91, *96*, *97*, 113, *120*
Pew, R.W. 39, *48*
Phillips, L. 90, *97*
Pickering, F. 111, *119*, 304, *309*
Pierman, B.C. 36, *46*, 110, *118*, 281, 289, 305, *309*
Pietrucha, M.T. 133, *145*
Pihlman, M. 72, *80*
Pittle, D. 258, *260*
Pittle, R. 256, *258*
Planek, T.W. 60, *80*
Plummer, R.W. 29, *50*
Politz, A. 73, *80*
Polzella, D.J. 36, 42, *50*
Ponsi, K.A. 34, *50*, 57, *80*, 155, *186*, 197, *217*, 226, *242*
Poon, L.W. 163, *185*
Popper, E.T. 107, *120*
Porter, R.F. 43, *49*
Portnoy, B. *260*
Poston, J. 114, *118*
Powell, K.B. 34, *50*
Precht, T. 30, *48*, 70, *79*, 139, *145*, 248, *259*
Preusser, D.F. 76, *80*
Prevey, M.L. 74, *78*
Public Health Service (PHS) 272, *290*
Purdon, S.E. 222, *242*
Purswell, J.L. 44, *47*, 70, 77, *78*, *80*, 102, *118*, 209, *217*, 236, *241*, 300, *310*
Pyrczak, F. 34, *50*

Rabinowitz, V. 113, *120*
Rachwal, G. 215
Racicot, B.M. 17, 30, 32, *51*, 56, 63, 67, *80*, *81*, 94, *97*, 108, 110, *121*, 124, 128, 140, *146*, *148*, 165, 170, *186*, *187*, 205, 211, *217*, 219, 229, 232, 235, *241*, 243, 247, 248, 254, *260*, *261*
Raghubir, P. 114, *120*
Rahneswar, S. *97*
Ramond, C. 73, *80*
Rao, K.V.N. 35, *48*, 154, *184*
Rashid, R. 30, *51*, 67, *81*, 127, 142, *146*, *148*, 173, 175, *186*, *187*, 248, *261*

Rasmussen, J. 139, *146*, 213, *217*
Rathneswar, S. 91, *97*
Raven, B. 88, *96*
Read, S. 41, *47*
Reddy, D.M. 107, *117*
Redish, J.C. 111, *119*, 304, *309*
Reisinger, K.S. 72, *80*
Rethans, A.J. 41, 42, *50*, *217*, 235, *241*
Rhoades, T.P. 6, *50*, 70, *79*, 128, 139, 153, *185*, 201, 205, *215*, 230, *240*, 247, *259*, 296, *309*, *310*, 330
Ricco, D. 113, *120*
Richards, J.W. 29, *47*, 105, 107, *119*, *120*
Richardson, S.L. 181, *184*
Riley, M.W. 107, 110, *118*, *120*, 173, *186*, 196, *217*, 230, *241*
Ringseis, E.L. 159, *186*
Roberts, D.S. 59, *80*
Robertson, E.K. *260*
Robertson, L.S. 43, *50*, 76, *80*
Robinson, J.N. 213
Rodgers, W. 113, *119*
Rodriguez, M.A. 67, *80*, 127, *146*
Rogers, R.W. 224, *241*
Rogers, W.A. 131, *147*, 252, *260*
Rojas, T.H. 107, 108, *119*
Rokeach, M 189, *217*
Roland, H.E. 294, *310*
Rosen, B. 89, *97*
Rosenstock, I.M. 201, *218*
Ross, K. 302, *310*
Ross, L. 230, *241*
Ross-Degnan, D. 87, *97*
Roth, D.H. 34, *50*
Rothman, A.J. 238, *241*
Rothschild, M.L. 105, 106, *119*
Rothstein, P.R. 56, *81*, 108, 110, *119*, *121*, *147*, *187*, 200, *219*, *243*, 253, *259*, *261*
Rousseau, G.K. 131, 135, 142, *147*
Rowe, A.L. *49*, 57, *79*, 196, *216*, 227, *240*
Rowe-Hallbert, A.L. 196, *216*, 227, *240*, 249, *260*
Roy, D. 255, *259*
Rumelhart, D.E. 151, *186*
Russ, F.A. 106, *117*
Russo, J.E. 101, *120*
Ryan, J.P. 248, *260*

Salop, S.C. 100, *118*
Salvendy, P. 5, *13*, 238, *241*, 294, *310*
Samet, M.G. 29, *50*
Sanders, M.S. 4, *13*, 124, 125, 126, 130, 132, 134, 138, *147*, 160, 161, *186*
Scammon, D.L. 61, *80*, 109, *120*, 204, *217*
Scancorelli, L.F. 71, *81*, 250, *261*
Schaeffer, M.A. 107, *117*
Schank, R.C. 205, *217*, 234, *241*
Scheiner, E. 133, *146*
Scheyer, R.D. 74, *78*
Schlegel, R.E. 77, *80*, 209, *217*, 236, *241*
Schmidt, R.A. 296, *309*
Schneider, T. 177, *185*
Schneider, K.C. 250, *260*
Schnell, T. 126, *148*
Schommer, J.C. 112, *119*
Schucker, R.E. 75, *80*

Schumann, D. 113, *120*
Schupack, S.A. 63, *80*
Schwarz, N. 112, 114, *120*
Scoggins, J.A. 37
Scott, K. 126, *148*
Sears, D.O. 199, *215*
Seiden, R.M. 294, 296, *311*
Senders, J.W. 346, *349*
Shaffer, D.R. 231, 232, *242*
Shean, R. 108, *118*
Sherrell, D.L. 87, 95, *97*
Shinar, D. 39, 43, 45, *50*, 109, *120*
Showers, L. 202, *214*
Shuv-Ami, A. 92, 93, *97*
Siegel, A.I. 34, *50*
Silver, N.C. 18, 34, *50*, 54, 57, 58, 67, 70, *78–80*, 155, 156, 166, 172, 177, 178, *182*, *186*, *187*, 195, 197, 207, 208, 210, *214*, *217*, *219*, 226, 230, *242*, 249, 257, *258*, *262*
Simpson, C.A. 29, *50*
Simpson, S.N. 29, 30, *51*, 63, *81*, 110, *121*, 140, *148*, 195, 197, 205, *219*, 228, 235, *243*, 246, *261*
Singer, R. 228, *241*
Six, B. 55, *78*
Sklar, D.L. 130, *147*
Slater, A.D. 332, 334, *349*
Sless, D. 39, *46*, 163, *183*
Sloan, J.J. 75, *79*
Slovic, P. 6, 41, 45–7, *49*, *50*, 192, 193, 197, 198, *215–17*, 226, 227, *242*, 251, 252, *260*
Smith, K.R. 61, *80*, 109, *120*, 204, *217*
Smith, S.J. 106, *120*
Smith, S.L. 131, *147*
Smith, V.L. 41, *48*, 109, *119*, 141, *145*, *180*, *184*, 204, 206, *215*, 226, *240*, 251, *259*
Smither, J.A. 252, *261*
Snyder, H.L. 29, *50*
Snyder, L.B. *218*
Sojourner, R.J. *51*, 128, *147*, 163, 171, *186*, *187*
Sokol, R.J. 75, *79*
Sorkin, R.D. 130, *147*, 300, *310*
Soumerai, S.B. 87, 95, *97*
Spira Kahn, J. 87, *97*
Spunar, M.E. 127, *146*
Stacy, A.W. 38, *49*, *80*, *216*
Staelin, R. 57, *80*, 100, 101, *118*
Stam, A. 107, *120*
Stankey, M.J. 105, 106, *119*
Stanton, N. 124, *145*, *147*, *184*
Stanush, J.A. 155, *185*
Starr, C. 210, *218*, 237, *242*
Sternthal, B. 90, *97*
Stewart, D.W. 109, *120*, 204, 205, *218*, *258*, *261*, 300, *311*
Stewart, M.L. 75, *80*
Stock, B.R. 208, *214*
Stokes, R.C. 75, *80*
Stokes, S.L. 58, *78*, 258, *258*
Strawbridge, J.A. 21, *23*, 32, 33, 37, 44, *51*, 69, *80*, 109, *120*, 126, 136, *147*, *218*, 234, *242*, 247, *261*
Strecher, V.J. 201, *218*
Stutts, M.A. 204, *217*
Sujan, M. 100, *121*

Summala, H. 72, *80*
Sumner, F.C. 126, *147*
Sutton, S.R. 224, *242*
Svarstad, B. 114, 115, *121*
Svenson, O. 125, *145*, 198, *218*, 235, *242*
Swasy, J.L. 108, 109, 113, *120*, *121*, 204, *217*
Swiernega, S.J. 127, *147*
Swindell, J.A. *148*

Tabak, E. 114, 115
Tabrizi, M.F. 41, *47*
Taylor, G.B. 29, *50*
Taylor, J.R. 106, *119*
Taylor, S.E. 55, 63, 74, *80*, 230, 242, 256, *261*
Tenney, J. *260*
Teret, S.P. *48*
Tesser, A. 231, 232, *242*
Testin, F.J. 29, *51*
Thompson, D.M. 170, *186*
Thompson, N.B. 36, *49*, 158, 170, *185*, 207, 216
Thompson, S.C. 230, *242*
Threlsall, S.M. 102, *121*, 180, *187*
Thuring, M. 112, *119*
Thyer, B.A. 246, *261*
Tinker, M.A. 132, *147*
Treasury Board of Canada 286, *290*
Treisman, A. 28, *51*
Trelfall, F.M. *148*
Trommelen, M. 302, *311*
Trucks, L.B. 29, *46*
Tuckermanty, E. *260*
Tulving, E. 170, *186*
Turk, D.C. 74, *80*
Tversky, A. 197, 199, *215*, *218*, 237, *240*
Twyman, M. 303, *311*

Udry, J. 256, *261*
Underwriters Laboratory (UL) 284, 285, *290*
Ulmer, R.G. 76, *80*
USPC 162, *186*
Ursic, M. 57, *80*, 155, 177, *186*, 195, *218*
USDL/OSHA *311*

Vainio, K. 114, *118*
Van der Molen, H.H. 256, *260*
van Dijk, T.A. 155, *184*
Van Dusen, L. 155, *182*
Vanderplas, J.H. 135, *147*
Vanderplas, J.M. 135, *147*
Vaubel, K.P. 35, 41, *48*, *49*, *51*, 57, *79*, 108, *119*, 127, *146*, 154, *186*, 196, *216*, *240*, 249, *260*
Vaught, C. 43, *50*, 87, *97*, 205, *216*
Venema, A. 71, *81*, 136, *147*, 250, *261*
Verry, S. *119*
Vigilante, W. 136, 140, *147*, *148*, 153, *186*
Vignali, R.M. 302, *311*
Viscusi, W.K. 109, *120*, *121*, 191, 195, 197, 206, *218*, 228, *242*
Voevodsky, J. 76, *81*
von Restorff, H. 181, *186*
Vredenburgh, A.G. 198, 206, 207, *218*, 235, *242*, 251, 253, *261*

Wagenaar, W.A. 209, *218*
Wallach, M. 237, *240*
Wallen, B.A. *48*
Wallenious, S. 114, *118*
Wallston, B.S. 190, *218*
Wallston, K.A. 222, *242*
Walster, E. 88, *96*
Wankling, J. 37, 41, *46*
Ward, L.M. 130, *144*, *183*
Ward, W.D. *146*, *147*
Ware Jr., J.E. 255, *259*
Warner, K.E. 75, *81*, *97*, 255, *261*
Watanabe, R.K. 133, *147*
Watzke, J.R. 252, *261*
Wax, Y. *260*
Webb, S. 108, *118*
Webster's New Universal Unabridged Dictionary 41, *51*
Weinstein, N.D. 190, 198, 201, 213, *218*, 222, 227, 235, *242*
Weiss, W. 87, *96*
Westinghouse Electric Corporation 127, *147*, 157, 160, 172, *187*, 196, *218*, 228, *242*, 279, *290*, 303, *311*
Wheale, J.L. 29, *51*
Wickens, C. 30, *51*
Wideman, M.V. 107, *117*
Wiklund, M.E. 196, 201, 211, *216*, 227, *241*, 299
Wilkie, W.L. 106, 116, *121*
Williams, A.F. 72, *80*
Williams, D. 29, *50*
Williams, R. 132, *147*
Wilson, D.K. 222, 224, *242*
Wilson, E.J. 87, 95, *97*
Winkler, J. 113, *119*
Witte, K. 224, 231, *242*
Wixom, C.W. 43, *50*, *260*
Wogalter, M.S. 9, 15, 17, 18, 30–3, 36–41, 44–8, *50–2*, 56–8, 63–7, 70–3, 77–*81*, 94, 95, *97*, 106–10, *117*, *121*, 124–9, 134, 136, 137, 139–42, *145–8*, 152, 153, 155, 160, 162, 163, 165–8, 170–3, 175, 177, *178–83*, *185–7*, 192, 193, 195–201, 203, 205–8, 211, *217–19*, 226–30, 232, 233, 235, 236, *240–3*, 246–*62*, 283, *290*, 305, *310*
Wolff, J.S. 36, *51*, 162, 166–8, *185*, *187*, 283, *290*
Wood, S.E. 230, *239*
Wood, C.T. 213, 296, *309*
Wood, P. 255, *260*
Wreggit, S.S. 41, *47*, 69, *78*, 139, *145*, 200, *214*, 229, *240*, 253, *258*
Wright, P. 94, *97*, 102, 107, *121*, 137, 157, 180, *187*, *219*

Yellin, A.K. 107, *117*
Young, S.L. 27–30, 32, 35, 38, 39, 39, 41, *48–52*, 54, 57, 67, 71, *79*, *81*, 108, *119*, 124, 126–9, 133, 136, 139, 142, *146*, *148*, 163, 165, 166, *187*, 192, 194–6, 198, 205, 206, *214*, *216*, *219*, 226, 227, 235, *239*, *243*, 249, 251, *260*, *262*, 305, *310*

Zeitlin, L.R. 70, *81*, 251, *262*
Zlotnik, M.A. 248, *262*
Zohar, D. 43, *52*, 72, *81*
Zuccollo, G. 135, *148*, 209, *219*

Zuckerman, M. 209, *219*, 236, *243*
Zwaga, H. 39, 40, *47*, 110, *118*, 162, 168, *183*, *187*
Zwahlen, H.T. 126, *148*